Enzyme
Structure and
Mechanism

Enzyme Structure and Mechanism

Second Edition

ALAN FERSHT
*Imperial College of Science
and Technology, London*

W. H. FREEMAN AND COMPANY
New York

Cover: Phosphofructokinase, R state, from *Bacillus stearothermophilus*. Structure determination by P. R. Evans and P. J. Hudson. Photo by Arthur M. Lesk, Medical Research Council, University Medical School, Cambridge.

Library of Congress Cataloging in Publication Data

Fersht, Alan.
   Enzyme structure and mechanism.

   Includes bibliographical references and index
   1. Enzymes. I. Title. [DNLM: 1. Enzymes. QU 135
F399e]
QP601.F42   1984   574.19'25   84-4172
ISBN 0-7167-1614-3
ISBN 0-7167-1615-1 (pbk.)

Printed in the United States of America

11 12 13 14 15 16 17 18 19 20   VB   9 9 8 7 6 5 4 3

To P. J. FERSHT *in memoriam*
W. P. JENCKS
M. F. PERUTZ

Preface

We are now entering a new golden age of enzymology. Just as x-ray protein crystallography transformed the 1960s and 1970s, so recombinant DNA technology is changing the whole perspective in the 1980s. The cloning of enzyme genes for overproduction has facilitated the study of well-known enzymes and has enabled previously inaccessible enzymes to be characterized. Even more exciting are the prospects opened up by our ability to tailor the structure and activity of enzymes by manipulation of their genes. Specific amino acid residues may be changed by site-directed mutagenesis. Even whole genomes may be synthesized and their gene products expressed.

The second edition contains a chapter on protein engineering (Chapter 14) to summarize these developments. Two further new chapters (8 and 9) have been added to cover the recent advances in stereochemical methods and the continued progress in the design of highly specific irreversible inhibitors. Elsewhere, the book has been generally brought up to date, and certain chapters have been reorganized. Chapter 1, in particular, has been expanded to include discussion of the dynamic aspects of protein structure and recent ideas on protein evolution. Also, the chapter on allosteric proteins (Chapters 10) has been expanded to include specific examples of enzymes that regulate metabolic pathways.

I am indebted to several colleagues for their useful comments, especially those of Sir John Cornforth on Chapter 8.

A. F.
London, 1984

Preface to the First Edition

During the past two decades the advances in x-ray crystallography, transient kinetics, and the study of chemical catalysis have revolutionized our ideas on enzyme catalysis and mechanism. It is the intention of this text to provide a brief account of these developments for senior undergraduate students and postgraduates who have attended courses in chemistry and biochemistry. The philosophical and theoretical aspects of this book center upon how the interactions of an enzyme with its substrates lead to enzyme catalysis and specificity, and upon the relationship between structure and mechanism. The experimental approaches emphasized are those involving the direct study of enzymes as molecules. As such, there is a strong emphasis on pre-steady state kinetics where enzymes are handled in substrate quantities and enzyme-bound intermediates observed directly. The steady state kinetics of multisubstrate enzymes and the detailed chemistry of coenzymes and cofactors are discussed only in a cursory manner.

There have been two guiding rules in the preparation of this book. The first is to discuss general principles and ideas using specific enzymes as examples. [Although to avoid overloading the more theoretical chapters on kinetics, most of the illustrative examples are presented in a separate chapter (7).] The second is to stick closely to examples where hard evidence is available and to avoid speculation and woolly evidence. In consequence, the discussion of detailed chemical mechanisms is generally restricted to enzymes whose tertiary structures have been solved by x-ray crystallography. Similarly, the discussion of the theoretical aspects of allosteric proteins is very much restricted to hemoglobin because it is the only example where good (or any) evidence is available on the nature of the interactions of positive cooperativity.

The references cited tend to be those of the most recent reviews or papers where more extensive bibliographies are given, and also those of the original papers in order to maintain a historical perspective. Illustrative examples have been taken were possible from the files of the MRC Laboratory of Molecular Biology because of their ready availability and uniform quality of presentation. In this context I must thank Annette Lenton both for the illustations she has prepared especially for this book and also for those prepared for other members of the laboratory whose files I have shamelessly raided.

I am particularly indebted to W. P. Jencks, H. B. F. Dixon, H. Gut-freund, K. F. Tipton, and R. S. Mulvey for their critical comments on the entire manuscript, and also M. F. Perutz and D. M. Blow for their comments on individual chapters. I wish to thank The Royal Society, the American Chemical Society, the Cornell University Press, Academic Press, John Wiley, and Alan R. Liss for permission to reproduce illustrations.

A. F.
Cambridge, 1977

Contents

1

The three-dimensional structure
of enzymes

In 1930, J. B. S. Haldane wrote a book on enzymes that is still worth reading today.[1] The most striking feature of this book is that so much was then known about the properties and action of enzymes, yet so little was known about the enzymes themselves: the question of whether or not enzymes are proteins was still the subject of raging controversy. The knowledge was so one-sided because there was no means of studying enzymes directly. All the information had been deduced indirectly from the effects of enzymes on their substrates. Nevertheless, the foundations of modern steady state kinetics had been laid in a little over thirty years, the first cell-free enzyme extract having been prepared by E. Büchner in 1897.

In order to proceed further, it was necessary to isolate purified enzymes in *substrate* quantities and examine them directly. This was accomplished in 1926, when J. B. Sumner crystallized urease from jack bean extracts. Soon afterwards (1930–36), J. H. Northrop and M. Kunitz crystallized pepsin, trypsin, and chymotrypsin. This provided the material to prove finally that enzymes are proteins, and to allow the development of the techniques of modern protein chemistry: the sequencing of proteins pioneered by F. Sanger; the solution of the three-dimensional structure of proteins pioneered by M. F. Perutz and J. C. Kendrew; and the use of rapid-reaction kinetics, which had been initiated by F. J. W. Roughton in 1923.

The major part of the present book deals with the direct study of enzymes in substrate quantities, taking up the story from where Haldane was forced to stop. This first chapter discusses the general features of the most significant advance in our knowledge of enzymes since then— their three-dimensional structure, from the basics of the peptide bond and the various elements of protein folding to macromolecular assemblies. The chapter also describes the evolution of protein structure and function, and considers the dynamic aspects of proteins. To set the scene for later chapters, there is a preview of chymotrypsin and lysozyme, since these enzymes have been the testing grounds of so much theory and experiment. The structures and mechanisms of individual enzymes are discussed at greater length in Chapter 15.

A. The primary structure of proteins

The major constituent of proteins is an unbranched polypeptide chain consisting of L-α-amino acids linked together by amide bonds between the α-carboxyl of one residue and the α-amino group of the next. Usually only the 20 amino acids listed in Table 1·1 are involved. The primary structure is defined by the sequence in which the amino acids form the polymer. By convention, the sequence is written as follows, beginning with the N-terminus on the left:

$$H_2NCH(R)CO-NHCH(R')CO-NHCH(R'')CO-NHCH(R''')CO-NH\cdots$$

$$(1\cdot1)$$

Although the primary structures of almost all intracellular proteins consist of linear polypeptide chains, many extracellular proteins contain covalent —S—S— cross-bridges in which two cysteine residues are linked by their thiols. This either creates loops in the main polypeptide chain due to intrachain bridges, or links different chains together (Figure 1·1). In these latter cases, the multiple chains are derived from a single-chain precursor that has become covalently modified by proteolysis—examples

TABLE 1·1. *The common amino acids*

Amino acid (three- and one-letter codes, M_r)	Side chain R in $RCH(NH_3{}^+)CO_2{}^-$	pK_a's[a]
Glycine (Gly, G, 75)	H—	2.35, 9.78
Alanine (Ala, A, 89)	CH_3—	2.35, 9.87
Valine (Val, V, 117)	H_3C $\diagdown$ CH— $\diagup$ H_3C	2.29, 9.74
Leucine (Leu, L, 131)	H_3C $\diagdown$ $CHCH_2$— $\diagup$ H_3C	2.33, 9.74
Isoleucine (Ile, I, 131)	CH_3CH_2 $\diagdown$ CH— $\diagup$ CH_3	2.32, 9.76

TABLE 1·1. *The common amino acids* (continued)

Amino acid (three- and one-letter codes, M_r)	Side chain R in $RCH(NH_3{}^+)CO_2{}^-$	pK_a's
Phenylalanine (Phe, F, 165)	⟨benzene ring⟩—CH_2—	2.16, 9.18
Tyrosine (Tyr, Y, 181)	HO—⟨benzene ring⟩—CH_2—	2.20, 9.11, 10.13
Tryptophan (Trp, W, 204)	⟨indole ring⟩—CH_2—	2.43, 9.44
Serine (Ser, S, 105)	$HOCH_2$—	2.19, 9.21
Threonine (Thr, T, 119)	HO, H_3C⟩CH—	2.09, 9.11
Cysteine (Cys, C, 121)	$HSCH_2$—	1.92, 8.35, 10.46
Methionine (Met, M, 149)	$CH_3SCH_2CH_2$—	2.13, 9.28
Asparagine (Asn, N, 132)	$H_2NC(=O)CH_2$—	2.1, 8.84
Glutamine (Gln, Q, 146)	$H_2NC(=O)CH_2CH_2$—	2.17, 9.13
Aspartic acid (Asp, D, 133)	$^-O_2CCH_2$—	1.99, 3.90, 9.90
Glutamic acid (Glu, E, 147)	$^-O_2CCH_2CH_2$—	2.10, 4.07, 9.47
Lysine (Lys, K, 146)	$H_3N^+(CH_2)_4$—	2.16, 9.18, 10.79
Arginine (Arg, R, 174)	H_2N^+, H_2N⟩C—$NH(CH_2)_3$—	1.82, 8.99, 12.48
Histidine (His, H, 155)	⟨imidazole ring⟩—CH_2—	1.80, 6.04, 9.33
Proline (Pro, P, 115)	⟨pyrrolidine ring with $CO_2{}^-$, C, H, N^+H_2⟩	1.95, 10.64

[a] From R. M. C. Dawson, D. C. Elliott, W. H. Elliott, and K. M. Jones, *Data for biochemical research*, 2nd ed., Oxford University Press (1969).

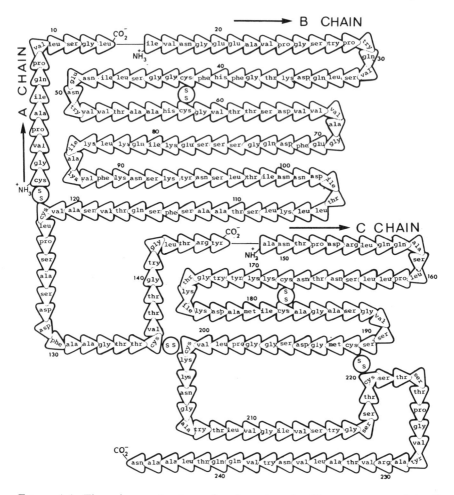

FIGURE 1·1. The primary structure of α-chymotrypsin. The enzyme consists of three chains linked together by —S—S— bridges. However, the chains are derived from the single polypeptide chain of chymotrypsinogen by the excision of residues Ser-14, Arg-15, Thr-147, and Asn-148. (Note that the code used here for tryptophan is "try" instead of "trp.") (Courtesy of D. M. Blow.)

being insulin from proinsulin and chymotrypsin from chymotrypsinogen. These bridges may be cleaved by reduction with thiols. It is of interest that single polypeptide chains may often be reduced and denatured, and then reversibly oxidized and renatured. But reduction and denaturation of proteins derived from the covalent modifications of precursors is generally irreversible.

The spontaneous refolding of single polypeptide chains that have not been covalently modified leads to an important principle: *The genetically encoded sequence of a protein determines its three-dimensional structure* (in a particular environment).

B. The three-dimensional structure of enzymes

1. X-ray diffraction methods

The importance of x-ray diffraction methods in enzymology cannot be overstated. Not only have they provided the experimental basis of our present knowledge of the structure of proteins, but they have proved to be the single most important factor in the investigation of enzyme mechanisms. This section briefly discusses the type of information that can be derived from x-ray diffraction experiments.

The way x-rays are scattered when they strike the electrons of atoms is similar to the way light waves are scattered by the engraved lines of a diffraction grating. The regular lattice of a crystal acts like a three-dimensional diffraction grating in scattering a monochromatic beam of x-rays, giving a pattern where the diffracted rays reinforce and do not destructively interfere. The structure of the crystal, or more precisely the distribution of its electron density, may be calculated from the diffraction pattern by Fourier transformation. This requires knowledge of the intensities and directions of the diffracted rays—which are easily measured on photographic film as a pattern of spots or by a diffractometer—and also of their *phases*. Determination of the phase of each diffracted ray (an essential requirement for the Fourier transformation) is the most difficult problem. The ways in which this problem was circumvented in analysis of simple crystal structures could not be applied to proteins. The solution of the *phase problem* was the stumbling block that held up protein crystallography until Perutz and his coworkers applied the method of *isomorphous replacement* in 1954.[2] In this procedure, a heavy metal atom is bound at specific sites in the crystal without disturbing its structure or packing. The metal scatters x-rays more than the atoms of the protein and its scattering is added to every diffracted ray. Information about the phases of the diffracted rays from the protein can then be obtained from the changes in intensity, depending on whether they are reinforced or diminished by the scattering from the heavy atom. Several different isomorphous substitutions are needed to give an accurate determination of the phases.

Once the phase and the amplitude of every diffracted ray are known, the electron density of the protein may be calculated. The structure of the protein is obtained by matching this density to the amino acid residues of the primary structure. This can be done mechanically with wire models; nowadays, however, a computer display system is more frequently used. The structure may then be refined by computer methods.

a. Accuracy and resolution

The degree of accuracy that is attained depends on both the quality of the data and the *resolution*. The term resolution is best illustrated for our purposes by the data in Table 1·2. At low resolution, 4 to 6 Å (0.4 to 0.6 nm), the electron density map reveals little more in most cases than the overall shape of the molecule. At 3.5 Å, it is often possible to follow the course of the polypeptide backbone, but there may be ambiguities. At 3.0 Å, it is possible in favorable cases to begin to resolve the amino acid side chains, and, with some uncertainty, to fit the sequence to the electron density. At 2.5 Å, the positions of atoms may often be fitted with an accuracy of ± 0.4 Å. In order to locate atoms to 0.2 Å, a resolution of about 1.9 Å and very well ordered crystals are necessary. What this means in practical terms is illustrated in Figure 1·2. The reflections required for high-resolution analysis are those that have been diffracted through the greater angles. But it is seen that the intensities rapidly decrease at higher resolution. Some crystals diffract better than the one shown in Figure 1·2; the majority do worse. A further point is that the number of reflections to be analyzed increases as the third power of the resolution; an increase from 3 to 1.5 Å increases the amount of data to be collected by a factor of 8, and the total effort required by an even larger factor due to the poorer quality of the data.

The isomorphous replacement method becomes ineffective for protein crystals beyond a resolution of 2 to 2.5 Å, since the addition of the heavy atom may cause some alteration in structure and since it is so difficult to observe small changes in intensities that are already very weak. However, computers can now refine a structure by using the measured intensities at higher resolution and the model structure that has been determined at 2- to 2.5-Å resolution.

TABLE 1·2. *Resolution and structural information*

Resolution (Å) (1 Å = 0.1 nm)	Structural features observable in a good map
5.5	Overall shape of the molecule. Helixes as rods of strong intensity.
3.5	The main chain (usually with some ambiguities).
3.0	The side chains partly resolved.
2.5	Side chains well resolved. The plane of the peptide bond resolved. Atoms located to about ± 0.4 Å.
1.5	Atoms located to about ± 0.1 Å (the present limit of protein crystallography).
0.77	Bond lengths in small crystals measured to 0.005 Å.

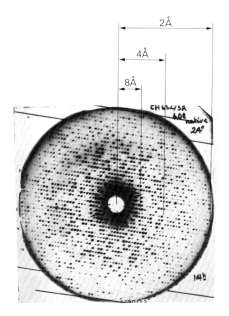

FIGURE 1·2. X-ray diffraction pattern from a crystal of α-chymotrypsin, showing the data required for various resolutions.

b. Neutron diffraction

It is not possible to locate by x-ray diffraction the positions of hydrogen atoms in proteins because the electron density of the H atoms is so weak. They may be observed, however, by neutron diffraction, because the scattering of neutrons by the hydrogen nucleus (the proton) is appreciable even compared with larger nuclei. The method is time-consuming and requires the use of specialized facilities. The structures of crystalline myoglobin[3] and trypsin[4] have been determined from neutron diffraction by using the structures from earlier x-ray diffraction studies as a basis for calculation. Important protons were located.

2. The structural building blocks

Even before the structure of any crystalline protein had been solved by x-ray crystallography, L. Pauling and R. B. Corey had worked out the structures of the units that have subsequently been found to be the basic building blocks of the architecture of proteins.[5] They first solved the structures of crystals of small peptides to find the dimensions and geometry of the peptide bond. Then, by constructing very precise models, they found structures that could fit the x-ray diffraction patterns of fibrous proteins. The diffraction patterns of fibers do not consist of the lattice of

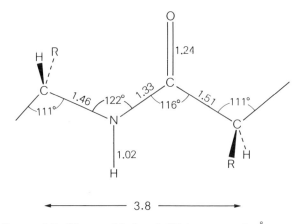

FIGURE 1·3. The peptide bond. Distances are in Å.

points found from crystals, but of a series of lines corresponding to the distances between constantly recurring elements of structure. This method of building models was later used by J. D. Watson and F. H. C. Crick to solve the double-helical nature of DNA.

a. The peptide bond[5]

The x-ray diffraction studies of the crystals of small peptides showed that the peptide bond is planar and trans (anti) (Figure 1·3). The same structure has been found for all peptide bonds in proteins, with a few rare exceptions. This planarity is due to a considerable delocalization of the lone pair of electrons of the nitrogen onto the carbonyl oxygen. The C—N bond is consequently shortened, and it has double-bond character (equation 1·2). Twisting of the bond breaks it and loses the 75 to 88 kJ/mol (18 to 21 kcal/mol) of delocalization energy.[5,6]

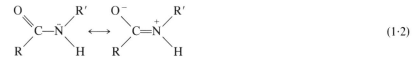

$$(1·2)$$

Proline is an exceptional amino acid residue in that the cis-trans equilibrium only slightly favors the trans form. Small proline-containing peptides in solution contain some 20 to 30% of the cis (syn) isomer,[7,8] as opposed to less than 0.1% of the cis isomers of the other amino acids.[9] The cis form is even found in native proteins: two of the four prolines in ribonuclease are cis. In native proteins, the overall structure of the molecule determines the isomeric form of the amino acid.[10] The slow cis-trans isomerization of proline residues causes complications in kinetic experiments on protein folding.[7,11]

b. The α helix and the β sheet[5]

The polypeptide chains of fibrous proteins are found to be organized into hydrogen-bonded structures. In these ordered regions, any buried carbonyl oxygen forms a hydrogen bond with an amido NH group. This is done in two major ways: by forming α helixes (found from the fiber dif-

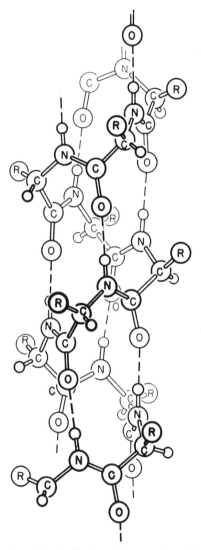

FIGURE 1·4. The right-handed α helix found in proteins. [Reprinted and modified from L. Pauling, *The nature of the chemical bond,* Cornell University (original edition, 1939; third edition, 1960). Used by permission of the Cornell University Press.]

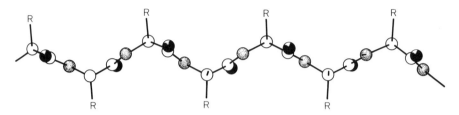

FIGURE 1·5. An extended polypeptide chain. The hydrogen bonds are made perpendicular to the plane of the paper so that the sheet made from successive parallel chains is pleated.

fraction studies on α-keratin) or by forming β sheets (found in β-keratin). The polypeptide chain of a globular protein also folds upon itself to create local regions of similar structures.

The α helix is illustrated in Figure 1·4. It is a stable structure, each amide group making a hydrogen bond with the third amide group away from it in either direction. The C=O groups are parallel to the axis of the helix and point almost straight at the NH groups to which they are hydrogen-bonded. The side chains of the amino acids point away from the axis. There are 3.6 amino acids in each turn of the helix. The rise of the helix per turn—the pitch—is about 5.4 Å.

An extended polypeptide chain (Figure 1·5) can make complementary hydrogen bonds with a parallel extended chain. This in turn can match up with another extended chain to build up a sheet. There are two stable arrangements: the parallel β-pleated sheet in which all the chains are

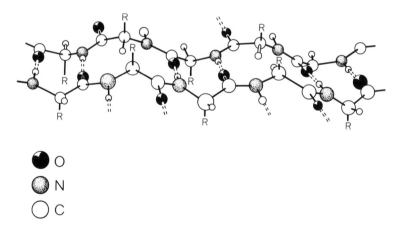

● O

◉ N

◯ C

FIGURE 1·6. Two strands of antiparallel β sheet. Other strands may be added in a similar manner.

aligned in the same direction, and the antiparallel β-pleated sheet in which the chains alternate in direction (Figure 1·6). The repeating unit of a planar peptide bond linked to a tetrahedral carbon produces a pleated structure. The parallel β sheet is not found in fibrous proteins but only in globular proteins. Mixed β sheets are also found in globular proteins that contain a mixture of parallel and antiparallel strands.

c. Bends in the main chain[12–15]

Another frequently observed structural unit occurs when the main chain sharply changes direction via a "β bend" composed of four successive residues. In these, the C=O group of residue i is hydrogen-bonded to the NH of residue $i + 3$ instead of $i + 4$ in the α helix. A 180° change of direction, which can link two antiparallel strands of pleated sheet, can be

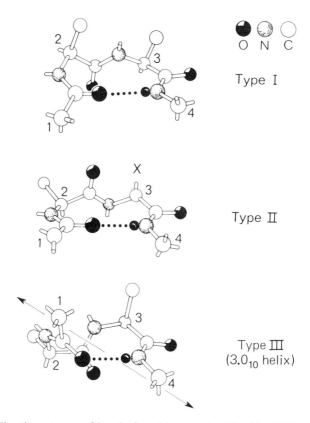

FIGURE 1·7. The three types of bends found in proteins. Residue X in type II is always glycine. [From J. J. Birktoft and D. M. Blow, *J. Molec. Biol.* **68**, 187 (1972).]

achieved in two ways (types I and II, Figure 1·7). Type II has the restriction that glycine must be the third residue in the sequence. A distortion of type I gives a bend, type III, which can be repeated indefinitely to form a helix known as the 3.0_{10} helix (a tighter, less stable, helix than the α, with 3.0 residues per turn, forming hydrogen-bonded loops of 10 atoms).

3. Assembly of proteins from the building blocks

There was great excitement when the structures of the first two proteins were solved. The single polypeptide chain of myoglobin was found to consist of a series of α-helical rods connected together without any β sheet. Then hemoglobin was found to be just like four myoglobin chains (Figure 1·8). But, as more protein structures were solved, the story rapidly complicated. Lysozyme and carboxypeptidase, the first enzymes whose structures were known, consist of α helixes and a β sheet. Soon afterwards, α-chymotrypsin was found to consist almost entirely of β sheet, with only two regions of α helix. It became apparent that general rules of protein folding would be hard to find. However, now that over 100 structures have been solved, the first steps in this direction are being taken. Certain recurring features have been noted, and methods of classifying structures are being developed.

M. Levitt and C. Chothia[16] have suggested that most proteins can be considered crudely as layered sandwich structures, with each layer consisting of either α helixes or a β sheet. They have identified four classes of protein molecules based upon different combinations of these structures: α/α (all α), β/β (all β), α/β, and $\alpha + \beta$. These are illustrated in Figure 1·9. The helixes and sheets pack by stacking their amino acid side chains. The internal packing is good, with the side chains of one piece of secondary structure fitting between those of another to give a hydrophobic region of similar density to a hydrocarbon liquid or wax.[17] There are few hydrogen bonds in these regions: in carboxypeptidase A there are only 17 hydrogen bonds connecting the 8 helixes to each other and to the sheet—some 2 per helix. The buried hydrogen-bond donors are invariably paired. Similarly, any buried charged group is paired with an interacting complementary charge to form a salt bridge, except in a few rare examples where a charged group is buried for purposes of catalysis. The irregularly folded segments of polypeptide chain that connect the helixes and sheets are usually exposed to solvent and are often short. Globular proteins thus have hydrophobic cores with their charged groups on the outside: they are "waxy" on the inside and "soapy" on the outside. The major driving force for the folding of proteins appears to be the burying and clustering

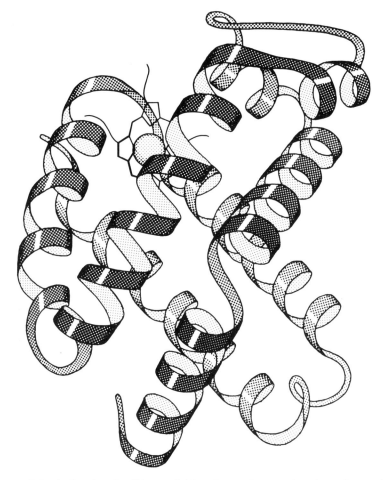

FIGURE 1·8. A β subunit of hemoglobin. An analogous drawing of myoglobin would be indistinguishable from this. (Courtesy of Jane Richardson.)

of hydrophobic side chains to minimize their contact with water (the "hy-drophobic effect" discussed in Chapter 11).

Proteins are not loose and floppy structures; their residues are packed as tightly as crystalline amino acids.[18–20] The packing density (the fraction of the total space occupied by the atoms) is about 0.75 for proteins, compared with values of 0.7 to 0.78 found for crystals. Close-packed spheres have a packing density of 0.74, and infinite cylinders, 0.91.

The following terms are used to describe the various aspects of protein structure.

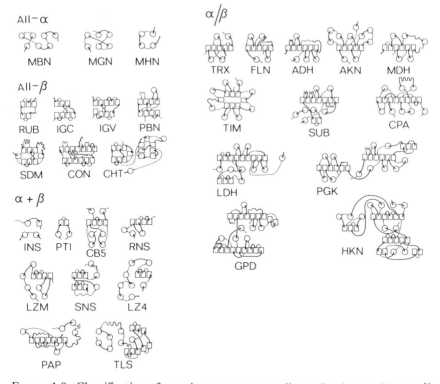

FIGURE 1·9. Classification of protein structure according to Levitt and Chothia.[16] The protein is viewed from such an angle that most segments of secondary structure are seen end on. Each strand of a β sheet is represented by a square. The front end of each α helix is represented by a circle. The segments that are close in space are close together in the diagram. The segments are connected by bold or thin arrows (from the N- to the C-terminus) to indicate whether the connection is at the near or the far end. The approximate scale is: diameter of α helix = 5 Å, β strand = 5 × 4 Å, separation of helixes = 10 Å, separation of hydrogen-bonded β strands = 5 Å, separation of non-hydrogen-bonded strands = 10 Å.

Abbreviations: MBN, myoglobin; MGN, myogen; MHN, myohemerythrin; RUB, rubredoxin; IGC, immunoglobulin constant region; IGV, immunoglobulin variable region; PBN, prealbumin; SDM, superoxide dismutase; CON, concanavalin A; CHT, chymotrypsin; INS, insulin; PTI, pancreatic trypsin inhibitor; CB5, cytochrome b_5; RNS, ribonuclease; LZM, lysozyme; SNS, staphylococcal nuclease; LZ4, T4 lysozyme; PAP, papain; TLS, thermolysin; TRX, thioredoxin; FLN, flavodoxin; ADH, alcohol dehydrogenase coenzyme domain; AKN, adenyl kinase; MDH, malate dehydrogenase; TIM, triosephosphate isomerase; SUB, subtilisin; CPA, carboxypeptidase; LDH, lactate dehydrogenase; PGK, phosphoglycerate kinase; GPD, glyceraldehyde 3-phosphate dehydrogenase; HKN, hexokinase.

14

1. *Secondary structure.* The secondary structure of a segment of polypeptide chain is the local spatial arrangement of its main-chain atoms, without regard to the conformation of its side chains or to its relationship with other segments. Sometimes the secondary structure is periodic in nature, giving rise to well-defined structures such as α helixes or β sheets.

2. *Tertiary structure.* The arrangement in space of all the atoms in a single polypeptide chain or in covalently linked chains is termed the tertiary structure.

3. *Quaternary structure.* Many proteins are oligomers composed of subunits that are not covalently linked together. The overall organization of the subunits is known as the quaternary structure. This refers to the arrangement of the subunits in space and the ensemble of subunit contacts and interactions, without regard to the internal geometry of the subunits. The three-dimensional structure of each subunit is still referred to as its tertiary structure. A change in quaternary structure means that the subunits move relative to each other.

The subunits may be identical polypeptide chains or they may be chemically different. As a general rule, at the contact interfaces the subunits are as closely packed as the interiors of proteins and hydrogen-bonding groups, and ions are paired on these surfaces.[21] The change of quaternary structure in hemoglobin involves a shift from one set of close-packed interfaces to another.

4. Prediction of three-dimensional structure from primary structure

One of the goals of enzymology is the prediction of the three-dimensional structure and the catalytic activity of an enzyme from its amino acid sequence. Now that we are able to synthesize proteins via their genes (Chapter 14), the production of new enzymes with specified catalytic activities is a challenging prospect.

The first step, the prediction of structure, involves two distinct tasks: the description of (1) the final folded state and (2) the kinetic pathway of folding. In theory, the calculation of both is possible from a knowledge of all the interaction energies involved. In practice, this is beyond present methods of computation.[22]

Direct computation of protein structure from interaction energies is feasible for small perturbations from a *known* three-dimensional structure. If a structure is known with a fair degree of precision from protein crystallography, then the coordinates of the atoms may be refined by calculation using energy functions. Further, fluctuations in the structure caused by the vibrations of the atoms about their mean positions may be calculated from the energy functions and Newton's equations of motion: this is the technique of *molecular dynamics*.[23] A possible strategy for structure

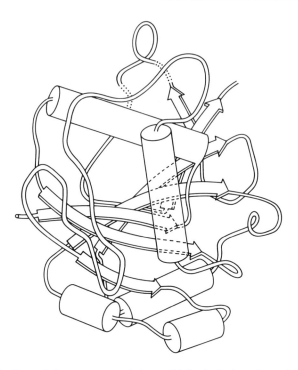

FIGURE 1·10. Part of the structure of glyceraldehyde 3-phosphate dehydrogenase from *Bacillus stearothermophilus,* showing the assembly from the helixes (cylinders) and parallel β sheet (arrows). Note the twist in the β sheet. (Courtesy of G. Biesecker and A. Wonacott.)

prediction is thus the derivation of methods for predicting approximate three-dimensional structure, followed by computer refinement.

One approach for deriving the approximate structure has been the computer simulation of folding by using simplified potential energy functions for the interactions. The renaturation of a very small protein (pancreatic trypsin inhibitor) from its unfolded state has been simulated by minimizing its energy in successive stages of folding.[24] This approach is still very much in its infancy.

Another method is to avoid any explicit calculation of energy terms or of the pathway of folding, and to concentrate just on describing elements of structure collated from the many known examples from protein crystallography. A productive approach has been based on the observation described in section 3, that proteins consist of stacked α helixes and β sheets (see Figure 1·9).[16] Rules are being established for how these groups of secondary structure interact. For example, one helix may pack on another by the mutual intercalation of their side chains, so that a "ridge" of projecting side chains on one helix fits into a "groove" between rows of side chains in the other.[25] The stacking of an α helix on a β sheet to give an α/β structure nearly always involves a parallel β sheet.[26] The main region of contact is an even, nonpolar surface made up of side chains of valine, leucine, and isoleucine residues in the β sheet. Antiparallel sheets can pack face to face.[27] There are few polar interactions across the surface. The β sheet is invariably twisted (Figure 1·10).[28]

Certain recurring structural motifs have been observed and a new hierarchy of substructures has been developed.[29,30] For example, the structures of most larger proteins ($M_r > 20\,000$) appear to be composed of independently folded globular units called *domains*. The domains often contain smaller, frequently recurring substructures or folds collectively termed *supersecondary structure*.[31] These are combinations of α and/or β structure.

A complementary approach is the prediction of secondary structure based on statistical analyses of known structures and on experiments with synthetic polymers.[32] Certain amino acids have a propensity for forming α helixes; others, for destabilizing them. Proline cannot be incorporated into a helix without destabilizing it. Despite much effort, however, the algorithms have been developed only to the stage where about 60% of the residues in a protein may be successfully predicted to belong to a particular secondary structure that forms an α helix, a β sheet, or a β bend.[33]

C. Enzyme diversity

There are over 1500 known enzymes. This raises some questions: How many different types of structure do we have to consider, and how did such a large number evolve on earth in a relatively short period of time? Some astrophysicists have queried the idea of evolution on the grounds that the chances of so many structures arising at random are vanishingly small. Fortunately, there are simplifying features that aid in the classification of structures and point to ways that different proteins could have been generated in nature via short cuts.

1. Divergent evolution of families of enzymes

a. The serine proteases[34–40]

The initial excitement of the discovery that the two oxygen-binding proteins hemoglobin and myoglobin have a common tertiary structure as well as a common function was rekindled when the same was found to be true of the mammalian serine proteases. The enzymes are so called because they have a uniquely reactive serine residue that reacts irreversibly with organophosphates such as diisopropyl fluorophosphate. The major pancreatic enzymes—trypsin, chymotrypsin, and elastase—are kinetically

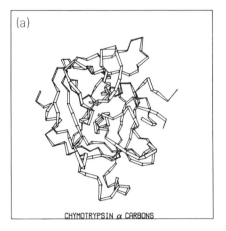

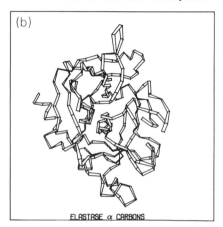

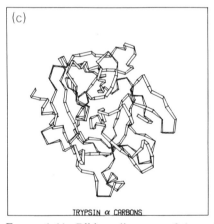

FIGURE 1·11. Ribbon diagrams of the polypeptide chains of (a) chymotrypsin, (b) elastase, and (c) trypsin. The α carbons are at the pleats in the ribbon. The small differences among the three structures occur in the external loops. (Courtesy of J. Smith.)

very similar, catalyzing the hydrolysis of peptides and synthetic ester substrates. Their activities peak at around pH 7.8 and fall off at low pH with a pK_a of about 6.8. In all three cases, the reaction forms an "acyl-enzyme" through esterification of the hydroxyl of the reactive serine by the carboxyl portion of the substrate.

The major difference between the three enzymes is specificity. Trypsin is specific for the peptides and esters of the amino acids lysine and arginine; chymotrypsin for the large hydrophobic side chains of phenylalanine, tyrosine, and tryptophan; elastase for the small hydrophobics, such as alanine. When the structures of the crystalline enzymes were solved, it was found that the polypeptide backbones of all three are essentially superimposable (Figure 1.11), apart from some small additions and deletions in the chain. The difference in their specificities is due to just a few changes in a pocket that binds the amino acid side chain. There is a well-defined binding pocket in chymotrypsin for the large hydrophobic side chains.[41] In trypsin, the residue at the bottom of the pocket is an aspartate instead of the Ser-189 of chymotrypsin.[42] The negatively charged carboxylate of Asp-189 forms a salt linkage with the positively charged ammonium or guanidinium at the end of the side chain of lysine or arginine. In elastase, the two glycines at the mouth of the pocket in chymotrypsin are replaced by a bulky valine (Val-216) and threonine (Thr-226).[43] This prevents the entry of large side chains into the pocket, and provides a way of binding the small side chain of alanine (Figure 1·12).

The remarkable similarity of all three tertiary structures could not have been guessed in advance from a comparison of their sequences. There is extensive homology among their primary structures, but only about 50% of the sequences of elastase and chymotrypsin are composed of amino acids that are chemically identical or similar to those in trypsin.[40] The crystallographic results now tell us, of course, that this level is highly significant. By this token, it is expected that the plasma serine proteases that occur in the blood clotting cascade system also have very similar tertiary structures (Table 1·3). Closer examination of the sequence homologies shows that 60% of the amino acids in the interior are conserved, but only 10% of the surface residues. The major differences occur in exposed areas and external loops.

A totally unexpected feature was found in the crystal structure of chymotrypsin. It was known from solution studies that the imidazole ring

TABLE 1·3. *Sequence homologies in mammalian serine proteases*[a]

Enzyme	% homology
Pancreas	
Trypsin	100
Chymotrypsin-A	53
Chymotrypsin-B	49
Elastase	48
Plasma	
Thrombin	38
Factor Xa	50

[a] % chemical similarity of residues in sequence compared with bovine trypsin. The enzymes are all bovine except for the elastase, which is porcine. [From B. S. Hartley, *Symp. Soc. Gen. Microbiol.* **24**, 152 (1974).]

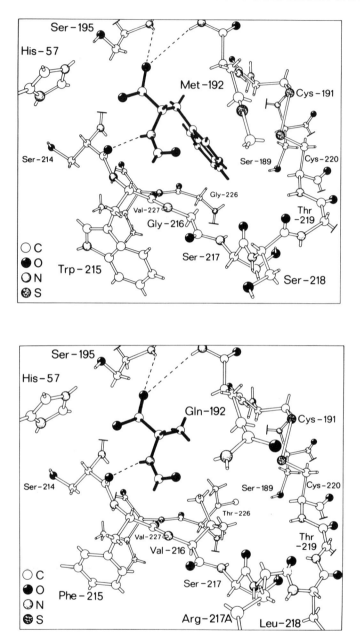

FIGURE 1·12. Comparison of the binding pockets in (a) chymotrypsin, with *N*-formyl-L-tryptophan bound, and (b) elastase, with *N*-formyl-L-alanine bound. The binding pocket in trypsin is very similar to that in chymotrypsin, except that residue 189 is an aspartate to bind positively charged side chains. Note the hydrogen bonds between the substrate and the backbone of the enzyme.

of His-57 increases the reactivity of Ser-195. But there was no hint that the histidine is hydrogen-bonded to the carboxylate of Asp-102 to form a catalytic triad now called the "charge relay system":[44]

$$\begin{array}{cc} & (1\cdot3) \end{array}$$

This structure has subsequently been found in all serine proteases.

The mammalian serine proteases appear to represent a classic case of *divergent evolution*. All were presumably derived from a common ancestral serine protease.

Subsequently, some nonmammalian serine proteases were shown to be 20 to 50% homologous with their mammalian counterparts (Table 1·4). We now know that this suggests very similar tertiary structure. The crystal structure of the elastase-like protease from *Streptomyces griseus* has been solved, and—despite its having only 186 amino acids in its sequence, compared with 245 in α-chymotrypsin—it is found to have two-thirds of the residues in a conformation similar to those in the mammalian enzymes.[45] Possibly, these bacterial enzymes and the pancreatic ones have evolved from a common precursor. But the evolutionary relationships are not clear.

b. The carboxypeptidases[46,47]

Carboxypeptidase A catalyzes the hydrolysis of the C-termini of proteins. It is specific for the hydrophobic side chains of phenylalanine, tyrosine, and tryptophan. Carboxypeptidase B is specific for the positively charged side chains of lysine and arginine. The two carboxypeptidases are related in the same way as chymotrypsin and trypsin. Forty-nine percent of their sequences are identical. Differences in their tertiary structures are con-

TABLE 1·4. *Species differences in serine proteases*[a]

Enzyme	Species	% homology
Trypsin	Cow	100
	Dogfish	69
	Streptomyces griseus	43
Elastase	Pig	48
	Myxobacter sorangium	26
	S. griseus	~20
Subtilisin	*Bacillus subtilis*	0
	Bacillus amyloliquifaciens	0

[a] From Table 1·3 and L. T. J. Delbaere, W. L. B. Hutcheon, M. N. G. James, and W. E. Thiessen, *Nature, Lond.* **257**, 758 (1975).

fined to the external regions. The major difference is that Ile-255 in the binding pocket of the A form becomes Asp-255 in the B form, to bind the positively charged side chains of the basic substrates.

2. Convergent evolution

The first crystal structure of a bacterial serine protease to be solved, subtilisin from *Bacillus amyloliquifaciens,* revealed an enzyme of apparently totally different construction from the mammalian serine proteases.[48] This was not unexpected, since there is no sequence homology between them. But closer examination shows that they are functionally identical as far as substrate binding and catalysis are concerned. Subtilisin has the same charge relay system, the same system of hydrogen bonds for binding the carbonyl oxygen and the acetamido NH of the substrate, and the same series of subsites for binding the acyl portion of the substrate as have the mammalian enzymes (see Figure 1·12 and section D2). This appears to be a case of *convergent evolution.* Different organisms, starting from different tertiary structures, have evolved a common mechanism.

Another example of convergent evolution is that of the endopeptidase thermolysin from *Bacillus thermoproteolyticus* and the carboxypeptidases.[49,50] There are no sequence or structural homologies except that the active sites are very similar, containing in each case a catalytically important Zn^{2+} ion. The enzymes consequently appear to have similar catalytic mechanisms.[50]

3. Convergence or divergence?

Our intuition tells us that the mammalian serine proteases have evolved through divergence, but that their common catalytic mechanism with subtilisin has developed through convergence. Other cases are not so clear cut. The accepted procedure for distinguishing between convergence and divergence is to count the number of common characteristics.[51] If there are many, then divergence is more likely; if there are few, convergence is the logical interpretation. This has been translated into molecular terms to provide the following six criteria for testing whether two proteins have evolved from a common precursor:[51]

1. The DNA sequences of their genes are homologous.
2. Their amino acid sequences are homologous.
3. Their three-dimensional structures are homologous.
4. Their enzyme–substrate interactions are homologous.
5. Their catalytic mechanisms are similar.
6. The segments of polypeptide chain essential for catalysis are in the same sequence (i.e., not transposed).

These criteria are in descending order of strength. If 1 and 2 hold, the rest will follow, in most—but not all—cases. Note that in the serine proteases, tertiary structure is more conserved than primary.

Lysozyme (section D3) from hen egg white and lysozyme produced by the bacteriophage T4 have no detectable homologies in their amino acid sequences. Yet, by showing that criteria 3 to 6 hold, a strong case has been made for their divergent evolution from a common precursor.[51,52]

Sometimes structure has been conserved through evolution but function has changed; that is, criteria 3 and 4 do not hold. For example, the binding protein haptoglobin appears to have diverged from the serine proteases.[53,54]

4. Dehydrogenases and domains[55]

One would expect that the NAD^+-dependent dehydrogenases, a class of enzymes that have the same chemical function and that bind the same cofactor, would form a structurally related family. They appear to do this, but not in the same clear-cut way as do the serine proteases. The tertiary structures of the first two crystal structures to be solved, dogfish lactate dehydrogenase and soluble porcine malate dehydrogenase, were found to be almost superimposable, if the first 20 residues of lactate dehydrogenase are discounted. It seems likely that the two dehydrogenases have evolved from a common precursor dehydrogenase. But the subsequent solution of the structures of horse liver alcohol dehydrogenase and lobster glyceraldehyde 3-phosphate dehydrogenase complicated the picture, because they are extensively different. However, it was noticed that the structure of each of the four enzymes consists of two domains, one of which is similar in all four. This domain binds the NAD^+ (Figure 1·13).

The nucleotide-binding fold is a complicated structure which differs in detail from one dehydrogenase to another. In its idealized form it consists of six strands of parallel β sheet with four parallel helixes running antiparallel to the sheet. This structure occurs in other nucleotide-binding proteins, such as phosphoglycerate kinase. But it is also found in proteins, such as flavodoxin, that are not involved in nucleotide binding. It is not known whether these similarities are evidence for a common evolutionary precursor protein, or whether they are caused by there being only a limited number of ways of folding a polypeptide chain.

5. Evolution of proteins by fusion of gene fragments?

The evolution of proteins by a random choice of individual amino acids seems a most unlikely process. For one thing, the number of different sequences possible for a protein containing 250 amino acid residues (e.g.,

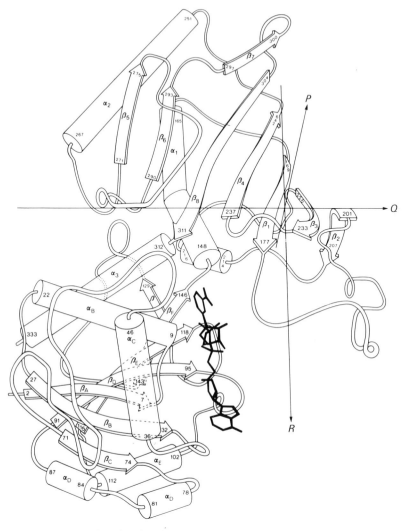

FIGURE 1·13. The domains in glyceraldehyde 3-phosphate dehydrogenase from *B. stearothermophilus*. [From G. Biesecker, J. I. Harris, J. C. Thierry, J. E. Walker, and A. Wonacott, *Nature, Lond.* **266**, 328 (1977).]

trypsin or chymotrypsin) is 20^{250} or 2×10^{325}. However, the domain structure of the dehydrogenases points to a very simple means of generating a family of enzymes with diverse specificity. Suppose that there is a gene coding for the dinucleotide-binding domain. Then the fusion of this gene with a series of genes each coding for a separate "catalytic domain" could have generated the family of dehydrogenases.[56]

This theory was reinforced by the discovery that the structural genes of eukaryotic (e.g., higher) organisms are often interspersed with extraneous segments of DNA that do not code for amino acid residues in the protein. The genes of prokaryotes (e.g., bacteria) consist of DNA containing contiguous sequences of nucleotide triplets, each coding for an amino acid (Chapter 14). In eukaryotes, the coding sections, termed *exons*, may be separated by insertion elements, termed *introns*. When the DNA is transcribed, the introns are excised from the mRNA to give a contiguous sequence of exons that is translated into the protein. It has been suggested that the exons could correspond to functional units of proteins, and that new proteins could have evolved by the recombination of different exons.[57] Circumstantial evidence is consistent with this idea.[58] For example, hen egg white lysozyme DNA contains four exons. Exon 2 has the catalytic center of the enzyme, and could perhaps have been a primitive glycosidase. Exon 3 codes for substrate-orienting residues. The lysozyme from bacteriophage T4 (section C3) is prokaryotic and is thus not derived from DNA containing introns. The catalytic center of the molecule appears, however, to be equivalent to the fusion of exons 2 and 3 of the eukaryotic enzyme.[59]

M. Go has used a diagonal plot to search for compact structural units, which she calls "modules," in proteins. There is a good correspondence between these units in hen egg white lysozyme and hemoglobin and the exons in their genes.[60] The modules are linked in lysozyme by disulfide bridges, but Go has suggested that the folding of each module is independent and that the bridges are not essential for the individual folding.

A related phenomenon is the repeat of a domain along the same polypeptide chain, as found in the immunoglobulins. Such repeats could have evolved by the fusion of two copies of the same gene.

The evolution of proteins in a modular fashion by fusion of segments of genes, each coding for a module of a compact structural unit of polypeptide, is thus a credible and attractive hypothesis for explaining the rapid generation of enzyme diversity.

D. The structure of enzyme–substrate complexes

The outstanding characteristic of enzyme catalysis is that the enzyme specifically binds its substrates, with the reactions taking place in the confines of the enzyme–substrate complex. Thus, to understand how an enzyme works, we need to know not only the structure of the native enzyme, but also the structures of the complexes of the enzyme with its substrates, intermediates, and products. Once these have been determined, we can see how the substrate is bound, what catalytic groups are close to the substrate, and what structural changes occur in the substrate and the enzyme on binding. There is one obvious difficulty: enzyme–

substrate complexes react to give products in fractions of a second, whereas the acquisition of x-ray data usually takes several hours. For this reason, it is usual to determine the structures of the complexes of enzymes with the reaction products, inhibitors, or substrate analogues.

1. Methods for examining enzyme–substrate complexes

a. The difference Fourier method

Protein crystals generally contain about 50% solvent, and never less than 30%. Often there are channels of solvent leading from the exterior of the crystal to the active site, so that substrates may be diffused into the crystals and bound to the enzyme. Alternatively, it may be possible to co-crystallize the enzyme and substrates. Provided that there are no drastic changes in the structure or packing of the enzyme when it binds the ligand, it is possible to solve the structure of the complex by the difference Fourier technique. This involves measuring the *differences* between the diffraction patterns of the native crystals and those soaked in a solution of the ligand. The electron density of the bound ligand and any minor changes in the structure of the enzyme may be obtained without solving the whole structure from scratch.

b. Production of stable complexes

The first attempts to determine the structure of a productively bound enzyme–substrate complex were based on extrapolation from the structures of stable enzyme–inhibitor complexes. (The classic example of this, lysozyme, is discussed in section D3.) Such extrapolation may be done in several ways. For example, a portion of the substrate may be bound to the enzyme, and the structure of the remainder determined by model building. An alternative method is to use a substrate analogue that is unreactive because its reactive bond is modified. A good example of this is the binding of a phosphonate, rather than a phosphate, to ribonuclease (Chapter 15).

Methods are also available in some cases for the direct study of productively bound enzyme–substrate complexes under unreactive conditions. These include the use of a substrate that is weakly reactive, either because it is at a pH at which the enzyme is largely inactive due to a residue being in the wrong ionic state (e.g., indolylacryloyl-chymotrypsin, section D2), or because it is at a very low temperature (e.g., an elastase acylenzyme, section D2; and ribonuclease, Chapter 15). A chemically modified enzyme may also be used: an example that has yet to be studied is lysozyme in which the —CO_2H of the catalytically important residue Asp-52 is converted to a —CH_2OH (section D3 and Chapter 15).

It is also possible on occasion to examine a productively bound enzyme–substrate complex directly under rapid reaction conditions. This happens when an equilibrium may be set up between substrates and products in such a way that the substrates are favored. Adding the enzyme to the equilibrium mixture cannot alter the position of the solution equilibrium, so it should be possible to obtain a stable enzyme–substrate complex. However, there is no guarantee that the equilibrium position is the same for the enzyme-bound reagents as it is for the solution (Chapter 3). One such equilibrium occurs in the hydrolysis of a dipeptide, which *can* be forced to favor synthesis by adding excess amine:[61]

$$RCONHR' \rightleftharpoons RCO_2^- + H^+ + R'NH_2 \qquad (1\cdot4)$$

This reaction has also been used for NMR experiments.[62] Other examples are the binding of dihydroxyacetone phosphate to triosephosphate isomerase, and the formation of ternary complexes of alcohol dehydrogenase (Chapter 15). These enzymes catalyze a simple equilibrium between two reagents. In this context it should be remembered that an enzyme–product complex is also an enzyme–substrate complex: that for the reverse reaction.

These methods are discussed in more detail in Chapter 15. However, the serine proteases and lysozyme are also discussed below because of their historical importance in the development of the ideas and techniques involved, and because they are referred to extensively in this text.

2. Example 1: The serine proteases

Peptide and synthetic ester substrates are hydrolyzed by the serine proteases by the acylenzyme mechanism (equation 1·5).[63] The enzyme and substrate first associate to form a noncovalent *enzyme–substrate* complex held together by physical forces of attraction. This is followed by the attack of the hydroxyl of Ser-195 on the substrate to give the *tetrahedral intermediate*. The intermediate then collapses to give the *acylenzyme*, releasing the amine or alcohol. The acylenzyme then hydrolyzes to form the *enzyme–product* complex. Crystallographic studies have been performed to give the structures of most of the complexes. The rest can be

$$
E{-}OH + RCONHR' \rightleftharpoons E{-}OH{\cdot}RCONHR' \rightleftharpoons E{-}O{-}\overset{\overset{\displaystyle O^-}{|}}{\underset{\underset{\displaystyle NHR'}{|}}{C}}{-}R \rightleftharpoons
$$

$$
E{-}OCOR \rightleftharpoons E{-}OH{\cdot}RCO_2H \rightleftharpoons E{-}OH + RCO_2H \qquad (1\cdot5)
$$
$$
\underset{\displaystyle NH_2R'}{+}
$$

obtained by model building. Working backwards, the structure of the enzyme–product complex, N-formyl-L-tryptophan chymotrypsin (see Figure 1·12), has been solved by diffusing the product into the crystal and then using the difference Fourier technique.[64] The structure of the nonspecific acylenzyme indolylacryloyl-chymotrypsin was solved at pH 4, where it is deacylated only very slowly.[65] The substrate, indolylacryloylimidazole, an activated derivative of the acid, was diffused into the crystal, where it acylated Ser-195. The structure of carbobenzoxyalanyl-elastase has been solved at $-55°C$.[66,67]

Nature has provided a rare opportunity for determining the structures of the enzyme–substrate complexes of trypsin and chymotrypsin with polypeptides. There are many naturally occurring polypeptide inhibitors that bind to trypsin and chymotrypsin very tightly because they are locked into the conformation that a normal flexible substrate takes upon binding.[68] They do not hydrolyze under normal physiological conditions because the amino group that is released on the cleavage of the peptide is constrained and cannot diffuse away from the active site of the enzyme. On removing the constraints in the pancreatic trypsin inhibitor by reducing an —S—S— bridge in its polypeptide chain, the peptide bond between Lys-15 and Ala-16 is readily cleaved by trypsin.[69] The structures of trypsin, its complex with the basic pancreatic trypsin inhibitor complex, and the free inhibitor (Figure 1·14) have been solved at resolutions of 1.4, 1.9, and 1.7 Å, respectively.[70] These are among the most accurate determinations yet done, so critical atomic positions are known to about 0.1 to 0.2 Å. Such studies have given the following information about the binding of substrates.[71-81]

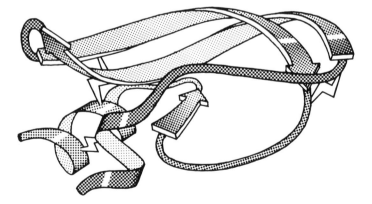

FIGURE 1·14. Pancreatic trypsin inhibitor. (Courtesy of Jane Richardson.)

a. The binding site

The binding site for a polypeptide substrate consists of a series of *subsites* across the surface of the enzyme. By convention they are labeled as in structure 1·6. The substrate residues are called P (for peptide); the subsites, S. Except at the primary binding site S_1 for the side chains of the aromatic substrates of chymotrypsin or the basic amino acid substrates of trypsin, there is no obvious well-defined cleft or groove for substrate binding. The subsites run along the surface of the protein.

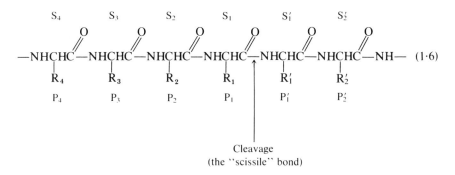

Cleavage
(the "scissile" bond)

b. The primary binding site (S_1)

The binding pocket for the aromatic side chains of the specific substrates of chymotrypsin is a well-defined slit in the enzyme 10 to 12 Å deep and 3.5 to 4 by 5.5 to 6.5 Å in cross section.[64] This gives a very snug fit, since an aromatic ring is about 6 Å wide and 3.5 Å thick. A methylene group is about 4 Å in diameter (Chapter 11), so the side chain of lysine or arginine is bound nicely by the same-shaped slit in trypsin. Of course, as mentioned in section C1a, there is a carboxylate at the bottom of the pocket in trypsin to bind the positive charge on the end of the side chain, and also the mouth of the pocket is blocked somewhat in elastase.

The binding pocket in chymotrypsin may be described as a "hydrophobic pocket," since it is lined with the nonpolar side chains of amino acids. It provides a suitable environment for the binding of the nonpolar or hydrophobic side chains of the substrates. The physical causes of hydrophobic bonding and its strength are discussed in Chapter 11.

The hydrophobic binding site in subtilisin is not a well-defined slit as in the pancreatic enzymes, but more like a shallow groove. However, certain hydrogen bonds are found for all the enzymes (see Figure 1·12): The carbonyl oxygen of the reactive bond has a binding site between the backbone NH groups of Ser-195 and Gly-193. The hydrogen bonds made there are very important because the oxygen becomes negatively charged during the reaction. There is also a hydrogen bond between the NH part of the *N*-acylamino group of the substrate and the C=O of Ser-214.

c. Sites S_1—S_2—S_3

The hydrogen bond between the *N*-acylamino NH and the carbonyl of Ser-214 initiates a short region of antiparallel β sheet between the residues Ser-214, Trp-215, and Gly-216 of the enzyme and the amino acids P_1, P_2, and P_3 of the substrate (structure 1·7).

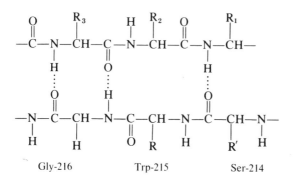

$$(1 \cdot 7)$$

d. Site S_1'—The leaving group site

There is a leaving group site that is constructed to fit L-amino acids.[81] The contacts with the enzyme are predominantly hydrophobic, which accounts for the lack of exopeptidase activity with the enzyme, since this would require binding a —CO_2^- in a nonpolar region. ("Leaving group" is chemists' jargon for a group displaced from a molecule—in this case from an acyl group.)

3. Example 2: Lysozyme

Lysozyme catalyzes the hydrolysis of a polysaccharide that is the major constituent of the cell wall of certain bacteria. The polymer is formed from β(1 → 4)-linked alternating units of *N*-acetylglucosamine (NAG) and *N*-acetylmuramic acid (NAM) (Figure 1·15). The solution of the structure of lysozyme is one of the triumphs of x-ray crystallography.[82–84] Whereas solving the crystal structures of the serine proteases represented the culmination of a long series of studies stretching back through the history of enzymology, the structure of the little-known enzyme lysozyme stimulated the solution studies. Also, the mechanism of the enzyme, which was previously unknown, was guessed from examining the crystal structure of the native enzyme and the complexes with inhibitors.

Unlike chymotrypsin, lysozyme has a well-defined deep cleft, running down one side of the ellipsoidal molecule, for binding the substrate. This cleft is partly lined with nonpolar side chains of amino acids for binding the nonpolar regions of the substrate, and it also has hydrogen-bonding sites for the acylamino and hydroxyl groups. The cleft is divided into six sites, A, B, C, D, E, and F. NAM residues can bind only in B, D, and

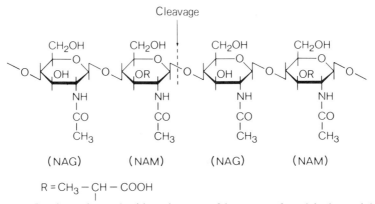

FIGURE 1·15. The polysaccharide substrate of lysozyme found in bacterial cell walls.

F, whereas NAG residues of synthetic substrates may bind in all sites. The bond that is cleaved lies between sites D and E. There is no means at present of solving the structure of a productively bound enzyme–substrate complex in the same way as for chymotrypsin and trypsin, since there are no similar natural inhibitors. The method of determining the structure of lysozyme complexes was based on model building, which has been a typical approach. The structure of the enzyme and the inhibitor (NAG)₃ was solved first.[83] This is a very poor substrate of the enzyme, since the structural studies show that it is bound nonproductively in the A, B, and C sites, avoiding the D and E sites where cleavage takes place. The structure of a productively bound complex was then obtained by building a wire model of the complex of (NAG)₃ with the enzyme and extending the polysaccharide chain by adding further NAG units, using chemical intuition about the contacts made with the enzyme. Nowadays, this procedure is done by using a computer program to optimize the fit between the enzyme and the substrate. It was found that the bond that is cleaved is located between the carboxyl groups of Glu-35 and Asp-52 (which were later proven to be in the un-ionized and ionized forms, respectively). How these contribute to the mechanism will be discussed in Chapter 15.

Small distortions occur in the enzyme structure at the active-site binding cleft when the inhibitor is bound. The enzyme is said to undergo a small *conformational change*.

To summarize, the binding sites of lysozyme and the serine proteases are approximately complementary in structure to the structures of the substrates: the nonpolar parts of the substrate match up with nonpolar side chains of the amino acids; the hydrogen-bonding sites on the substrate bind to the backbone NH and CO groups of the protein and, in the case of lysozyme, to the polar side chains of amino acids. The reactive part of the substrate is firmly held by this binding next to acidic, basic, or nucleophilic groups on the enzyme.

E. Flexibility and conformational mobility of proteins

Although globular proteins are generally close-packed, they do have certain degrees of flexibility. Some enzymes are known to undergo conformational changes on binding ligands or substrates. These changes may be small, as with lysozyme, or they may involve large movements. It will be seen in Chapter 10 that such changes are important in a certain class of enzymes (allosteric) for modulating activity: certain ligands (allosteric effectors) can alter the shape of the protein. This clearly affects the status of crystal structures, since, although protein crystals contain about 50% water, the lattice forces constrain the structure of the protein. The following question is thus often raised.

1. Are the crystal and solution structures of an enzyme essentially identical?

The answer is, in general, yes. The identity of the crystal and solution structures is supported by the following evidence:

1. Some enzymes (e.g., subtilisin[85,86] and lysozyme[87]) have been crystallized from different solvents and in different forms, but their structures remain essentially identical.
2. Enzymes in families (e.g., the serine proteases) have structures that are similar.
3. Areas of α-chymotrypsin that are in contact within the dimer in solution are also in contact within the dimer in the crystal.[88,89]
4. In many cases (e.g., ribonuclease[90,91] and carboxypeptidase[92]), the crystal retains enzymatic activity.
5. In some cases, the activity of the enzyme in the crystal is the same as that in solution. It is hard to compare activities in solution and in the crystalline state because of the difficulty of diffusing the substrate into the crystal. But in one ingenious example this problem was overcome by crystallizing the acylenzyme indolylacryloyl-chymotrypsin at a pH at which it is stable. On changing the pH to increase the reactivity, the intermediate was found to hydrolyze with the same first-order rate constant as occurs in solution.[93] Glyceraldehyde 3-phosphate dehydrogenase has also been shown to have identical reaction rates in solution and in the crystal, under certain conditions.[94]

NMR studies of globular proteins in solution give data consistent with the crystal structure, as has been amply demonstrated for lysozyme.[95,96]

However, where there is an equilibrium among two or more conformations of the enzyme in solution, crystallization may select out only one of the conformations. α-Chymotrypsin has a substantial fraction of an inactive conformation present under the conditions of crystallization, but only the active form of the enzyme crystallizes.[97] An allosteric effector molecule that changes the conformation of the protein in solution may have no effect on the crystalline protein, as has been found, for example, with phosphorylase b.[98] The enzyme is frozen in one conformation, with the crystal lattice forces preventing any conformational change. On the other hand, the addition of an effector to phosphorylase a causes the crystals first to crack and then to anneal, giving crystals of the enzyme in a second conformation.[99]

2. Modes of motion and flexibility observed in proteins[100,101]

a. Molecular tumbling

Globular proteins are found to rotate in solution at frequencies close to those calculated for rigid spheres. The frequencies are usually expressed in terms of a *rotational correlation time*, ϕ, which is the reciprocal of the rate constant for the randomization of the orientation of the molecule by Brownian motion. For a rigid sphere, ϕ is given by

$$\phi = \frac{V\eta}{kT} \tag{1.8}$$

where V is the molecular volume, η is the viscosity of the medium, k is Boltzmann's constant, and T is the absolute temperature. Substituting the values of η, k, and T, and using an approximate relationship between the relative molecular mass M_r of a globular protein and V, gives an approximation that holds at ambient temperatures:

$$\phi \approx \frac{M_r}{2000} \tag{1.9}$$

The rate of molecular tumbling affects the shapes of lines observed in EPR and NMR spectra. Slow tumbling causes the line widths to broaden. In conventional EPR spectroscopy, the lines undergo a transition from narrow to broad as the correlation time increases through the nanosecond time region. This enables correlation times of proteins of appropriate size to be calculated from the shape of the lines. For example, chymotrypsin ($M_r = 25\,000$) that has been spin-labeled has a measured correlation time of 12 ns,[102] which is close to that expected from equations 1.8 and 1.9. In conventional NMR spectroscopy, on the other hand, the lines broaden indefinitely as the correlation time increases. The consequence is that for correlation times greater than 20 ns or so (i.e., $M_r > \sim40\,000$), the lines begin to broaden to such an extent that they cannot be readily analyzed. A useful corollary of this is that the presence of sharp NMR (or EPR) lines in the spectra of large molecules with long correlation times indicates that the groups being observed have their own independent mobility.

Molecular tumbling also affects fluorescence polarization. Fluorescence (Chapter 6) occurs when a photon is absorbed by a molecule and re-emitted at a longer wavelength. The lifetime of the excited state varies from 1 to 20 ns or so, depending on the nature of the fluorophore. If a fluorescent molecule is excited by a pulse of plane-polarized light, then the degree of polarization of the emitted light (the *anisotropy*) will decay exponentially according to the rotational correlation time. Through use of time-resolved fluorescence polarization spectroscopy and excitation of the tryptophan residue in staphylococcal nuclease B ($M_r = 20\,000$) and in serum albumin ($M_r = 69\,000$), rotational correlation times of 9.9 ns and 31.4 ns, respectively, were measured;[103] these are close to those calculated for the rigid spheres. Just as the presence of sharp lines in NMR spectra implies the existence of independent mobility, so the finding of a correlation time for a fluorophore shorter than that expected for the overall molecular tumbling implies that the fluorophore has additional mobility.

b. *"Breathing"*
The compact globular regions of proteins have structural fluctuations that have been observed by a variety of techniques. The accessibility of backbone NH groups to solvent has been traditionally measured by rates of isotopic exchange. The amide hydrogen atoms exchange with tritium from ^3H_2O or deuterium from D_2O in an acid- or base-catalyzed reaction. The rates were originally measured by radioactive incorporation or by analysis of infrared spectra.[104,105] These methods have been superseded by ^1H NMR, which can measure the rate of exchange of *individual* protons of assigned groups in the structure. The exposed backbone NH hydrogen atoms exchange rapidly. The majority of the buried NH groups also exchange, albeit at lower rates, despite their apparently not being exposed to solvent. There is strong evidence that exchange occurs in native structures by a "breathing" of the protein molecule that allows solvent to penetrate without a mandatory unfolding of the protein (although unfolding does provide a further pathway). Perhaps the nicest demonstration of this comes from measurements of the exchange of H by D in crystalline myoglobin in neutron diffraction studies. Soaking the crystals in D_2O leads to the exchange of 95% of the amide hydrogen atoms under conditions in which the protein cannot possibly unfold because of the constraints of the crystal lattice.[106]

Another useful probe of exposure is the quenching of fluorescence by the direct collision of the excited fluorophore with a solute molecule.

It has been found that all the tryptophan residues, both exposed and buried, in a series of proteins are quenched by dissolved oxygen. The protein matrix is penetrated by O_2 at 25 to 50% of the rate of diffusion of O_2 in water.[107] This implies fluctuations in the protein structure on the nanosecond time scale. Computer simulations of the vibrational modes of the basic trypsin inhibitor calculated by molecular dynamics predict such modes.[108]

c. Segmental flexibility and "hinge bending"
Some segments of proteins are flexibly attached to the bulk of the molecule. The higher mobility of such segments compared with the overall structure has been shown by time-resolved fluorescence polarization spectroscopy. For example, in an immunoglobulin molecule, a part that is attached to the major portion by a "hinge" of polypeptide (the Fab fragment) has a measured correlation time of 33 ns, compared with 168 ns for the overall structure.[109]

X-ray diffraction studies of crystalline hexokinase (see Figure 12·10) show that two lobes of the protein move together on binding a substrate in a hinge bending motion.[110] Part of the pyruvate dehydrogenase complex (section F2) is very flexible, as shown by sharp NMR lines.[111]

d. Rotation of side chains
As well as providing a means of measuring H/D exchange in proteins, NMR is a most powerful technique for studying the mobility of individual amino acids. For example, the rotational freedom of the aromatic side chains of tyrosine and phenylalanine about the C^β—C^γ bond is readily studied by various NMR methods. ^{1}H NMR can detect whether or not the aromatic ring is constrained in an anisotropic environment.[112-114] In an isotropic environment or where there is rapid rotation on the NMR time scale, the 3 and 5 protons of phenylalanine and tyrosine are symmetrically related, as are the 2 and 6 (structures 1·10). The resultant spectrum is of the AA′BB′ type, containing two pairs of closely separated

(1·10)

doublets. However, if there is slow rotation in an anisotropic environment, the symmetry breaks down to give four separate resonances (an ABCD spectrum), since the 5 and 6 protons are in different states from the 2 and 3. At an intermediate time range for rotation, the two spectra coalesce and the lines are said to undergo *exchange broadening*. This occurs when the rate constant k for the exchange between the two states is approximately equal to $\nu_A - \nu_B$, where ν_A is the NMR resonance frequency in one state and ν_B is its counterpart in the other state. Exchange broadening may be observed and analyzed to give k when $\nu_A - \nu_B$ is in the region 10 to 10^4 Hz, i.e., for $k = 10$ to 10^4 s^{-1}. Thus, "slow" rotation means less than 1 to 10 s^{-1}, and "fast" rotation means greater than 10^4 to 10^5 s^{-1}. By this means, it has been shown that three of the four phenylalanine and three of the four tyrosine residues in the basic pancreatic trypsin inhibitor are rotating at greater than 100 Hz.[115,116] Yet, all four phenylalanine and three of the tyrosine residues are buried in the interior of the protein. The remaining one tyrosine and one phenylalanine ring in the trypsin inhibitor are immobilized. This distribution of immobilized and rotating aromatic rings is quite typical. The measured rotation frequencies for the trypsin inhibitor are in accord with calculations of the rotation barriers from molecular dynamics.[117]

More recently, ^{15}N, ^{13}C, and ^{2}H quadrupole-interaction NMR of immobilized samples of a protein (the coat protein of bacteriophage fd) have confirmed that the motion of the aromatic side chains is best described as a series of *jumps* or 180° flips at greater than 10^6 Hz, rather than as a continuous rotation.[118,119] In contrast to the frequent flipping of the aromatic side chains of buried tyrosine residues, the side chains of buried tryptophan residues are usually immobile.[101,119] The difference between them may be rationalized when it is realized that the benzene ring is close to being cylindrical because of the thickness of the π-electron cloud, whereas the indole ring of tryptophan is quite asymmetric along the axis of the C^β—C^γ bond. The tryptophan side chain can act as a rigid platform for the construction of hydrophobic regions.

Surface amino acids are more mobile than interior ones, and many surface side chains have no unique conformation.[95]

e. Static disorder and real motion in crystals from x-ray diffraction studies
Although protein crystallography has traditionally been used to give only a static picture of protein structure, it is now proving to be a powerful technique for studying the mobility of every residue in a protein chain. This is because in addition to locating the positions of atoms from their electron density, it is also possible to measure their *mean square displacement* from the smearing out of the electron density. It has been noticed in certain structures, such as in trypsinogen, that some regions of the electron density are very weak.[120] In other words, the structure in that region is not uniquely defined in space during the time of the experiment. This can be due to either *static disorder* (i.e., the atoms are occupying a number of different conformations in space) or *real motion* (i.e., the atoms are rapidly vibrating about their mean positions). It is

possible in theory to distinguish between these modes by studying the dependence of the mean displacement on temperature. The apparent mean amplitudes of true vibrations increase with increasing temperature in accordance with normal thermal motions, whereas those of static disorder do not change. A pioneering study has been conducted on sperm whale myoglobin at four temperatures between 200 and 300 K.[121] As shown in Figure 1·16, the greatest displacement of the backbone occurs at the C-terminus. The greatest average root mean square displacements of side chains are about 0.4 to 0.5 Å, for the charged chains on the surface. Buried nonpolar side chains have average values of 0.2 to 0.25 Å. The core of the protein is, in general, more rigid than the outer regions. Examination of the periodicities of the amplitudes suggests that not only do the helixes move as rigid units, but they also experience breathing or rippling modes of vibration. The experimental data are in broad agreement with predictions from molecular dynamics calculations.[25] A study of lysozyme crystals has revealed that the residues of highest apparent motion are in the active-site region where the conformational change is observed on substrate binding: specifically, in the lips of the binding crevice.[122]

In summary, then, small movements and distortions of proteins are possible, especially at the surface, because of their inherent flexibility.

f. Protein mobility and enzyme mechanism
A major question is: Are the modes of mobility observed in enzymes just incidental? That is, are they simply an inherent property of proteins that must always be borne in mind, or are they essential for catalysis? A crude analogy of the problems involved may be made by comparing the properties of a grandfather clock and a digital quartz watch. In the grandfather clock, the large swing of the pendulum is all-important for regulating the time, while the vibrations of the atoms in the pendulum rod are incidental. However, in that the vibrations of the atoms are responsible for the thermal expansion of the rod, they do exert a second-order effect on the accuracy of the clock. On the other hand, in the quartz watch, the vibrations are the central mechanism. Enzymes often resemble the grandfather clock. The larger movements, such as hinge bending, are important for specific biological processes such as muscle contraction (Chapter 7), for specific purposes such as the "swinging-arm" mechanism for the pyruvate dehydrogenase complex (section F2), and for the structural changes in allosteric enzymes. The importance of the vibrational and breathing modes and the internal rotations is not clear. It has not been demonstrated to date that the vibrational modes can be coupled with the chemical steps of catalysis to enhance reaction rates. However, these modes do contribute to the flexibility of the protein as regards distortion of its structure. This is relevant to theories of enzyme catalysis such as "strain" (Chapter 12) and the limits of "lock and key" specificity (Chap-

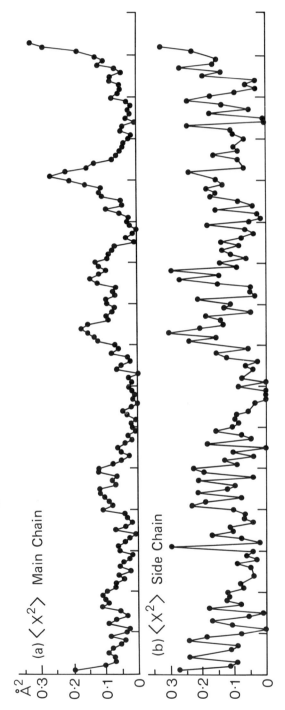

FIGURE 1·16. Sum of the conformational and vibrational mean square displacements against the residue number of myoglobin. (a) Average displacement of main chain N, C, and Cα. (b) Largest displacement in each side chain. [From H. Frauenfelder, G. A. Petsko, and D. Tsernoglou, *Nature, Lond.* **280**, 558 (1979).]

ter 13)—theories that depend on whether an enzyme can distort a substrate or *vice versa*.

Flexibility could be useful in aiding the access of ligands to active sites. The binding of glucose to hexokinase may be a good example: in order for the enzyme to bind the whole molecule, the protein structure has to open and close. The binding of O_2 to myoglobin and hemoglobin appears to require some movement of the protein to allow access to the heme. Flexibility of binding sites could also affect the rate and equilibrium constants for the binding of ligands.

One set of examples in which some flexibility of protein structure does appear necessary is in electron transfer, where the catalyzing protein—for example, a cytochrome—is reversibly oxidized and reduced. It is essential for the structure to be able to relax in order to be able to accommodate the changes of geometry around the iron atom as it changes its oxidation state. In this case, the protein is one of the reactants. The protein could not be rigid, since the intermediates have different geometries. This latter point must apply to some extent to any enzymatic reaction involving covalent intermediates with the enzyme.

F. Higher levels of organization: Multienzyme complexes

Certain enzymes, notably those involved in sequential steps in a biosynthetic pathway or those involved in a complex biochemical process, are organized into physical aggregates. These vary in size and complexity from the simple association of two enzymes, through multienzyme complexes, to large units such as the ribosome. Well-documented examples are tryptophan synthetase and the 2-oxoacid dehydrogenase complexes discussed below, and the fatty acid synthetases.[123] There is evidence that the glycolytic enzymes in *Escherichia coli* are physically associated into a large complex.[124] Eight of the twenty aminoacyl-tRNA synthetases are isolated as a large complex from rat liver.[125] The group of diverse enzymes responsible for the priming of DNA replication in *E. coli* appears to be organized into a "primosome," with each enzyme having its defined role in the team. This example has been likened to a train. One member is the "engine," which consumes ATP as fuel.[126] Some of the enzymes of the citric acid cycle appear to associate.[127]

1. Double-headed enzymes and the noncovalent association of different activities

The enzyme tryptophan synthetase is an $\alpha_2\beta_2$ tetramer (i.e., it contains two pairs of identical chains).[128] In DNA, the α chain (M_r = 29 000) and the β chain (M_r = 44 000) are the two most distal genes of the tryptophan operon. (An operon is a set of consecutive genes that may be induced or repressed as a group. The entire set of enzymes of a biosynthetic pathway is frequently encoded in this way so that the pathway can be turned on or off as a whole.) The tetrameric enzyme catalyzes the following reaction:

$$\text{Indole 3-glycerol phosphate} + \text{serine} \xrightarrow{\alpha_2\beta_2}$$
$$\text{tryptophan} + \text{glyceraldehyde 3-phosphate} \quad (1\cdot11)$$

The enzyme can be resolved into active α and β_2 subunits that separately catalyze the partial reactions $1\cdot12$ and $1\cdot13$, respectively. The overall reaction $(1\cdot11)$ is the sum of the two:

$$\text{Indole 3-glycerol phosphate} \xrightarrow{\alpha}$$
$$\text{indole} + \text{glyceraldehyde 3-phosphate} \quad (1\cdot12)$$

$$\text{Indole} + \text{serine} \xrightarrow{\beta_2} \text{tryptophan} \quad (1\cdot13)$$

There is evidence that the intermediate (indole) is shuttled directly between the subunits in the complex and is not released into solution.[129]

Besides having a noncovalent association of subunits as in tryptophan synthetase, some enzymes are *double-headed,* in that they contain two distinct activities in a single polypeptide chain. A good example of this is the aspartokinase I-homoserine dehydrogenase I enzyme, which catalyzes two steps in the biosynthesis of threonine from aspartic acid. The enzyme is an α_4 tetramer of M_r $4 \times 86\ 000$. Each polypeptide chain contains both activities, the aspartokinase at the N-terminus portion and the homoserine dehydrogenase at the C-terminus. The activities may be separated by proteolysis. Digestion with subtilisin gives a globular fragment ($M_r = 55\ 000$) that contains the dehydrogenase activity. The N-terminus, containing the aspartokinase activity, may be isolated from a mutant that produces only a portion of the chain ($M_r = 47\ 000$). The enzyme thus appears to consist of two globular, independently folded portions that are covalently joined.[130,131] This could have arisen by the fusion of two genes that were originally independent.

The *arom* complex of *Neurospora crassa* has five enzymatic activities on a single polypeptide chain. These are responsible for successive steps in the skikimic acid pathway.[132]

2. The pyruvate dehydrogenase complex

There are two 2-oxoacid dehydrogenase multienzyme complexes in *E. coli.* One is specific for pyruvate, the other for 2-oxoglutarate.[123,133-135] Each complex is about the size of a ribosome, about 300 Å across. The pyruvate dehydrogenase is composed of three types of polypeptide chains: E1, the pyruvate decarboxylase (a dimer of $M_r = 2 \times 100\ 000$); E2, lipoate acetyltransferase ($M_r = 80\ 000$); and E3, lipoamide dehydrogenase (a dimer of $M_r = 2 \times 56\ 000$). These catalyze the oxidative decarboxylation of pyruvate via reactions $1\cdot14$, $1\cdot15$, and $1\cdot16$. [The relevant chemistry of the reactions of thiamine pyrophosphate (TPP), hydroxyethylthiamine pyrophosphate (HETPP), and lipoic acid (lip-S_2) is discussed in detail in Chapter 2, section C3.]

$$\text{Pyruvate} + \text{TPP}-\text{E1} \longrightarrow \text{HETPP}-\text{E1} + CO_2 \tag{1·14}$$

$$\text{HETPP}-\text{E1} + (\text{lip-}S_2)-\text{E2} \longrightarrow \text{TPP}-\text{E1} + (\text{acetyl-S-lip-SH})-\text{E2} \tag{1·15}$$

$$(\text{Acetyl-S-lip-SH})-\text{E2} + \text{CoA} \longrightarrow [\text{lip-}(\text{SH})_2]-\text{E2} + \text{acetyl-CoA} \tag{1·16}$$

The complex is constructed around a core of 24 E2 molecules arranged in octahedral symmetry. The lipoic acid residues that transfer the important acetyl group are attached to E2 on the ϵ-NH_2 groups of two lysine residues per chain. This enables the lipoates to be on a swinging arm of radius 14 Å.[133,136] The correlation time of an arm that has been spin-labeled is 0.2 ns, compared with 0.01 ns for the residue in solution and 10 μs for the rotational correlation time of the total complex,[137] indicating the free rotation of an acetylated chain. ^{1}H NMR measurements show that the polypeptide chain of E2 is itself very mobile.[111,135] This increases the spatial freedom of the swinging arm and so facilitates the rapid transacetylation from one E2 chain to another.[133,138] By this mechanism, a single E1 dimer can acetylate the lipoic acid residues of possibly 12 E2 chains.

3. Reasons for multiple activities and multienzyme complexes

In some cases there are fairly obvious and specific reasons for the association of different activities of the complex structures. For example, as discussed in Chapters 13 and 14, certain DNA polymerases have a 3' → 5' hydrolytic activity in addition to the polymerization activity. This is an error-correcting mechanism in which hydrolysis must be coupled with synthesis. For a complex series of reactions as in the "primosome," there are so many different activities that they require an organized system—rather than random pathways—of association and dissociation of the constituent enzymes from the point of priming of DNA replication.

The following list of possible advantages of multienzyme complexes over individual activities in solution has also been proposed:

1. *Catalytic enhancement:* The reduction of diffusion time of an intermediate from one enzyme to the next.
2. *Substrate channeling:* The control over which biosynthetic route an intermediate should follow, by directing it to a specified enzyme rather than allowing competition from other enzymes in solution.
3. *Sequestration of reactive intermediates:* The protection of chemically unstable intermediates from aqueous solution.
4. *"Servicing"*: The rapid intramolecular acetyl transfer reactions that are observed in the pyruvate dehydrogenase complex.

References

1 J. B. S. Haldane, *Enzymes*, Longmans, Green and Co. (1930). M.I.T. Press (1965).
2 D. W. Green, V. Ingram, and M. F. Perutz, *Proc. R. Soc.* **A225**, 287 (1954).
3 S. E. V. Phillips and B. P. Schoenborn, *Nature, Lond.* **292**, 81 (1981).
4 A. A. Kossiakoff and S. A. Spencer, *Biochemistry* **20**, 642 (1981).
5 L. Pauling, *The nature of the chemical bond*, Cornell University Press (1960).
6 A. R. Fersht, *J. Am. Chem. Soc.* **93**, 3504 (1971).
7 J. F. Brandts, H. R. Halvorson, and M. Brennan, *Biochemistry* **14**, 4953 (1975).
8 C. Grathwohl and K. Wütrich, *Biopolymers* **15**, 2025 (1976).
9 G. N. Ramachandran and A. K. Mitra, *J. Molec. Biol.* **107**, 85 (1976).
10 M. Levitt, *J. Molec. Biol.* **145**, 251 (1981).
11 M. Jullien and R. L. Baldwin, *J. Molec. Biol.* **145**, 265 (1981).
12 C. M. Venkatachalam, *Biopolymers* **6**, 1425 (1968).
13 P. N. Lewis, F. A. Momany, and H. A. Scheraga, *Proc. Natn. Acad. Sci. USA* **68**, 2293 (1971).
14 I. D. Kuntz, *J. Am. Chem. Soc.* **94**, 4009 (1972).
15 J. L. Crawford, W. N. Lipscomb, and C. G. Schellman, *Proc. Natn. Acad. Sci. USA* **70**, 538 (1973).
16 M. Levitt and C. Chothia, *Nature, Lond.* **261**, 552 (1976).
17 I. D. Kuntz, *J. Am. Chem. Soc.* **94**, 8568 (1972).
18 M. H. Klapper, *Biochim. Biophys. Acta* **229**, 557 (1971).
19 F. M. Richards, *J. Molec. Biol.* **82**, 1 (1974).
20 C. Chothia, *Nature, Lond.* **254**, 304 (1975).
21 C. Chothia and J. Janin, *Nature, Lond.* **256**, 705 (1975).
22 G. Némethy and H. A. Scheraga, *Quart. Rev. Biophys.* **10**, 289 (1977).
23 J. A. McCammon, B. R. Gelin, and M. Karplus, *Nature, Lond.* **267**, 585 (1977).
24 M. Levitt and A. Warshel, *Nature, Lond.* **253**, 694 (1975).
25 C. Chothia, M. Levitt, and D. Richardson, *J. Molec. Biol.* **145**, 215 (1981).
26 J. Janin and C. Chothia, *J. Molec. Biol.* **143**, 95 (1980).
27 F. E. Cohen, M. J. E. Sternberg, and W. R. Taylor, *J. Molec. Biol.* **148**, 253 (1981).
28 C. Chothia, *J. Molec. Biol.* **75**, 295 (1973).
29 M. J. E. Sternberg, F. E. Cohen, W. R. Taylor, and R. J. Feldman, *Phil. Trans. R. Soc.* **B293**, 177 (1981).
30 M. G. Rossman and P. Argos, *Ann. Rev. Biochem.* **50**, 497 (1981).
31 S. T. Rao and M. G. Rossman, *J. Molec. Biol.* **76**, 241 (1973).
32 P. Y. Chou and G. D. Fasman, *Ann. Rev. Biochem.* **47**, 251 (1978).
33 W. Kabsch and C. Sander, *FEBS Lett.* **155**, 179 (1983).
34 G. P. Hess, *The Enzymes* **3**, 213 (1971).
35 B. Keil, *The Enzymes* **3**, 249 (1971).
36 S. Magnusson, *The Enzymes* **3**, 277 (1971).
37 B. S. Hartley and S. Magnusson, *The Enzymes* **3**, 323 (1971).
38 D. M. Blow, *The Enzymes* **3**, 185 (1971).
39 J. Kraut, *The Enzymes* **3**, 547 (1971).
40 B. S. Hartley, *Symp. Soc. Gen. Microbiol.* **24**, 151 (1974).
41 B. W. Matthews, P. B. Sigler, R. Henderson, and D. M. Blow, *Nature, Lond.* **214**, 652 (1967).

42 R. M. Stroud, L. M. Kay, and R. E. Dickerson, *J. Molec. Biol.* **83**, 185 (1974).
43 D. M. Shotton and H. C. Watson, *Nature, Lond.* **225**, 811 (1970).
44 D. M. Blow, J. J. Birktoft, and B. S. Hartley, *Nature, Lond.* **221**, 337 (1969).
45 L. T. J. Delbaere, W. L. B. Hutcheon, M. N. G. James, and W. E. Thiessen, *Nature, Lond.* **257**, 758 (1975).
46 F. A. Quiocho and W. N. Lipscomb, *Adv. Protein Chem.* **25**, 1 (1971).
47 M. F. Schmid and J. R. Herriott, *J. Molec. Biol.* **103**, 175 (1976).
48 C. S. Wright, R. A. Alden, and J. Kraut, *Nature, Lond.* **221**, 235 (1969).
49 W. R. Kester and B. W. Matthews, *Biochemistry* **16**, 2506 (1977); *J. Biol. Chem.* **252**, 7704 (1977).
50 P. Argos, R. M. Garavito, W. Eventoff, M. G. Rossman, and C.-I. Brändén, *J. Molec. Biol.* **126**, 141 (1978).
51 B. W. Matthews, S. J. Remington, M. G. Grutter, and W. F. Anderson, *J. Molec. Biol.* **147**, 545 (1981).
52 M. G. Rossman and P. Argos, *J. Molec. Biol.* **105**, 75 (1976).
53 A. Kurosky, D. R. Barnett, T. H. Lee, B. Touchstone, R. E. Hay, M. S. Arnott, B. H. Bowman, and W. M. Fitch, *Proc. Natn. Acad. Sci. USA* **77**, 3388 (1980).
54 J. Greer, *Proc. Natn. Acad. Sci. USA* **77**, 3393 (1980).
55 M. G. Rossman, A. Liljas, C.-I. Brändén, and L. J. Banaszak, *The Enzymes*, **11**, 61 (1975).
56 M. G. Rossman, D. Moras, and K. W. Olsen, *Nature, Lond.* **250**, 194 (1974).
57 W. Gilbert, *Nature, Lond.* **271**, 501 (1978).
58 C. C. F. Blake, *Trends Biochem. Sci.* **8**, 11 (1983).
59 P. J. Artymiuk, C. C. F. Blake, and A. E. Sippel, *Nature, Lond.* **290**, 287 (1981).
60 M. Go, *Proc. Natn. Acad. Sci. USA* **80**, 1964 (1983).
61 A. R. Fersht and M. Renard, *Biochemistry* **13**, 1416 (1974).
62 G. Robillard, E. Shaw, and R. G. Shulman, *Proc. Natn. Acad. Sci. USA* **71**, 2623 (1974).
63 J. Fastrez and A. R. Fersht, *Biochemistry* **12**, 2025 (1973).
64 T. A. Steitz, R. Henderson, and D. M. Blow, *J. Molec. Biol.* **46**, 337 (1969).
65 R. Henderson, *J. Molec. Biol.* **54**, 341 (1970).
66 A. L. Fink and A. I. Ahmed, *Nature, Lond.* **263**, 294 (1976).
67 T. Alber, G. A. Petsko, and D. Tsernoglou, *Nature, Lond.* **263**, 297 (1976).
68 M. Laskowski, Jr. and R. W. Sealock, *The Enzymes* **3**, 375 (1971).
69 K. A. Wilson and M. Laskowski, Sr., *J. Biol. Chem.* **246**, 3555 (1971).
70 M. Rigbi, *Proceedings of the International Conference on Proteinase Inhibitors* (H. Fritz and H. Tschesche, Eds.), Walter de Gruyter, Berlin, p. 117 (1971).
71 W. Bode, P. Schwager, and R. Huber, *Proceedings of the Tenth FEBS Meeting* **40**, 3 (1975).
72 R. M. Sweet, H. T. Wright, J. Janin, C. M. Chothia, and D. M. Blow, *Biochemistry* **13**, 4212 (1974).
73 J. D. Robertus, J. Kraut, R. A. Alden, and J. J. Birktoft, *Biochemistry* **11**, 4293 (1972).
74 J. Kraut, J. D. Robertus, J. J. Birktoft, R. A. Alden, P. E. Wilcox, and J. C. Powers, *Cold Spring Harb. Symp. Quant. Biol.* **36**, 117 (1971).
75 D. M. Shotton, N. J. White, and H. C. Watson, *Cold Spring Harb. Symp. Quant. Biol.* **36**, 91 (1971).

76   D. M. Segal, G. H. Cohen, D. R. Davies, J. C. Powers, and P. E. Wilcox, *Cold Spring Harb. Symp. Quant. Biol.* **36**, 85 (1971).

77   D. Atlas, S. Levit, I. Schechter, and A. Berger, *FEBS Lett.* **11**, 281 (1970).

78   A. Gertler and T. Hofmann, *Can. J. Biochem.* **48**, 384 (1970).

79   R. C. Thompson, *Biochemistry* **13**, 5495 (1974).

80   S. A. Bizzozero, W. K. Baumann, and H. Dutler, *Eur. J. Biochem.* **122**, 251 (1982).

81   A. R. Fersht, D. M. Blow, and J. Fastrez, *Biochemistry* **12**, 2035 (1973).

82   C. C. F. Blake, D. F. Koenig, G. A. Mair, A. C. T. North, D. C. Phillips, and V. R. Sarma, *Nature, Lond.* **206**, 757 (1965).

83   D. C. Phillips, *Proc. Natn. Acad. Sci. USA* **57**, 484 (1967).

84   C. C. F. Blake, L. N. Johnson, G. A. Mair, A. C. T. North, D. C. Phillips, and V. R. Sarma, *Proc. R. Soc.* **B167**, 378 (1967).

85   J. D. Robertus, R. A. Alden, and J. Kraut, *Biochem. Biophys. Res. Commun.* **42**, 334 (1971).

86   J. Drenth, W. G. T. Hol, J. N. Jansonius, and R. Koekoek, *Cold Spring Harb. Symp. Quant. Biol.* **36**, 107 (1971).

87   J. Moult, A. Yonath, W. Traub, A. Smilansky, A. Podjarny, D. Rabinovich, and A. Saya, *J. Molec. Biol.* **100**, 179 (1976).

88   C. S. Hexter and F. H. Westheimer, *J. Biol. Chem.* **246**, 3928 (1971).

89   A. R. Fersht and J. Sperling, *J. Molec. Biol.* **74**, 137 (1973).

90   M. S. Doscher and F. M. Richards, *J. Biol. Chem.* **238**, 2399 (1963).

91   J. Bello and E. F. Nowoswiat, *Biochim. Biophys. Acta* **105**, 325 (1965).

92   C. A. Spilburg, J. L. Bethune, and B. L. Vallee, *Proc. Natn. Acad. Sci. USA* **71**, 3922 (1974).

93   G. L. Rossi and S. A. Bernhard, *J. Molec. Biol.* **49**, 85 (1970).

94   M. Vas, R. Berni, A. Mozzarelli, M. Tegoni, and G. L. Rossi, *J. Biol. Chem.* **254**, 8480 (1979).

95   C. M. Dobson, in *NMR in biology* (R. A. Dwek, I. D. Campbell, R. E. Richards, and R. J. P. Williams, Eds.), Academic Press, London, p. 63 (1977).

96   C. C. F. Blake, R. Cassels, C. M. Dobson, F. M. Poulsen, R. J. P. Williams, and K. S. Wilson, *J. Molec. Biol.* **147**, 73 (1981).

97   A. R. Fersht and Y. Requena, *J. Molec. Biol.* **60**, 279 (1971).

98   L. N. Johnson, E. A. Stura, K. S. Wilson, M. S. P. Sansom, and I. T. Weber, *J. Molec. Biol.* **134**, 639 (1979).

99   N. B. Madsden, P. J. Kasvinsky, and R. J. Fletterick, *J. Biol. Chem.* **253**, 9097 (1978).

100   F. R. N. Gurd and T. M. Rothgeb, *Adv. Protein Chem.* **33**, 74 (1979).

101   R. J. P. Williams, *Biol. Rev.* **54**, 389 (1979).

102   E. J. Shimshick and H. M. McConnell, *Biochem. Biophys. Res. Commun.* **46**, 321 (1972).

103   I. Munro, I. Pecht, and L. Stryer, *Proc. Natn. Acad. Sci. USA* **76**, 56 (1979).

104   K. Linderström-Lang, *Chem. Soc. Spec. Publ.* **2**, 1 (1955).

105   S. W. Englander, N. W. Downder, and H. Teitelbaum, *Ann. Rev. Biochem.* **41**, 903 (1972).

106   J. C. Norvell, A. C. Nunes, and B. P. Schoenborn, *Science* **190**, 568 (1975).

107   J. R. Lakowicz and G. Weber, *Biochemistry* **12**, 4161, 4171 (1973).

108   M. Karplus and J. A. McCammon, *C.R.C. Crit. Rev. Biochem.* **9**, 293 (1981).

109   J. Yguerabide, H. F. Epstein, and L. Stryer, *J. Molec. Biol.* **51**, 573 (1970).

110   T. A. Steitz, M. Shaoham, and W. S. Bennett, Jr., *Phil. Trans. R. Soc.* **B293**, 43 (1981).

111 R. N. Perham and G. C. K. Roberts, *Biochem. J.* **199**, 733 (1981).
112 I. D. Campbell, C. M. Dobson, and R. J. P. Williams, *Proc. R. Soc.* **B189**, 503 (1975).
113 G. H. Snyder, R. Rowan III, S. Karplus, and B. D. Sykes, *Biochemistry* **14**, 3765 (1975).
114 K. Wütrich and G. Wagner, *FEBS Lett.* **50**, 265 (1975).
115 G. Wagner, H. Tschesche, and K. Wütrich, *Eur. J. Biochem.* **95**, 239 (1979).
116 W. H. Hull and B. D. Sykes, *J. Molec. Biol.* **98**, 121 (1975).
117 B. R. Gelin and M. Karplus, *Proc. Natn. Acad. Sci. USA* **72**, 2002 (1978).
118 C. M. Gall, J. A. Diverdi, and S. J. Opella, *J. Am. Chem. Soc.* **103**, 5039 (1981).
119 C. M. Gall, T. A. Cross, J. A. Diverdi, and S. J. Opella, *Proc. Natn. Acad. Sci. USA* **79**, 101 (1982).
120 H. Fehlhammer, W. Bode, and R. Huber, *J. Molec. Biol.* **111**, 415 (1977).
121 H. Frauenfelder, G. A. Petsko, and D. Tsernoglou, *Nature, Lond.* **280**, 558 (1979).
122 P. J. Artymiuk, C. C. F. Blake, D. E. P. Grace, S. J. Oatley, D. C. Phillips, and M. J. E. Sternberg, *Nature, Lond.* **280,** 563 (1979).
123 L. J. Reed and D. J. Cox, *The Enzymes* **1**, 213 (1970).
124 D. M. Gorringe and V. Moses, *Internatn. J. Biol. Macromol.* **2**, 161 (1980).
125 C. V. Dang and D. C. H. Yang, *J. Biol. Chem.* **254**, 5350 (1979).
126 K.-I. Arai, R. L. Low, and A. Kornberg, *Proc. Natn. Acad. Sci. USA* **78**, 707 (1981).
127 J. P. Weber and S. A. Bernhard, *Biochemistry* **21**, 4189 (1982).
128 C. Yanofsky and I. P. Crawford, *The Enzymes* **7**, 1 (1972).
129 W. H. Matchett, *J. Biol. Chem.* **249**, 4041 (1974).
130 M. Veron, F. Falcoz-Kelly, and G. N. Cohen, *Eur. J. Biochem.* **28**, 520 (1972).
131 M. Veron, J. C. Saari, C. Villar-Palasi, and G. N. Cohen, *Eur. J. Biochem.* **38**, 325 (1973).
132 J. Lumsden and J. Coggins, *Biochem. J.* **161**, 599 (1977); **169**, 441 (1978).
133 L. J. Reed, *Accts. Chem. Res.* **7**, 40 (1974).
134 G. G. Hammes in *Mobility and migration of biological molecules* (P. B. Garland and R. J. P. Williams, Eds.), The Biochemical Society, London (*Biochem. Soc. Symp.* **46**), p. 73 (1981).
135 G. C. K. Roberts, H. W. Duckworth, L. C. Packman, and R. N. Perham, Ciba Foundation No. 93 (1982).
136 J. H. Collins and L. J. Reed, *Proc. Natn. Acad. Sci. USA* **74**, 4223 (1977).
137 M. C. Ambrose and R. N. Perham, *Biochem. J.* **155**, 425 (1976).
138 M. J. Danson, A. R. Fersht, and R. N. Perham, *Proc. Natn. Acad. Sci. USA* **75**, 5386 (1978).

Further reading

Physical methods

C. R. Cantor and P. R. Schimmel, *Biophysical chemistry*, W. H. Freeman and Company (1980).
O. Jardetzky and G. C. K. Roberts, *NMR in molecular biology*, Academic Press (1981).

Protein structure

J. S. Richardson, "The anatomy and taxonomy of protein structure," *Adv. Protein Chem.* **34**, 167 (1981).

M. G. Rossman and P. Argos, "Protein folding," *Ann. Rev. Biochem.* **50**, 497 (1981).

F. R. N. Gurd and T. M. Rothgeb, "Motions in proteins," *Adv. Protein Chem.* **33**, 74 (1979).

M. J. E. Sternberg, "The analysis and prediction of protein structure," in *Computing in biological science* (M. J. Geisow and A. N. Barrett, Eds.), Elsevier (1983).

D. C. Phillips, M. J. E. Sternberg, and B. J. Sutton, "Intimations of evolution from the three-dimensional structures of proteins," in *Evolution from molecules to men* (D. S. Bendall, Ed.), Cambridge University Press (1983).

2

Chemical catalysis

We know from structural and kinetic studies that enzymes have well-defined binding sites for their substrates, that they sometimes form co-valent intermediates, and that they generally involve acidic, basic, and nucleophilic groups. In this chapter we shall see why these features are necessary for catalysis and how they are used. The importance of the enzyme–substrate binding energy is dealt with in Chapter 12.

Many of the concepts in catalysis are based on *transition state theory*. Because an elementary knowledge of the theory greatly simplifies the understanding of some ideas and is essential for others, we begin with a discussion of its principles and applications. This is followed by an introduction to the basic principles of chemical catalysis and the factors responsible for the magnitude of enzyme catalysis. Subsequent sections deal with progressively more advanced topics in kinetics and solution catalysis, starting with the factors that determine chemical reactivity, such as what is a good nucleophile and what is a good leaving group in a particular reaction, and the analysis of structure–reactivity relationships.

A. Transition state theory[1-4]

There are several theories to account for chemical kinetics. The simplest is the collision theory, which will be used in Chapter 4 to calculate the rate constants for the collision of molecules in solution. A more sophisticated theory, one that is particularly useful for analyzing structure–reactivity relationships, is the transition state theory. The processes by which the reagents collide are ignored: the only physical entities considered are the reagents, or ground state, and the most unstable species on the reaction pathway, the *transition state*. The transition state occurs at the peak in the reaction coordinate diagram (Figure 2·1), in which the energy of the reagents is plotted as the reaction proceeds. In the transition state, chemical bonds are in the process of being made and broken. In contrast, *intermediates*, whose bonds are fully formed, occupy the troughs in the diagram. A simple way of deriving the rate of the reaction is to consider that the transition state and the ground state are in thermodynamic equilibrium, so that the concentration of the transition state may be calculated from the difference in their energies. The overall reaction rate is then obtained by multiplying the concentration of the transition state by the rate constant for its decomposition. This process is simpler than it sounds because the energy difference between the ground state and the transition state is used only qualitatively, and it may be shown that all transition states decompose at the same frequency for a given temperature.

The analysis for a unimolecular reaction is as follows. Suppose that the difference in Gibbs energy between the transition state, $X^{\ddagger}$, and the ground state, X, is $\Delta G^{\ddagger}$. Then, using a well-known relationship from equilibrium thermodynamics,[†] it is derived that

$$[X^{\ddagger}] = [X] \exp \left(- \frac{\Delta G^{\ddagger}}{RT} \right) \tag{2·1}$$

The frequency at which the transition state decomposes is the same as the vibrational frequency v of the bond that is breaking. This frequency is obtained from the equivalence of the energies of an excited oscillator calculated from quantum theory ($E = hv$) and classical physics ($E = kT$); that is,

$$v = \frac{kT}{h} \tag{2·2}$$

where k is the Boltzmann constant and h is the Planck constant. At 25°C, $v = 6.212 \times 10^{12} \ \mathrm{s}^{-1}$.

The rate of decomposition of X is thus given by

$$\frac{-d[X]}{dt} = v[X^{\ddagger}] \tag{2·3}$$

$$= [X]\left(\frac{kT}{h}\right) \exp \left(\frac{-\Delta G^{\ddagger}}{RT} \right) \tag{2·4}$$

[†] The two thermodynamic equations that are most useful for simple kinetic and binding experiments are: (1) the relationship between the Gibbs energy change and the equilibrium constant of a reaction,

$$\Delta G = -RT \ln K$$

where R is the gas constant and T is the absolute temperature, and (2) the relationship between the Gibbs energy change and the changes in the enthalpy and entropy,

$$\Delta G = \Delta H - T \Delta S$$

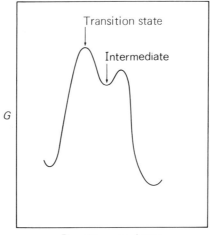

FIGURE 2·1. Transition states occur at the peaks of the energy profile of a reaction, and intermediates occupy the troughs.

The first-order rate constant for the decomposition of X is given by

$$k_1 = \left(\frac{kT}{h}\right) \exp\left(\frac{-\Delta G^{\ddagger}}{RT}\right) \qquad (2\cdot5)$$

The Gibbs energy of activation, $\Delta G^{\ddagger}$, may be separated into enthalpic and entropic terms, if required, by using another relationship from equilibrium thermodynamics,

$$\Delta G^{\ddagger} = \Delta H^{\ddagger} - T\Delta S^{\ddagger} \qquad (2\cdot6)$$

(where $\Delta H^{\ddagger}$ is the enthalpy, and $\Delta S^{\ddagger}$ the entropy, of activation). The rate constant becomes

$$k_1 = \left(\frac{kT}{h}\right) \exp\left(\frac{\Delta S^{\ddagger}}{R}\right) \exp\left(\frac{-\Delta H^{\ddagger}}{RT}\right) \qquad (2\cdot7)$$

(A more rigorous approach includes a factor known as the transmission coefficient, but this is generally close to 1 so it may be ignored.)

1. The significance and the application of transition state theory

The importance of transition state theory is that it relates the rate of a reaction to the difference in Gibbs energy between the transition state and the ground state. This is especially important for comparing the relative reactivities of pairs of substrates, or the rates of a given reaction under different sets of conditions. Under some circumstances the ratio of rates may be calculated; or, more generally, the trends in reactivity may be estimated qualitatively. For example, the alkaline hydrolysis of an ester, such as phenyl acetate, involves the attack of the negatively charged hydroxide ion on the neutral ground state. This means that in the transition state of the reaction some negative charge must be transferred to the ester. We can predict that *p*-nitrophenyl acetate will be more reactive than phenyl acetate, since the nitro group is electron-withdrawing and will stabilize the negatively charged transition state with respect to the neutral ground state. Consider also the spontaneous decomposition of *tert*-butyl bromide into the *tert*-butyl carbonium ion and the bromide ion. The transition state of this reaction must be dipolar. Therefore a polar solvent, such as water, will stabilize the transition state, and a nonpolar solvent, such as diethyl ether, will destabilize the transition state with respect to the ground state.

In Chapters 12 and 13 it will be seen how the transition state theory may be used quantitatively in enzymatic reactions to analyze structure reactivity and specificity relationships involving discrete changes in the structure of the substrate.

2. The Hammond postulate[5]

A useful guide in the application of transition state theory or in the analysis of structure–reactivity data is the Hammond postulate, which states that if there is an unstable intermediate on the reaction pathway, the transition state for the reaction will resemble the structure of this intermediate. The reasoning is that the unstable intermediate will be in a small dip at the top of the reaction coordinate diagram. This is a useful way of guessing the structure of the transition state for predicting the types of stabilization it requires. For example, a carboxonium ion is an intermediate in the

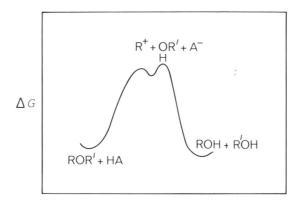

FIGURE 2·2. The transition state for the general-acid-catalyzed cleavage of an acetal resembles the carbonium ion intermediate that occupies the small "dip" at the top of the energy diagram.

reaction catalyzed by lysozyme (Figure 2·2). Since these are known to be unstable high-energy compounds, the transition state is assumed to resemble the structure of the carboxonium ion. One cannot really apply the Hammond postulate to bimolecular reactions. Because these involve two molecules condensing to form one transition state, a large part of the Gibbs energy change is caused by the loss in entropy (section B4). The Hammond postulate applies mainly to energy differences so it works best with unimolecular reactions. The postulate is sometimes extended to cases in which there are large energy differences between the reagents and products. If the products are very unstable, the transition state is presumed to resemble them; the same applies if the reagents are very unstable. However, the postulate is less reliable in these situations.

B. Principles of catalysis

1. Where, why, and how catalysis is required

In order to understand why enzymes are such efficient catalysts, it is necessary to understand first why uncatalyzed reactions in solution are so slow. As illustrations, we consider the reactions that may be catalyzed by chymotrypsin or lysozyme.

The uncatalyzed attack of water on an ester leads to a transition state in which a positive charge develops on the attacking water molecule, and a negative charge on the carbonyl oxygen (equation 2·8). The uncatalyzed

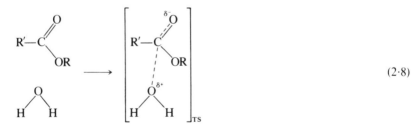

$$(2·8)$$

hydrolysis of an acetal involves a transition state that is close in structure to a carbonium ion and an alkoxide ion (equation 2·9).

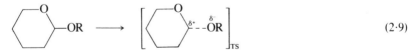

$$(2·9)$$

In both reactions the transition state is very unfavorable because of the unstable positive and negative charges that are developed. Stabilization of these charges catalyzes the reaction by lowering the energy of the transition state. Such stabilization can be achieved in the case of the positive charge developing on the attacking water molecule by transferring one of the protons to a base during the reaction. This is known as *general-base catalysis* (equation 2·10). Similarly, the negative charge de-

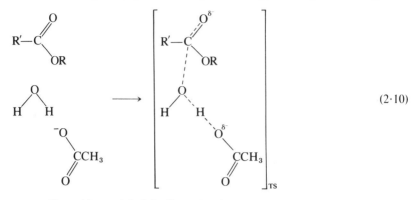

$$(2·10)$$

General-base catalysis by the acetate ion

veloping on the alcohol expelled from the acetal can be stabilized by proton transfer from an acid. This is known as *general-acid catalysis* (equation 2·11). The acid-base catalysis illustrated in these equations is

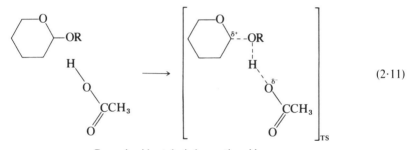

$$(2·11)$$

General-acid catalysis by acetic acid

termed *general* to distinguish it from *specific* acid or base catalysis in which the catalyst is the proton or hydroxide ion.

Positive and negative charges may also be stabilized by *electrostatic catalysis*. The positively charged carbonium ion cannot be stabilized by general-base catalysis because it does not ionize. But it can be stabilized by the electric field from a negatively charged carboxylate ion. The negative charge on an oxyanion may also be stabilized by a positively charged metal ion, such as Zn^{2+} or Mg^{2+}. The stabilization of a negative charge, i.e., an electron, is known as *electrophilic catalysis*.

The above types of catalysis function by stabilizing the transition state of the reaction without changing the mechanism. Catalysts may also involve a different reaction pathway. A typical example is *nucleophilic catalysis* in an acyl transfer or hydrolytic reaction. The hydrolysis of acetic anhydride is greatly enhanced by pyridine because of the rapid formation of the highly reactive acetylpyridinium ion (equation 2·12). For nucleo-

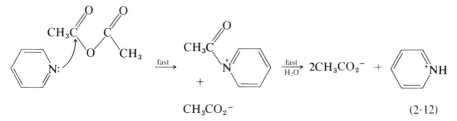

$$(2 \cdot 12)$$

philic catalysis to be efficient, the nucleophile must be more nucleophilic than the one it replaces, and the intermediate must be more reactive than the parent compound.

Nucleophilic catalysis is a specific example of *covalent catalysis*: the substrate is transiently modified by formation of a covalent bond with the catalyst to give a reactive intermediate. There are also many examples of electrophilic catalysis by covalent modification. It will be seen later that in the reactions of pyridoxal phosphate, Schiff base formation, and thiamine pyrophosphate, electrons are stabilized by delocalization.

There is a further important factor that is responsible for slowing down *multimolecular* reactions in solution. These require the bringing together of many molecules in the transition state. This is in itself an unfavorable event because it requires the right number of molecules simultaneously colliding in the correct orientation. The problem is exacerbated by acid-base or covalent catalysis, because even more molecules have to collide in the transition state. The magnitude of this factor is considered later, in the discussions on intramolecular catalysis and entropy (sections B3 and B4).

2. General-acid-base catalysis

a. Detection and measurement

The general-species catalysis of the hydrolysis of an ester is measured from the increase in the hydrolytic rate constant with increasing concentration of the acid or base. This is usually done at constant pH by maintaining a constant ratio of the acidic and basic forms of the catalyst. It is important to keep the ionic strength of the reaction medium constant because many reactions are sensitive to changes in salt concentration. In order to tell whether the catalysis is due to the acidic or basic form of the catalyst, it is necessary to repeat the measurements at a different buffer ratio. For ester hydrolysis, it is found that the increase in rate is generally proportional to the concentration of the basic form, so the catalysis is general-base. The slope of the plot of the rate constant against the concentration of base gives the second-order rate constant k_2 for the general-base catalysis (Figure 2·3).

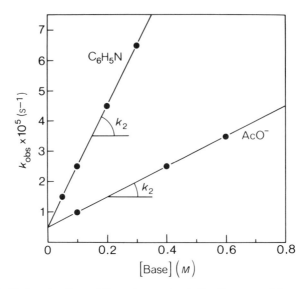

FIGURE 2·3. Determination of the rate constants for the general-base catalysis of the hydrolysis of ethyl dichloroacetate. The first-order rate constants for the hydrolysis are plotted against various concentrations of the base. The slope of the linear plot is the second-order rate constant (k_2). The intercept at zero buffer concentration is the "spontaneous" hydrolysis rate constant for the particular pH. A plot of the spontaneous rate constants against pH gives the rate constants for the H^+ and OH^- catalysis. It is seen that pyridine is a more effective catalyst than the weaker base acetate ion. [From W. P. Jencks and J. Carriuolo, *J. Am. Chem. Soc.* **83**, 1743 (1961).]

b. The efficiency of acid-base catalysis: The Brönsted equation[6]
It is found experimentally that the general-base catalysis of the hydrolysis of an ester is proportional to the basic strength of the catalyst (Figure 2·4).[7] The second-order rate constant k_2 for the dependence of the rate of hydrolysis on the concentration of the base is given by the equation

$$\log k_2 = A + \beta \, pK_a \tag{2·13}$$

Equation 2·13 is an example of the Brönsted equation. The β is known as the Brönsted β value. It measures the sensitivity of the reaction to the pK_a of the conjugate acid of the base. The A is a constant for the particular reaction.

    Brönsted equations are also common in general-acid catalysis, as, for example, in the hydrolysis of certain acetals.[8-10] In acid catalysis, α rather than β is used:

$$\log k_2 = A - \alpha \, pK_a \tag{2·14}$$

The values of α and β are always between 0 and 1 for acid-base catalysis

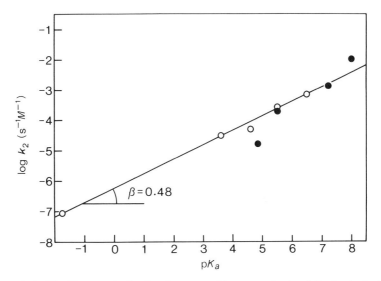

FIGURE 2·4. The Brönsted plot for the general-base catalysis of the hydrolysis of ethyl dichloroacetate. The logarithms of the second-order constants obtained from the plot of Figure 2·3 are plotted against the pK_a's of the conjugate acid of the catalytic base. The slope is the β value. Note that the points for amine bases (●) fall on the same line as those for oxyanion bases (○), showing that the catalysis depends primarily on the basic strength of the base and not on its chemical nature.

(except with some peculiar carbon acids),[11] because complete transfer of a proton gives a value of 1, and no transfer a value of 0. The usual values for ester hydrolysis are 0.3 to 0.5, and for acetal hydrolysis about 0.6.

Some values of the rate enhancement by general-base and general-acid catalysis are listed in Tables 2·1 and 2·2. These numbers are obtained from the Brönsted equation by using pK_a's of 15.74 and -1.74 for the ionization of H_2O and H_3O^+, respectively (i.e., from $[H^+][OH^-]$ =

TABLE 2·1. *Influence of β on general-base catalysis*

β	(Rate in 1-M solution of base) ÷ (rate in water)	
	$pK_a = 5$	$pK_a = 7$
0	1	1
0.3	2.9	8.6
0.5	44	427
0.7	951	2.4×10^4
0.85	9.7×10^3	4.9×10^5
1	1×10^5	1×10^7

TABLE 2·2. *Influence of α and state of ionization on general-acid catalysis*

α	(Rate in 1-*M* solution of acid) ÷ (rate in water)			(Rate in 1-*M* solution at pH 7) ÷ (rate in water)[a]		
	$pK_a = 5$	$pK_a = 7$	$pK_a = 9$	$pK_a = 5$	$pK_a = 7$	$pK_a = 9$
0	1	1	1	1	1	1
0.3	31	8.6	2.9	1.3	4.8	2.9
0.5	4.3×10^3	427	44	42	214	43.2
0.7	6×10^5	2.4×10^4	951	6×10^3	1.2×10^4	940
0.85	2.5×10^7	4.9×10^5	9.7×10^5	2.4×10^5	2.4×10^5	9.7×10^3
1	1×10^9	1×10^7	1×10^5	1×10^7	5×10^6	9.9×10^4

[a] The rate of a 1-*M* solution of both the acid and the base forms compared with the uncatalyzed water reaction. It should be noted that the proton becomes an efficient catalyst at higher values of α. For α = 0.3, 0.5, 0.7, 0.85, and 1.0, the reaction rate increases at pH 7 by factors of 1.0003, 2, 3×10^3, 1.3×10^6, and 5.5×10^8, respectively, thus swamping out the catalysis by other acids at the higher values.

10^{-14} M^2, and the concentration of water = 55 *M*). The magnitude of the catalysis depends strongly on α and β and the pK_a of the catalyst. In simple terms: *The stronger the base, the better the general-base catalysis; the stronger the acid, the better the general-acid catalysis.*

c. The efficiency of acid-base catalysis: The ionization state of the catalyst

A crucial factor in whether or not an acid-base catalyst is effective is whether or not it is in the correct ionization state under the reaction conditions: an acid has to be in acidic form to be an acid catalyst, and a base in its basic form. For example, an acid of pK_a 5 is a much better general-acid catalyst than one of pK_a 7, but at pH 7 only 1% of an acid of pK_a 5 is in the active acid form while the remaining 99% is ionized. An acid of pK_a 7 is only 50% ionized at pH 7 and still 50% active. Table 2·2 shows that for α = 0.85 or less, an acid of pK_a 7 is a better catalyst at pH 7 than is an acid of pK_a 5. Similarly, a base of pK_a 7 is a more effective catalyst than one of pK_a 9 at pH 7 (at β = 0.85 or less), due to the inherently more reactive base being mainly protonated at the pH below its pK_a. The most effective acid-base catalysts at pH 7 are those whose pK_a's are about 7. This accounts for the widespread involvement of histidine, with an imidazole pK_a of 6 to 7, in enzyme catalysis.

3. Intramolecular catalysis: The "effective concentration" of a group on an enzyme

Acid-base catalysis is seen to be an effective way of catalyzing reactions. We should now like to know the contribution of this to enzyme catalysis, but there is a fundamental problem in directly applying the results of the last section to an enzyme. The crux of the matter is that the rate constants for the solution catalysis are second-order, the rate increasing with increasing concentration of catalyst; whereas reactions in an enzyme–substrate complex are first-order, the acids and bases being an integral part of the molecule. So what is the concentration of the acid or base that is to be used in the calculations? The experimental approach is to synthesize model compounds with the catalytic group as part of the substrate molecule, and to compare the reaction rates with the corresponding intermolecular reactions.

A typical example of an *intramolecularly catalyzed* reaction is the hydrolysis of aspirin (equation 2·15).[12] The hydrolysis of the ester bond

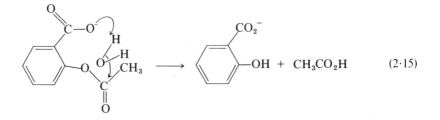

$$(2·15)$$

is achieved by intramolecular general-base catalysis. Comparison with the uncatalyzed hydrolysis rate of similar compounds gives a rate enhancement of some 100-fold due to the catalysis.[13] This may be extrapolated to a figure of 5000-fold if the pK_a of the base is 7 rather than the value of 3.7 in aspirin.

The intermolecular general-base catalysis of the hydrolysis may also be measured. Comparing the rate constants for this with those of the intramolecular reaction shows that a 13-M solution of an external base is required to give the same first-order rate as the intramolecular reaction has.[12] The "effective concentration" of the carboxylate ion in aspirin is therefore 13 M. This is a typical value for intramolecular general-acid-base catalysis.

The effective concentrations of nucleophiles in intramolecular reactions are often far higher than this. The examples that follow are for "unstrained" systems. The chemist can synthesize compounds that are strained; the relief of strain in the reaction then gives a large rate enhancement. In the succinate and aspirin derivatives that follow, the attacking nucleophile can rotate away from the ester bond to relieve any strain. The observed rate enhancements are due entirely to the high effective concentration of the neighboring group:

1. *Rates of acyl transfer in succinates:*[14]

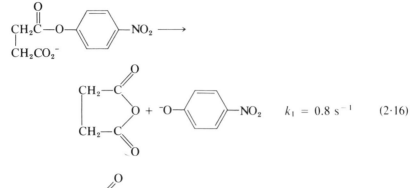

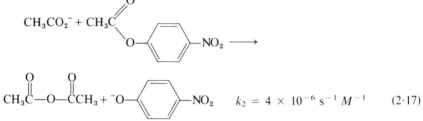

$k_1 = 0.8 \text{ s}^{-1}$ (2·16)

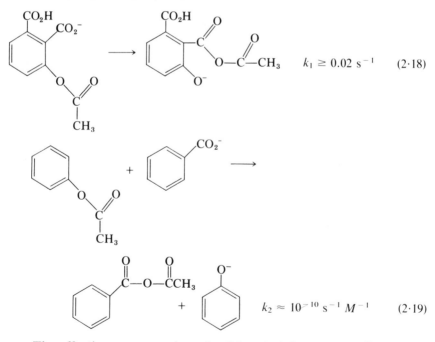

$k_2 = 4 \times 10^{-6} \text{ s}^{-1} M^{-1}$ (2·17)

The effective concentration of $-CO_2^-$ is $k_1/k_2 = 2 \times 10^5 \ M$.

2. *Rates of acyl transfer in aspirin derivatives:*[15]

$k_1 \geq 0.02 \text{ s}^{-1}$ (2·18)

$k_2 \approx 10^{-10} \text{ s}^{-1} M^{-1}$ (2·19)

The effective concentration of $-CO_2^-$ is $k_1/k_2 > 2 \times 10^7 \ M$.

3. *Equilibria for acyl transfer in succinates:*[16,17]

$$CH_2CO_2H \atop | \atop CH_2CO_2H \quad \rightleftharpoons \quad \begin{array}{c} O \\ \parallel \\ CH_2C-C \\ | \qquad\qquad O \\ CH_2C-C \\ \parallel \\ O \end{array} \qquad K_{eq} = 8 \times 10^{-7} \qquad (2\cdot20)$$

$$2CH_3CO_2H \rightleftharpoons CH_3\overset{O}{\overset{\parallel}{C}}-O-\overset{O}{\overset{\parallel}{C}}CH_3 \qquad K_{eq} = 3 \times 10^{-12} \, M \qquad (2\cdot21)$$

The effective concentration of $-CO_2H$ is $3 \times 10^5 \, M$.

These examples show that enormous rate enhancements come from intramolecular nucleophilic catalysis.

4. Entropy: The theoretical basis of intramolecular catalysis and effective concentration[18,19]

The high effective concentration of intramolecular groups is one of the most important reasons for the efficiency of enzyme catalysis. This can be explained theoretically by using transition state theory and examining the entropy term in the rate equation (2·7). It will be seen that effective concentrations may be calculated by substituting certain entropy contributions into the $\exp(\Delta S^{\ddagger}/R)$ term of equation 2·7.

a. The meaning of entropy
The most naive explanation of entropy, much frowned upon by purists but adequate for this discussion, is that entropy is a measure of the degree of randomness or disorder of a system. The more disordered the system, the more it is favored and the higher its entropy. Entropy is similarly associated with the spatial freedom of atoms and molecules.

The catalytic advantage of an intramolecular reaction over its intermolecular counterpart is due to entropy. The intermolecular reaction involves two or more molecules associating to form one. This leads to an increase in "order" and a consequent loss of entropy. An effective concentration may be calculated from the entropy loss.

b. The magnitude of entropy
The entropy of a molecule is composed of the sum of its translational, rotational, and internal entropies. The translational and rotational entropies may be precisely calculated for the molecule in the gas phase from its mass and geometry. The entropy of the vibrations may be calculated

from their frequencies, and the entropy of the internal rotations from the energy barriers to rotation.

The translational entropy is high: about 120 J/deg/mol (30 cal/deg/mol) for a 1-*M* solution of a small molecule. This is equivalent to about 40 kJ/mol (9 kcal/mol) at 25°C (298 K). The translational entropy is proportional to the volume occupied by the molecule; the smaller the volume, the more the molecule is restricted and the lower the entropy. Similarly, the entropy decreases with increasing concentration, since the average volume occupied by a molecule is inversely proportional to its concentration. It is important to note that the dependence on mass is low (Table 2·3). A 10-fold increase in mass on going from, say, a relative molecular mass of 20 to one of 200 leads to only a small increase in translational entropy.

The rotational entropy is also high: up to 120 J/deg/mol (30 cal/deg/mol) for a large organic molecule. It, too, increases only slowly with increasing mass, but it is independent of concentration.

Stiff vibrations, as found in most covalent bonds, make very low individual contributions to the entropy. Low-frequency vibrations, where

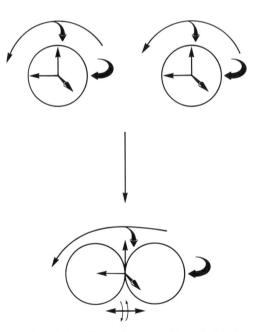

FIGURE 2·5. A free molecule has three degrees of translational entropy and three degrees of overall rotational entropy. When two molecules condense to form one, the resulting adduct has only three degrees of translational and three degrees of rotational entropy overall, a loss of three degrees of each. A compensating gain of internal vibrational and rotational entropy partly offsets this loss.

TABLE 2·3. *Entropy of translation, rotation, and vibration (298 K)*[a]

Motion	Entropy	
	(J/deg/mol)	(cal/deg/mol)
3 degrees of *translational freedom* for M_r 20–200 (standard state = 1 M)	120–150	29–36
3 degrees of *rotational freedom*:		
Water	44	10.5
n-Propane	90	21.5
endo-Dicyclopentadiene	114	27.2
Internal rotation	13–21	3–5
Vibrations (cm^{-1}):		
1000	0.4	0.1
800	0.8	0.2
400	4.2	1.0
200	9.2	2.2
100	14.2	3.4

[a] From M. I. Page and W. P. Jencks, *Proc. Natn. Acad. Sci. USA* **68**, 1678 (1971).

the atoms are less constrained, can contribute a few entropy units. Internal rotations have entropies in the range of 13 to 21 J/deg/mol (3 to 5 cal/deg/mol) (Table 2·3).

c. The loss of entropy when two molecules condense to form one
The combining of two molecules to form one leads to the loss of one set of rotational and translational entropies. The rotational and translational entropies of the adduct of the two molecules are only slightly larger than those of one of the original molecules, since these entropies increase only slightly with size (Table 2·3). The entropy loss is on the order of 190 J/deg/mol (45 cal/deg/mol) or 55 to 59 kJ/mol (13 to 14 kcal/mol) at 25°C. This may be offset somewhat by an increase in internal entropy due to new modes of internal rotation and vibration (Figure 2·5).

This loss is for a standard state of 1 M. If the solutions are more dilute, the loss will be correspondingly greater since the translational entropy is concentration-dependent.

d. The entropic advantage of a unimolecular over a bimolecular reaction
Let us compare the reaction of two molecules A and B combining together to form a third, with its intramolecular counterpart:

$$A + B \longrightarrow AB^{\ddagger} \longrightarrow AB \qquad (2·22)$$

$$\underset{\smile}{A\ B} \longrightarrow \underset{\smile}{AB^{\ddagger}} \longrightarrow \underset{\smile}{AB} \qquad (2·23)$$

The formation of the transition state $AB^{\ddagger}$ leads to the loss of translational and rotational entropy as described above, although there are some compensating gains in internal rotation and vibration. The intramolecular cyclization in equation 2·23 involves the loss of only some entropy of internal rotation.

Depending on the relative gains and losses in internal rotation, the intramolecular reaction is favored entropically by up to 190 J/deg/mol (45 cal/deg/mol) or 55 to 59 kJ/mol (13 to 14 kcal/mol) at 25°C. Substituting 190 J/deg/mol (45 cal/deg/mol) into the exp $(\Delta S^{\ddagger}/R)$ term of equation 2·7 gives a factor of 6×10^9. Taking into account the difference in molecularity between the second-order and first-order reactions, this may be considered as the maximum effective concentration of a neighboring group, i.e., $6 \times 10^9\ M$. In other words, for B in equation 2·22 to react with the same first-order rate constant as A⌣B in 2·23, the concentration of A would have to be $6 \times 10^9\ M$.

The loss of internal rotation lowers the effective concentration quite considerably. In the succinate case there are three internal rotations lost:

M. I. Page and W. P. Jencks suggest that the entropy is lowered by about 13 J/deg/mol (3 cal/deg/mol) by the loss of the rotation about the methylene group, and by 25 J/deg/mol (6 cal/deg/mol) by the loss of rotation about each methylene-carboxyl bond. This decrease is equivalent to a factor of 2×10^3 in effective concentration. If the free rotations in the succinate compounds were frozen out, the effective concentration of the neighboring carboxyl group would be increased to about $5 \times 10^8\ M$. This could be increased by a further factor of 10 by allowing for the unfavorable energy associated with the eclipse of the methylene hydrogens on formation of the five-membered ring. Considering these factors, the theoretical value for the maximum effective concentration appears to be quite realistic.

The theoretical basis of these calculations is completely valid for the gas phase. However, before the work of Page and Jencks it was not generally realized that reactions in solution involve similar entropy changes. Up till that point, the effective concentration of a neighboring group was thought to be no more than 55 M, the concentration of water in water.[20] This is the figure that is normally found for intramolecular general-base catalysis.

e. The dependence of effective concentration on "tightness" of transition state

The lower effective concentrations found in intramolecular base catalysis are due to the loose transition states of these reactions. In nucleophilic reactions, the nucleophile and the electrophile are fairly rigidly aligned so that there is a large entropy loss. In general-base or -acid catalysis, there is considerable spatial freedom in the transition state. The position of the catalyst is not as closely defined as in nucleophilic catalysis. There is consequently a smaller loss in entropy in general-base catalysis, so that the intramolecular reactions are not favored as much as their nucleophilic counterparts.

D. E. Koshland's original treatment of the problem of calculating effective concentrations was to consider the concentration of an intramolecular group to be approximately the same as that of water in aqueous solution, since a molecule in solution is completely surrounded by water.[20] This gives an upper limit of 55 M for effective concentration, equivalent to 34 J/deg/mol (8 cal/deg/mol) of entropy. That figure does represent the probability of two molecules being next to each other in solution. But, as soon as the two molecules are tightly linked, there is a large loss of entropy. A loose transition state may, perhaps, be interpreted as two molecules that are in close juxtaposition but that retain considerable entropic freedom.

In summary, one of the most important factors in enzyme catalysis is entropy. Catalyzed reactions in solution are slow because the bringing together of the catalysts and the substrate involves a considerable loss of entropy. Enzymatic reactions take place in the confines of the enzyme–substrate complex. The catalytic groups are part of the same molecule as the substrate, so there is no loss of translational or rotational entropy in the transition state. One way of looking at this is that the catalytic groups have very high effective concentrations compared with bimolecular reactions in solution. This advantage in entropy is "paid for" by the enzyme–substrate binding energy; the rotational and translational entropies of the substrate are lost on formation of the enzyme–substrate complex, and not during the chemical steps. (The loss of entropy on formation of the enzyme–substrate complex increases its dissociation constant.)

5. "Orbital steering"[21]

Attempts have been made to account for the rate enhancements in intramolecular catalysis on the basis of an effective concentration of 55 M combined with the requirement of very precise alignment of the electronic orbitals of the reacting atoms: "orbital steering." Although this treatment does have the merit of emphasizing the importance of correct orientation in the enzyme–substrate complex, it overestimates this importance, because, as we now know, the value of 55 M is an extreme underestimate of the contribution of translational entropy to effective concentration. The consensus is that although there are requirements for the satisfactory overlap of orbitals in the transition state, these amount to an accuracy of only 10° or so.[22,23] The distortion of even a fully formed carbon-carbon bond by 10° causes a strain of only 11 kJ/mol (2.7 kcal/mol). A distortion of 5° costs only 2.8 kJ/mol (0.68 kcal/mol).[19]

6. Electrostatic catalysis

a. Solvent water obscures electrostatic catalysis in aqueous solution
Chemical studies of model compounds do not show large effects due to electrostatic catalysis. This has led some chemists to reject the idea of electrostatic catalysis, but those who do so misunderstand the electrostatic forces involved. The electrostatic interaction energy between two point charges e_1 and e_2 separated by a distance r in a medium of dielectric constant D is given by

$$E = \frac{e_1 e_2}{Dr} \tag{2·24}$$

For a proton and an electron separated by 3.3 Å (0.33 nm) *in vacuo*, this gives -418 kJ/mol (-100 kcal/mol). But in water of dielectric constant 79, the value drops to -5.4 kJ/mol (-1.3 kcal/mol).

It should be noted that the water does not have to be inserted between the two charges to lower the interaction energy. This may be illustrated by the pK_a's of the two following amine bases in water:

$$\text{H}_2\text{NNH}_2 \underset{\text{H}^+}{\overset{pK_a=8}{\rightleftarrows}} \text{H}_2\text{NNH}_3{}^+ \underset{\text{H}^+}{\overset{pK_a=-1}{\rightleftarrows}} {}^+\text{H}_3\text{NNH}_3{}^+ \tag{2·25}$$

$$\text{N}{\Large\bigcirc}\text{N} \underset{\text{H}^+}{\overset{pK_a=8.8}{\rightleftarrows}} \text{N}{\Large\bigcirc}\text{NH}^+ \underset{\text{H}^+}{\overset{pK_a=3}{\rightleftarrows}} {}^+\text{HN}{\Large\bigcirc}\text{NH}^+ \tag{2·26}$$

In the protonation of hydrazine (equation 2·25), the juxtaposition of the two positive charges in the dication should destabilize it by 920 kJ/mol (220 kcal/mol), according to two positive charges separated by 1.5 Å *in vacuo*. Instead, the two pK_a's differ by only 9 units, reflecting a difference of only 50.2 kJ/mol (12 kcal/mol). In equation 2·26, the two nitrogen atoms of triethylenediamine are separated by 2.6 Å. The two positive charges in the dication would be expected to have an unfavorable interaction energy of 546 kJ/mol (130 kcal/mol) *in vacuo*, but the second pK_a is perturbed by only 33.4 kJ (8 kcal). In both cases there appears to be an effective dielectric constant of about 17 between the two nitrogen atoms.

This is due to the positively charged ions polarizing the solvent and inducing dipoles. The electrostatic field from these dipoles and from any counter ions partially neutralizes the positive field from the cations. For this reason, the surrounding of ions by a dielectric without interposing it between them lowers the interaction energy between the ions. Model studies in water greatly underestimate the importance of electrostatic catalysis in proteins.

b. Enzymes may stabilize polar transition states better
than water does
Electrostatic interactions are much stronger in organic solvents than in water because of lower dielectric constants. It has been felt for some time now that this property could be used by enzymes as a means of stabilizing polar transition states: positive or negative charges built into the low dielectric medium of the protein structure would have strong electrostatic interactions with charges on polar transition states.[24,25] The crux of the matter is that a protein is a very heterogeneous medium as far as its dielectric constant is concerned; apolar alkyl side chains are juxtaposed with polar backbone amide groups. Although this clearly opens up exciting possibilities of electrostatic catalysis, it renders calculation most difficult. Recently, A. Warshel has made calculations that take into account the local dielectric constant of the enzyme and the effects of the local permanent dipoles and the induced dipoles.[26,27] Warshel's conclusions are:

1. Two or three fixed dipoles in an enzyme (e.g., backbone $>$NH groups) can stabilize a charge as effectively as bulk water can.
2. An ion pair may be stabilized more effectively by fixed dipoles in an enzyme than by bulk water.

   It is instructive to consider the reasons for these statements. Water solvates an ion by forming a tight solvation shell in contact with the ion; this shell is surrounded by further shells that interact through their electrostatic dipoles. But these outer shells also have to interact with the dipoles and hydrogen bonds of the bulk solvent. This has the effect of randomizing the orientation of the dipoles in the outer shells and partly neutralizing them. In consequence, the calculations suggest that an ion has a higher solvation energy when it is surrounded by just 10 water molecules than when it is in bulk water. (Perhaps primitive enzymes stabilized charged transition states by simply enveloping a hydrated substrate in a hydrophobic pocket?) In a protein, on the other hand, the dipoles are rigidly held in a fixed orientation, pointing toward the substrate.

   Although the calculations need experimental testing, it is very pertinent, as illustrated in Figure 2·6, that enzymes always use parts of their own structure or bound ions to solvate transition states, and do not use bulk water. This means that enzymes contain parallel dipoles—that is,

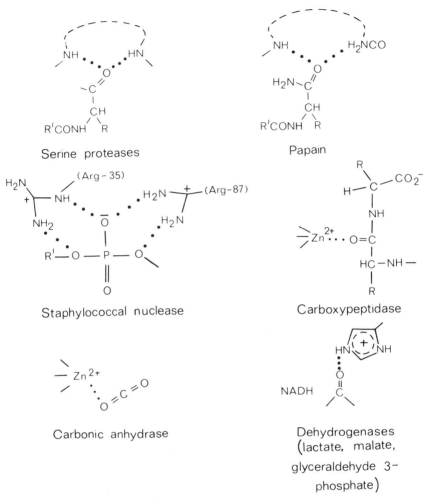

FIGURE 2·6. Solvation of substrates by enzymes.

dipoles in a high-energy geometry—and so have an "electrostatic strain" built in. A further source of electrostatic energy may come from the oriented dipoles of the hydrogen-bonded backbone groups of an α helix.[28]

Calculated electrostatic stabilization energies for the hydrolysis of a saccharide substrate in water and in catalysis by lysozyme are illustrated in Figure 2·7. It is estimated that the ionic triplet of (Asp-52)–oxocarbonium ion–(Glu-35) (i.e., − + −) is stabilized by some 25 to 42 kJ/mol (6 to 10 kcal/mol) by the environment of the enzyme compared with bulk water.

To summarize, it seems likely that enzymes can stabilize ion pairs and other charge distributions more effectively than water can because the enzyme has dipoles that are kept oriented toward the charge, whereas water dipoles are randomized by outer solvation shells interacting with bulk solvent.[26]

7. Metal ion catalysis[29]

a. Electrophilic catalysis

One obvious role for metals in metalloenzymes is to function as electrophilic catalysts, stabilizing the negative charges that are formed. In carboxypeptidase (Chapter 15), the carbonyl oxygen of the amide substrate is coordinated to the Zn^{2+} of the enzyme (Figure 2·6). The coordination polarizes the amide to nucleophilic attack and strongly stabilizes the tetrahedral intermediate (equation 2·27). This type of complex formation has been mimicked in model compounds to give rate enhancements of 10^4 to 10^6.[30,31] For example, the base-catalyzed hydrolysis of glycine ethyl ester

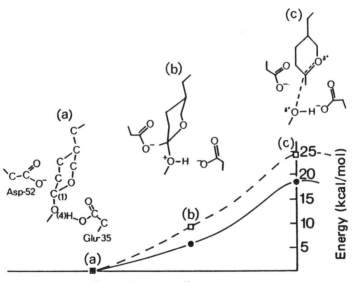

FIGURE 2·7. Energy diagram for the lysozyme reaction. The solid line traces the enzymatic reaction. The dashed line traces the same reaction but with the catalytic groups and the substrate surrounded by water. [From A. Warshel, *Proc. Natn. Acad. Sci. USA* **75**, 5250 (1978).]

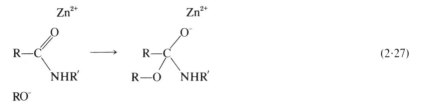

$$\text{(2·27)}$$

is increased 2×10^6-fold when the compound is coordinated to (ethylenediamine)$_2$Co^{3+} (equations 2·28 and 2·29).

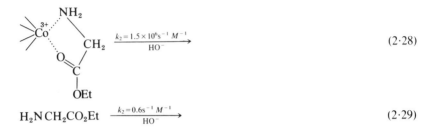

$$\text{(2·28)}$$

$$H_2NCH_2CO_2Et \xrightarrow[HO^-]{k_2 = 0.6s^{-1} M^{-1}} \text{(2·29)}$$

b. A source of hydroxyl ions at neutral pH

Another very important discovery from kinetic studies in inorganic chemistry is that metal-bound hydroxyl ions are potent nucleophiles.[32–34] The cobalt-bound water molecule in equation 2·30 ionizes with a pK_a of 6.6,

$$(NH_3)_5Co^{3+}OH_2 \rightleftharpoons (NH_3)_5Co^{2+}OH + H^+ \tag{2·30}$$

a value 9 units below the pK_a of free H$_2$O; yet the cobalt-bound hydroxyl group is only 40 times less reactive than the free hydroxide ion in catalyzing the hydration of carbon dioxide.[32] This insensitivity of the high reactivity to pK_a is quite general and independent of the metal involved.[34] Thus, metal-bound water molecules provide a source of nucleophilic hydroxyl groups at neutral pH. Just as a base of pK_a 7 is most effective in general-base catalysis (section B2b), so a metal-bound water molecule of pK_a 7 is most effective for nucleophilic attack because it combines a high reactivity with a high fraction in the correct ionization state.

This is of relevance to the mechanism of carbonic anhydrase. This enzyme, which catalyzes the hydration of CO$_2$, has at its active site a Zn^{2+} ion ligated to the imidazole rings of three of the histidines. The classic mechanism for the reaction is that the fourth ligand is a water molecule which ionizes with a pK_a of 7.[35] The reactive species is considered to be the zinc-bound hydroxyl. Chemical studies show that zinc-bound hydroxyls are no exception to the rule of high reactivity. The H$_2$O in structure 2·31 ionizes with a pK_a of 8.7 and catalyzes the hydration of carbon dioxide and acetaldehyde.[36] The carbonic anhydrase mechanism

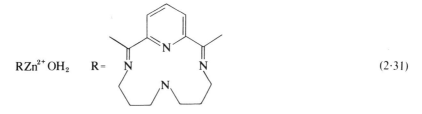

$$RZn^{2+}OH_2 \quad R = \qquad\qquad\qquad\qquad\qquad\qquad\qquad (2\cdot31)$$

probably involves the step in equation 2·32. (The Zn^{2+} ion possibly also polarizes the carbonyl oxygen atom.)[37]

$$E—Zn^{2+} \quad \underset{H}{\overset{O}{\underset{}{\overset{C}{\rightleftharpoons}}}} \quad E—Zn^{2+} + HCO_3^- \qquad\qquad (2\cdot32)$$

The combination of a metal-bound hydroxyl group and an intramolecular reaction provides some of the largest rate enhancements that can be found in a strain-free system. The complex of glycylglycine with (ethylenediamine)$_2$Co^{3+} and a *cis*-hydroxyl (equation 2·33) hydrolyzes at pH 7 nearly 10^{10} times faster than the free glycylglycine.[38]

$$\overset{NH_2CH_2CONHCH_2CO_2^-}{\underset{OH}{Co^{2+}}} \quad \xrightarrow[\text{pH 7, 25°C}]{k_1 = 5.5 \times 10^{-3}\,s^{-1}} \qquad (2\cdot33)$$

There are more complex examples of metal ion catalysis. Cobalt in vitamin B_{12} reactions forms covalent bonds with carbons of substrates.[39,40] Metals can also act as electron conduits in redox reactions. For example, in cytochrome *c* the iron in the heme is reversibly oxidized and reduced.

C. Covalent catalysis

1. Electrophilic catalysis by Schiff base formation[41]

A good example of how transient chemical modification can activate a substrate is Schiff base formation from the condensation of an amine with a carbonyl compound (equation 2·34). The Schiff base may be protonated

$$\underset{R''}{\overset{R'}{C}}{=}O \quad H_2\overset{.}{N}R \rightleftharpoons \underset{R'' + H_2O}{\overset{R'}{C}}{=}NR \overset{H^+}{\rightleftharpoons} \underset{R''}{\overset{R'}{C}}{=}\overset{+}{\underset{H}{N}}R \qquad (2\cdot34)$$

at neutral pH. This acts as an *electron sink* to stabilize the formation of a negative charge on one of the α carbons (equation 2·35). After tauto-

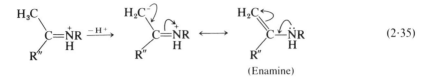

$$(2·35)$$

(Enamine)

merization to form the enamine, the methylene carbon is activated as a nucleophile. Another consequence of Schiff base formation is that the carbonyl carbon is activated toward nucleophilic attack because of the strong electron withdrawal of the protonated nitrogen.

a. Acetoacetate decarboxylase[42,43]
This enzyme catalyzes the decarboxylation of acetoacetate to acetone and carbon dioxide. The nonenzymatic reaction involves the expulsion of a highly basic enolate ion at neutral pH (equation 2·36), but the en-

$$CH_3C \longrightarrow CH_3C \quad + CO_2 \qquad (2·36)$$

zymatic reaction circumvents this by the prior formation of a Schiff base with a lysine residue. The protonated imine is then readily expelled. This process may be mimicked in solution by using aniline as a catalyst (equation 2·37). The evidence for the intermediate is that the enzyme is irre-

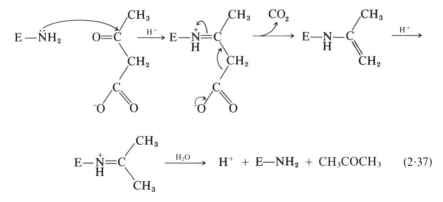

$$E—\overset{+}{\underset{H}{N}}=C \overset{CH_3}{\underset{CH_3}{\diagup}} \xrightarrow{H_2O} H^+ + E—NH_2 + CH_3COCH_3 \qquad (2·37)$$

versibly inhibited when sodium borohydride is added to the complex with the substrate. Borohydride is known to reduce Schiff bases, and the hydrolysate of the inhibited protein is found to contain isopropyllysine (equation 2·38). The carbon in the Schiff base is activated to the attack of an

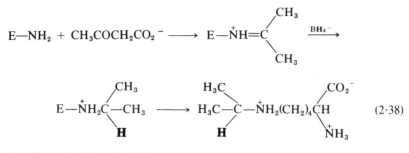

$$E-NH_2 + CH_3COCH_2CO_2^- \longrightarrow E-\overset{+}{N}H=C\overset{\displaystyle CH_3}{\underset{\displaystyle CH_3}{<}} \xrightarrow{BH_4^-}$$

$$E-\overset{+}{N}H_2\overset{|}{\underset{\textbf{H}}{C}}-CH_3 \longrightarrow H_3C-\overset{|}{\underset{\textbf{H}}{C}}-\overset{+}{N}H_2(CH_2)_4\overset{CO_2^-}{\underset{\overset{+}{N}H_3}{CH}} \tag{2·38}$$

H^- ion from the borohydride:

$$\underset{\textbf{H}}{\overset{+\frown}{-N\!=\!C}}\overset{\frown}{} \quad H\!-\!\bar{B}H_3 \tag{2·39}$$

b. Aldolase and transaldolase[44–47]

The aldol condensation and the reverse cleavage reaction catalyzed by these enzymes both involve a Schiff base. The cleavage reaction is similar to the acetoacetate decarboxylase mechanism, with the protonated imine being expelled. The condensation reaction illustrates the other function of a Schiff base, the activation of carbon via an enamine (equation 2·40).

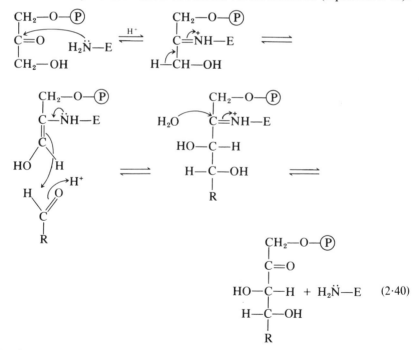

The intermediate may be trapped as before.

2. Pyridoxal phosphate—Electrophilic catalysis[41,48]

The principles of the above reactions form the basis of a series of important metabolic interconversions involving the coenzyme pyridoxal phosphate (equation 2·41). This condenses with amino acids to form a

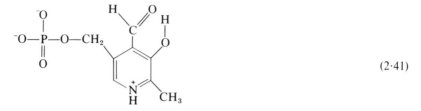

$$(2·41)$$

Schiff base (equation 2·42). The pyridine ring in the Schiff base acts as an

$$(2·42)$$

"electron sink" which very effectively stabilizes a negative charge. Each one of the groups around the chiral carbon of the amino acid may be cleaved, forming an anion that is stabilized by the Schiff base with the pyridine ring. (See Chapter 8, section F1, for a stereochemical explanation of why a particular bond is cleaved.)

a. Removal of the α hydrogen
The removal of the α hydrogen (equation 2·43) gives a key intermediate

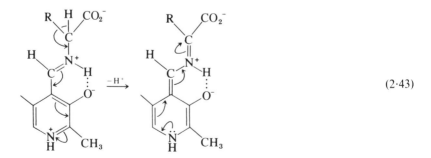

$$(2·43)$$

that may react in several different ways:

1. *Racemization*. Addition of the proton back to the amino acid will lead to racemization unless it is done stereospecifically.

2. *Transamination*. Addition of a proton to the carbonyl carbon of the pyridoxal leads to a compound that is the Schiff base of an α-keto acid and pyridoxamine. Hydrolysis of the Schiff base gives the α-keto acid and pyridoxamine, which may react with a different α-keto acid to reverse the sequence:

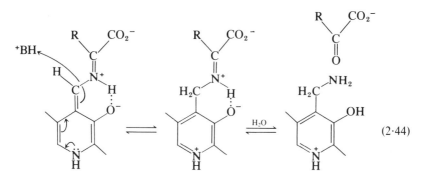

$$(2 \cdot 44)$$

The overall reaction is

$$R'CH(NH_3{}^+)CO_2{}^- + R''COCO_2{}^- \rightleftharpoons R'COCO_2{}^-$$
$$+ R''CH(NH_3{}^+)CO_2{}^- \qquad (2 \cdot 45)$$

3. *β-Decarboxylation*. When the amino acid is aspartate, the second compound in equation 2·44 is analogous to the Schiff base in the aceto-acetate decarboxylase reaction and may readily decarboxylate:

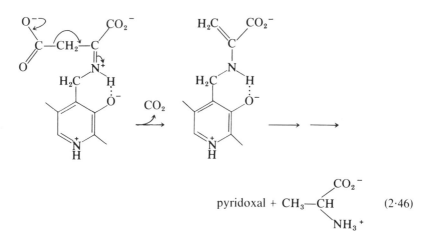

$$\text{pyridoxal} + CH_3{-}\underset{\underset{NH_3{}^+}{|}}{CH}{-}CO_2{}^- \qquad (2 \cdot 46)$$

4. *Interconversion of side chains.* When RX— is a good leaving group, it may be expelled as in equation 2·47. RX— may be a thiol, a

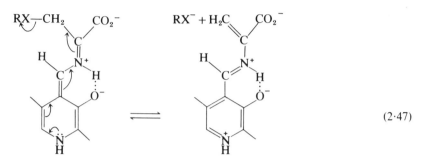

$$(2·47)$$

hydroxyl, or an indole group. In this way, serine, threonine, cysteine, tryptophan, cystathionine, and serine and threonine phosphates may be interconverted or degraded.

b. α-Decarboxylation

The "electron sink" allows facile decarboxylation (equation 2·48). The decarboxylated adduct will add a proton to the amino acid carbonyl car-

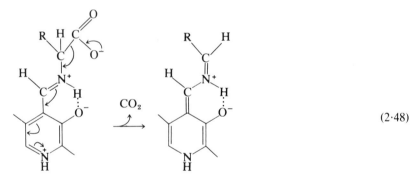

$$(2·48)$$

bon and then hydrolyze to give the amine and pyridoxal (equation 2·49), or else it will add the proton to the pyridoxal carbonyl carbon and then hydrolyze to give the aldehyde and pyridoxamine (equation 2·50).

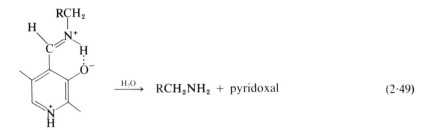

$$(2·49)$$

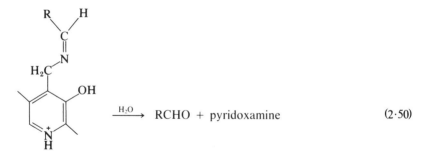

$$\xrightarrow{\text{H}_2\text{O}} \quad \text{RCHO + pyridoxamine} \qquad\qquad (2\cdot50)$$

3. Thiamine pyrophosphate—Electrophilic catalysis

Thiamine pyrophosphate (structure 2·51) is another coenzyme that co-valently bonds to a substrate and stabilizes a negative charge. The positive

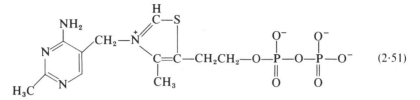

$$(2\cdot51)$$

charge on the nitrogen promotes the ionization of the C-2 carbon by elec-trostatic stabilization. The ionized carbon is a potent nucleophile (equa-tion 2·52). The nitrogen atom can also stabilize by delocalizing a negative

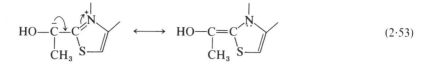

$$(2\cdot52)$$

charge on the adduct of thiamine with many compounds, as, for example, in *hydroxyethylthiamine pyrophosphate*, a form in which much of the coenzyme is found *in vivo* (equation 2·53). The combination of these re-

$$\text{HO}-\overset{\curvearrowleft}{\underset{\underset{\text{CH}_3}{|}}{\text{C}}}\overset{+}{\underset{\text{S}}{\overset{|}{\text{C}}}}\overset{+}{\underset{}{\text{N}}} \quad \longleftrightarrow \quad \text{HO}-\underset{\underset{\text{CH}_3}{|}}{\text{C}}=\overset{\overset{|}{\text{N}}}{\underset{\text{S}}{\text{C}}} \qquad (2\cdot53)$$

actions allows the decarboxylation of pyruvate by the route shown in equation 2·54. Other carbon-carbon bonds adjacent to a carbonyl group may be cleaved in the same manner.

The hydroxyethylthiamine pyrophosphates are potent nucleophiles and may add to carbonyl compounds to form carbon-carbon bonds. A good illustration of carbon-carbon bond making and breaking occurs in

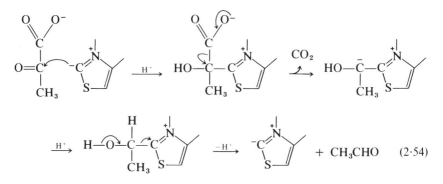

the reactions of transketolase. The enzyme contains tightly bound thiamine pyrophosphate and shuttles a dihydroxyethyl group between D-xylulose 5-phosphate and D-ribose 5-phosphate to form D-sedoheptulose 7-phosphate and D-glyceraldehyde 3-phosphate (equations 2·55 and 2·56).

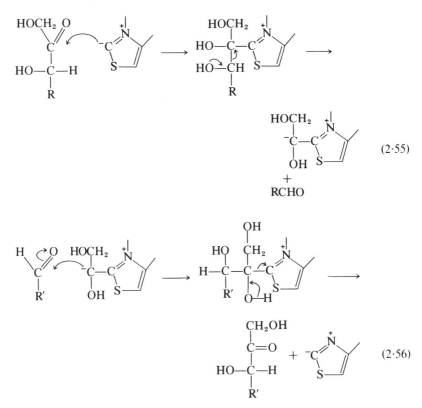

Hydroxyethylthiamine pyrophosphate is also nucleophilic toward a thiol of reduced lipoic acid. A hemithioacetal is formed, and this decomposes to give a thioester:

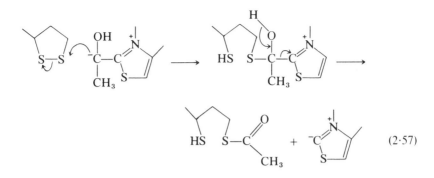

(2·57)

4. Nucleophilic catalysis

In enzymes, the most common nucleophilic groups that are functional in catalysis are the serine hydroxyl—which occurs in the serine proteases, cholinesterases, esterases, lipases, and acid and alkaline phosphatases—and the cysteine thiol—which occurs in the thiol proteases (papain, ficin, and bromelain), in glyceraldehyde 3-phosphate dehydrogenase, etc. The imidazole of histidine usually functions as an acid-base catalyst and enhances the nucleophilicity of hydroxyl and thiol groups, but it sometimes acts as a nucleophile with the phosphoryl group in phosphate transfer (Table 2·4).

The hydrolysis of peptides by these proteases represents classic nucleophilic catalysis. The relatively inert peptide is converted to the far more reactive ester or thioester acylenzyme, which is rapidly hydrolyzed. The use of the serine hydroxyl rather than the direct attack of a water molecule on the substrate is favored in several ways: alcohols are often better nucleophiles than the water molecule in both general-base-catalyzed and direct nucleophilic attack; the serine reaction is intramolecular and hence favored entropically; and the arrangement of groups is more ''rigid'' and defined for the serine hydroxyl compared with a bound water molecule.

5. A summary of factors responsible for enzyme catalysis

Uncatalyzed reactions in solution are often slow because they involve the formation of unstable positive and negative charges in the transition state, and frequently require that several molecules be brought together with the concomitant loss of entropy. These difficulties are lessened with enzymes: charges in transition states are stabilized electrostatically by strategically placed acids, bases, metal ions, or dipoles that are part of the structure of the enzyme; covalent catalysis is used to give reaction pathways of lower energy; and entropy losses are minimized because the necessary catalytic groups are part of the enzyme structure, with the

TABLE 2·4. *Nucleophilic groups in enzymes*

Nucleophile	Enzyme	Intermediate
—OH (serine)	Serine proteases	Acylenzyme
	Acid and alkaline phosphatases, phosphoglucomutase	Phosphorylenzyme
OH$^-$ (zinc-bound)	Carbonic anhydrase, liver alcohol dehydrogenase	—
—SH (cysteine)	Thiol proteases, glyceraldehyde 3-phosphate dehydrogenase	Acylenzyme
—CO$_2^-$ (aspartate)	ATPase (K$^+$/Na$^+$, Ca^{2+})	Phosphorylenzyme
—NH$_2$ (lysine)	Acetoacetate decarboxylase, aldolase, transaldolase, pyridoxal enzymes	Schiff base
	DNA ligase	Adenylenzyme (phosphoamide)
Imidazole (histidine)	Phosphoglycerate mutase, succinyl-CoA synthetase, nucleoside diphosphokinase, histone phosphokinase	Phosphorylenzyme
—OH (tyrosine)	Glutamine synthetase	Adenylenzyme
	Topoisomerases	Nucleotidylenzyme (phosphotyrosine)

reaction occurring in the confines of the enzyme–substrate complex. These features are paid for in two ways. (1) The original synthesis of the enzyme costs energy (though the enzyme can be used repeatedly). (2) The enzyme–substrate binding energy is used to immobilize the substrate at the active site and hold it next to the catalytic groups; this binding energy is inherently available for use but it is generally not utilized in uncatalyzed reactions. The utilization of binding energy has a profound effect on enzymatic reactions. Three later chapters, 11, 12, and 13, are devoted to this topic.

D. Structure–reactivity relationships

One of the most fruitful approaches in the study of organic reaction mechanisms has been to measure changes in reactivity with changes in the structures of the reagents. These studies have given considerable information on the electronic structures of transition states and on the features that determine reactivity, nucleophilicity, and leaving group ability. Structure–reactivity studies with enzymes tend to measure the effects of changes in the structure of the substrate on its interaction with the enzyme, rather than the effects of these changes on the distribution of electrons in the transition state. In general, it is difficult to obtain useful data on the electronic requirements of enzymatic reactions from structure–reactivity studies, both because the range of changes that can be made in the substrate is restricted and because the inductive effects of substituents are often obscured by the effects on binding. However, the lessons learned from the chemical studies have been invaluable to our understanding of the mechanisms of enzymatic reactions.

1. Nucleophilic attack at the carbonyl group

Structure–reactivity studies have been used to give information concerning the charge distribution in the transition state by noting the effects of electron-withdrawing and electron-donating substituents on the reaction rate. For example, it has been found that the rate of nucleophilic attack on esters increases with (1) electron withdrawal in the acyl portion ($CHCl_2CO_2Et$ is far more reactive than CH_3CO_2Et), (2) electron withdrawal in the leaving group (p-nitrophenyl acetate is more reactive than phenyl acetate), and (3) increasing basic strength of the nucleophile—that is, electron donation in the nucleophile (the hydroxide ion is far more reactive than the acetate ion). Using the idea from transition state theory that the reaction rate depends on the energy difference between the transition state and the ground state, we can deduce that the reaction involves an increase of negative charge on the substrate (since the reaction rate is increased by electron withdrawal), and a decrease in charge on the nucleophile (since the rate is increased by electron donation, i.e., electron repulsion). This is consistent either with mechanism 2·58, where the rate-

$$(2·58)$$

determining step is the formation of the tetrahedral intermediate, or with mechanism 2·59, where the rate-determining step is the breakdown.

$$(2·59)$$

a. Linear free energy relationships and the Brönsted and Hammett equations

A quantitative assessment of the sensitivity of the reaction to electron withdrawal and donation in the attacking nucleophile may be made by measuring the second-order rate constants for the attack of a series of nucleophiles on a particular ester. A plot of the logarithms of the rate constants against the pK_a's of the nucleophiles may be made in the same way as for general-base catalysis (Figure 2·8). Generally, when the measurements are restricted to nucleophiles with similar chemical natures and to a not-too-wide range of pK_a's, a straight-line relationship is found. The slope of the line is termed β, as for general-base catalysis.

These linear relationships between the logarithms of rate constants and the pK_a's are known as *linear free energy relationships*, since the logarithm of a rate constant is proportional to its Gibbs energy of activation, and the logarithm of an equilibrium constant (such as a pK_a) is proportional to the Gibbs energy change of a reaction (the Gibbs energy was formerly called the free energy). The relationship between the nucleophilicity of a nucleophile and its basic strength shows that the Gibbs energy of activation of bond formation with the carbonyl carbon is proportional to the Gibbs energy of transfer of a proton to the nucleophile.

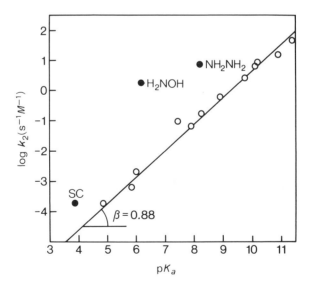

FIGURE 2·8. A Brönsted plot for the nucleophilic attack of primary and secondary amines on *p*-nitrophenyl acetate. Note that the "α"-effect nucleophiles—semicarbazide (SC), hydroxylamine, and hydrazine—are more reactive than would be expected from their pK_a's. [From W. P. Jencks and M. Gilchrist, *J. Am. Chem. Soc.* **90**, 2622 (1968).]

A similar Brönsted relationship is found when the rate constants for the attack of a particular nucleophile on a series of esters with differing leaving groups are plotted against the pK_a's of the leaving groups. An alternative way of plotting the data for aromatic compounds is to use the *Hammett* equation and the Hammett substituent constants (Table 2·5). The substituent constants, which measure the electron-donating or -withdrawing power of a substituent in a benzene ring, are derived empirically from the pK_a's of substituted benzoic acids by using the equation

$$(pK_a)_X = (pK_a)_0 - \sigma_X \tag{2·60}$$

where X is a substituent in the meta or para position of benzoic acid, and where $(pK_a)_X$ is the pK_a of the substituted acid, $(pK_a)_0$ that of the unsubstituted acid, and σ_X the substituent constant for X (different for the para and meta positions). It is found that the rates of other reactions of benzoic acid derivatives—such as the hydrolysis of benzoate esters—and also the equilibria of other reactions involving the benzene ring—such as the ionization of phenols—follow similar Hammett equations, except that a constant of proportionality, ρ, the reaction constant, is invoked to meas-

TABLE 2·5. *Substituent constants*[a]

Substituent	Aliphatic[b] (σ_I)	Aromatic[c]	
		σ_m	σ_p
—NH$_2$	0.10	−0.16	−0.66 (−0.17)[d]
—NH$_3$$^+$	0.60	0.63	—
—NHAc	0.28	0.21	0.00
—OH	0.25	0.121	−0.37 (−0.11)[d]
—OAc	0.39	0.39	0.31
—N(CH$_3$)$_3$$^+$	0.73	0.88	0.82
—CO$_2$$^-$	−0.17	−0.10	0.00
—CO$_2$H	0.34	0.37	0.45
—COCH$_3$	0.29	0.38	0.50 (0.87)[e]
—F	0.52	0.337	0.062
—Cl	0.47	0.373	0.227
—Br	0.45	0.391	0.232
—I	0.39	0.352	0.18
—NO$_2$	0.63	0.710	0.778 (1.24)[e]
—CN	0.58	0.56	0.66 (0.90)[e]
—CH$_3$	−0.05	−0.069	−0.17
—OCH$_3$	0.25	0.115	−0.268 (−0.11)[d]

[a] From M. Charton, *J. Org. Chem.* **29**, 1222 (1964); C. D. Ritchie and W. F. Sager, *Prog. Phys. Org. Chem.* **2**, 323 (1964).
[b] Based on $(\sigma_I)_X = [(pK_a)_{CH_3CO_2H} - (pK_a)_{X—CH_2CO_2H}]/3.95$.
[c] Based on $\sigma = (pK_a)_{PhCO_2H} - (pK_a)_{PhCO_2H}$; σ_m and σ_p are for meta and parasubstituents, respectively.
[d] For use with compounds other than benzoic acids and carbonyl compounds.
[e] For use with phenols, anilines, and thiophenols.

ure the sensitivity of the reaction toward changes in σ. For example, for the ionization of phenols, we have

$$(pK_a)_X = (pK_a)_0 - \rho\sigma_X \qquad (2\cdot61)$$

where $\rho = 2.1$. And for the alkaline hydrolysis of phenyl acetates, we have

$$\log k_X = \log k_0 + \rho\sigma_X \qquad (2\cdot62)$$

where $\rho = 0.8$.

The Brönsted and Hammett plots are equivalent. In the Hammett treatment the logarithms of the rate constants for, say, the alkaline hydrolysis of phenyl acetates are proportional to σ, and so are the pK_a's of the parent phenols. Hence, the logarithms of the rate constants for the hydrolysis are proportional to the pK_a's, as is found directly from the Brönsted plot of logarithm of rate constant against pK_a.

Brönsted and Hammett relationships are also found for the attack of a particular nucleophile on an ester of a particular leaving group as the acyl portion is varied.

b. Interpretation of Brönsted β values
The important parameter that is derived from the Brönsted plots is the β value. The sign and magnitude of this value are an indication of the charge developed in the transition state. Consider, for example, the attack of nucleophiles on esters. The β value for the equilibrium constants for the transfer of acetyl groups between oxanions and also between tertiary amines is 1.6 to 1.7. The value is greater than 1.0 because the acetyl group is more electron-withdrawing than the proton (for which $\beta = 1.0$ by definition), and is more sensitive to the pK_a of the alcohol or amine. Now it is found that the value of β for the attack of tertiary amines on esters of very basic alcohols is $+1.5$ for the variation of the nucleophile's pK_a, and -1.5 for the variation of the alcohol's.[49] This shows that the transition state for the reaction is very close to the structure of the products; that is, close to complete acyl transfer from the alcohol to the amine:

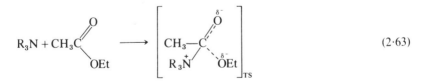

$$(2\cdot63)$$

At the other extreme, the reaction of basic nucleophiles with esters containing activated leaving groups exhibits β values of only $+0.1$ to 0.2 for the variation of the nucleophile's pK_a, and -0.1 to 0.2 for the variation of the leaving group's. This shows that the transition state involves little bond making and breaking and is close to the starting materials:

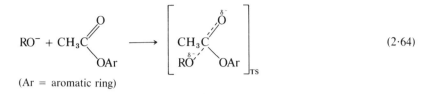

$$RO^- + CH_3C \overset{\displaystyle O}{\underset{\displaystyle OAr}{}} \longrightarrow \left[CH_3C \overset{\displaystyle \overset{\delta^-}{O}}{\underset{\displaystyle \underset{\delta^-}{RO}}{} OAr} \right]_{TS} \tag{2.64}$$

(Ar = aromatic ring)

The β value is a measure of the charge formed in the transition state rather than of the extent of bond formation. However, in equations 2·63 and 2·64 and in other examples where there is no acid-base catalysis, charge and bond formation are linked, so β does also give a measure of the extent of bond formation. But when there is also acid-base catalysis partly neutralizing the charges formed in the transition state, there is no relation between β and the extent of bond formation.[50]

2. Factors determining nucleophilicity and leaving group ability

The magnitude of general-acid-base catalysis by oxygen and nitrogen bases depends only on their pK_a's, and is independent of their chemical natures (apart from an enhanced activity of oximes in general-acid catalysis). Nucleophilic reactivity depends markedly on the nature of the reagents. These reactions may be divided into two broad classes: nucleophilic attack on *soft* and on *hard* electrophilic centers.[51]

a. Nucleophilic reactions with the carbonyl, phosphoryl, sulfuryl, and other hard groups
The attack of a nucleophile on an amide, ester, or carbonyl carbon involves the formation of a "real" chemical intermediate, and the valency of carbon is not extended beyond its normal value of 4. The attack on a phosphate ester is similar; a pentacovalent phosphate intermediate is formed. The transition state of the reaction involves the formation of a normal bond. This is a characteristic of hard centers. The dominant feature controlling nucleophilicity in these reactions is the basic strength of the nucleophile: the stronger the basic strength, the greater the nucleophilicity. There are, however, differences in reactivity among different classes of nucleophiles: amines and thiolate anions tend to be more nucleophilic than oxyanions.[52,53] Also, certain nucleophiles that have two electronegative atoms next to each other, such as NH_2OH, NH_2NH_2, $NH_2CONHNH_2$, HOO^-, and CH_3OO^-, are more reactive than would be expected from their pK_a's (Figure 2·9). This is known as the α effect. The relative reactivities of the different classes are often reflected in increased equilibrium constants for their addition to aldehydes and ketones.[54]

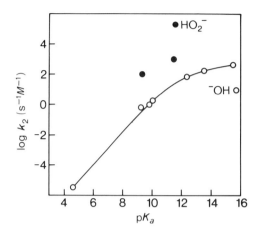

FIGURE 2·9. A Brönsted plot for the attack of oxyanion nucleophiles on *p*-nitro-phenyl acetate. As in Figure 2·8, the α-effect nucleophiles (●) are unusually re-active. Note how the linear plot breaks down with increasing pK_a for the more reactive nucleophiles. In general, the Brönsted relationships hold only over a limited range of pK_a's in these reactions. The curvature is not often seen in prac-tice because of the limited range of bases used.

The *ease of expulsion* of a group depends both on its pK_a and on its state of protonation. Basically, a "good" leaving group is one that is stable in solution. For example, the *p*-nitrophenolate ion is a good leaving group because it is weakly basic; the pK_a of *p*-nitrophenol is 7.0. The chemical cause of the stability of the ion is that the negative charge is delocalized around the aromatic ring and onto the nitro group (equation 2·65). The ion is readily and directly expelled from a tetrahedral intermediate (equa-tion 2·66). In this class of leaving group, the lower the basic strength, the

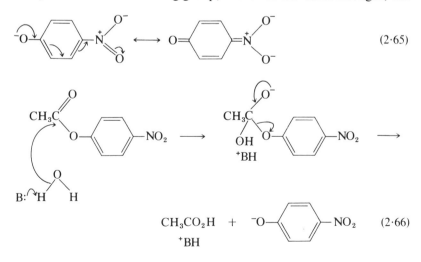

greater the ease of expulsion. Acetate (the pK_a of acetic acid = 4.76) is a better leaving group than *p*-nitrophenol and phosphate (with pK_a's ~ 7), which are better leaving groups than OH$^-$ (the pK_a of water = 15.8).

Alcoholate ions are difficult to expel because they are strongly basic; the pK_a's of simple alcohols are about 16. The expulsion of alcohols is aided by general-acid catalysis:

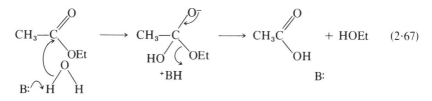

$$CH_3-C \longrightarrow CH_3-C \longrightarrow CH_3C \qquad + \text{ HOEt} \qquad (2\cdot67)$$

Nitrophenyl esters are often used as synthetic substrates for two reasons: (1) the nitrophenolate ion is a very good leaving group so it forms a reactive substrate; and (2) it has a characteristic absorption at 400 nm and is thus easily assayed spectrophotometrically. Both these factors are caused by the delocalization shown in equation 2·65.

Amines have to be protonated to be expelled from a molecule, since the amide ions, RNH$^-$, are far too unstable to be directly released into solution. (The exceptions to this rule are the highly activated derivatives such as 2,4-dinitroaniline.)

b. Nucleophilic reactions with saturated carbon
The attack of a nucleophile on saturated carbon—for example, the bimolecular attack of a thiol on the methyl carbon of *S*-adenosylmethionine—involves a transition state in which five groups surround the normally tetravalent carbon (equation 2·68). This is not a "normal" bond

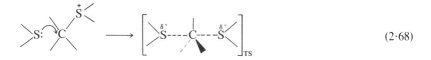

$$(2\cdot68)$$

with carbon, and it is peculiar to the transition state. The reaction is typical of a "soft" center. Large, polarizable atoms such as sulfur and iodine (i.e., "soft" ligands) react more rapidly in these reactions, whereas the small atoms of low polarizability, oxygen and nitrogen (i.e., "hard" ligands), are less reactive. The dominant factor in nucleophilicity toward alkyl groups and other soft centers is polarizability. Within any particular class of compounds, increasing basic strength increases the nucleophilicity, but between classes, polarizability is all important (Table 2·6).

As with reactions at the carbonyl group, weakly basic leaving groups are more readily displaced than strongly basic ones.

c. Leaving group activation

It was seen in the last section that highly basic groups are not readily displaced from carbonyl compounds and from saturated carbon. An extreme example of this is the esterification of an alcohol by a carboxylate ion. This would require the formation of a tetrahedral intermediate with two negatively charged oxygens, and the subsequent expulsion of O^{2-} (equation 2·69). When an aminoacyl-tRNA synthetase catalyzes the es-

$$RCO_2^- + R'OH \rightleftharpoons R\underset{\underset{OR'}{|}}{\overset{\overset{O^-}{|}}{C}}-O^- \xrightarrow{\quad} RCO_2R' + O^{2-} \tag{2·69}$$

$$+ H^+$$

terification of tRNA with an amino acid, the amino acid is activated by formation of an enzyme-bound mixed anhydride with AMP (equation 2·70)

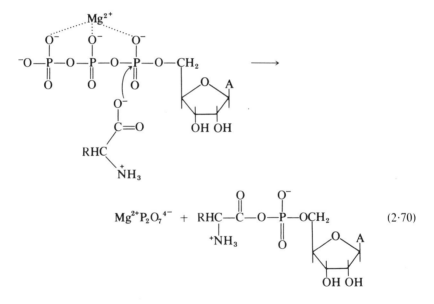

in the same way that an organic chemist activates a carboxylic acid by forming an acyl chloride or a mixed anhydride. (Note that in equation 2·70 the substrate is the magnesium complex of ATP, with the metal ion acting as an electrophilic catalyst. The Mg^{2+} binds primarily to the β,γ-phosphates.[55,56]) The carbonyl group of the amino acid is activated by equation 2·70 because it is bound to a good leaving group (the phosphate of AMP has a pK_a of about 6 or 7) and it may be readily attacked by one of the hydroxyl groups of the ribosyl ring of the terminal adenosine of the tRNA.

TABLE 2·6. *Nucleophilic reactivity toward saturated carbon*[a]

Nucleophile	pK_a	Relative reactivity toward CH_3Br
H_2O	−1.74	1.00
NO_3^-	−1.3	11
F^-	3.17	100
$CH_3CO_2^-$	4.76	525
Cl^-	−7.0	1.1×10^3
C_5H_5N	5.17	4.0×10^3
HPO_4^{2-}	7.21	6.3×10^3
Br^-	−9.0	7.8×10^3
OH^-	15.74	1.6×10^4
$C_6H_5NH_2$	4.62	3.1×10^4
I^-	−10.0	1.1×10^5
CN^-	9.40	1.3×10^5
SH^-	7.00	1.3×10^5

[a] From C. G. Swain and C. B. Scott, *J. Am. Chem. Soc.* **75**, 141 (1953).

Another example of leaving group activation is the utilization of *S*-adenosylmethionine rather than methionine in methylation reactions. A relatively basic thiolate anion has to be expelled from methionine, while the nonbasic neutral sulfur is displaced from the activated derivative:

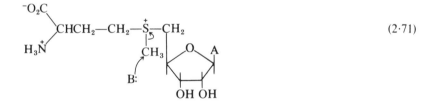

$$(2·71)$$

3. Application of linear free energy relationships to enzyme reactions

Although β values have been successfully obtained for some reactions, there are theoretical difficulties. There is first the problem of whether to examine the effects of substituents on k_{cat} or on k_{cat}/K_M. The difficulty with k_{cat}/K_M is that the binding energy terms come directly into this term, so the electronic effects of a substituent may be masked by its contribution to binding. On the other hand, use of k_{cat} is complicated in that effects such as the accumulation of intermediates, nonproductive binding, strain, and induced fit alter it (see Chapters 3 and 12). There is at least one example in which the effect of substituents on nonproductive binding has been mistakenly attributed to an inductive effect on the reaction rate.[57]

In many ways, k_{cat}/K_M is safer to use since all the above-mentioned artifacts that obscure k_{cat} cancel out in this parameter (Chapter 12). Nevertheless, some good structure–reactivity studies have been performed with enzymes. Earlier studies on the acylation of chymotrypsin by nonspecific substrates[58] and on the deacylation of nonspecific acylenzymes[59] showed that the reactions are very sensitive to electron withdrawal, indicating that there is the attack of a basic group on the substrate. A more recent study on the hydrolysis of substituted anilides by papain (Chapter 15, section B2) shows that there is a high negative value of ρ; that is, that the values of k_{cat} and k_{cat}/K_M are increased by electron-donating substituents in the aniline ring.[60] This is consistent with the rate-determining breakdown of a tetrahedral intermediate in which the aniline ring is pro-

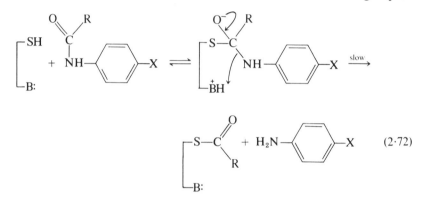

$$(2 \cdot 72)$$

tonated (equation 2·72). These examples are relatively straightforward to interpret, as the ρ values are large. Small values are more difficult to interpret, since they can be caused by a variety of different things: for example, a transition state in which there is very little bond making and breaking will be very insensitive to electron withdrawal, as will a transition state in which extensive bond making and breaking are compensated for by extensive proton transfer from acid-base catalysis.[50]

The reactions of yeast alcohol dehydrogenase have been analyzed by more sophisticated techniques.[61] Multiple linear regression analysis has been applied to separate the results of both the electronic effects of substituents and the contributions of their hydrophobic binding energy (see Chapter 11, section A). A similar analysis has been applied to the reactions of acetylcholinesterase.[62] In both examples, there is a feature that will be discussed at length in Chapter 12: binding energy is used to increase the value of k_{cat}.

Although it is sometimes thought that structure–reactivity relationships are more useful in teaching us the principles of catalysis in simple model systems than in being directly applicable to enzymes, there is no doubt that these examples from enzymology have provided useful information and ideas.

E. The principle of microscopic reversibility or detailed balance

The principle of microscopic reversibility or detailed balance is used in thermodynamics to place limitations on the nature of transitions between different quantum or other states. It applies also to chemical and enzymatic reactions: each chemical intermediate or conformation is considered as a "state." The principle requires that the transitions between any two states take place with equal frequency in either direction at equilibrium.[63] That is, the process A → B is exactly balanced by B → A, so equilibrium cannot be maintained by a cyclic process, with the reaction being A → B in one direction and B → C → A in the opposite. A useful way of restating the principle for reaction kinetics is that the reaction pathway for the reverse of a reaction at equilibrium is the exact opposite of the pathway for the forward direction. In other words, the transition states for the forward and reverse reactions are identical. This also holds for (nonchain) reactions in the steady state, under a given set of reaction conditions.[64]

The principle of microscopic reversibility is very useful for predicting the nature of a transition state from a knowledge of that for the reverse reaction. For example, as the attack of ethanol on acetic acid is general-base-catalyzed at low pH, the reverse reaction must involve the general-acid-catalyzed expulsion of ethoxide ion from the tetrahedral intermediate (equation 2·73). Similarly, since p-nitrophenol is ionized above pH 7, its

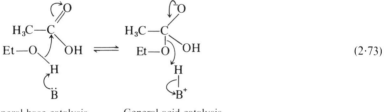

(2·73)

General-base catalysis   General-acid catalysis

attack on a carbonyl compound cannot be general-base-catalyzed: not only is the p-nitrophenolate ion carrying a full negative charge a more powerful nucleophile than p-nitrophenol, but the ion is present at a higher concentration. Therefore, the ion directly attacks a carbonyl compound, and, by the principle of microscopic reversibility, the expulsion of p-nitrophenolate ion from a tetrahedral intermediate is uncatalyzed (equation 2·74).

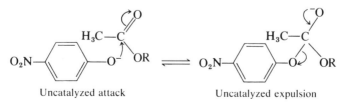

Uncatalyzed attack   Uncatalyzed expulsion

Similar arguments have been used to show that the attack of thiols upon esters is not general-base-catalyzed but involves the direct attack of the thiolate ion even at 10 pH units below the pK_a of the thiol, where only 1 part in 10^{10} is ionized.[50]

Care must always be taken to verify that a proposed reaction mechanism satisfies the principle of microscopic reversibility. Periodically, someone publishes a mechanism that contravenes the principle because the reverse reaction uses a different pathway from the forward reaction, under the same set of reaction conditions at equilibrium or in the steady state.

F. The principle of kinetic equivalence

There is an inherent ambiguity, which has been stated as the principle of kinetic equivalence, in interpreting the pH dependence of chemical reactions. When a rate law shows, for example, that the reaction rate is proportional to the concentration of an acid HA, it means that the net ionic charge of the acid appears in the transition state of the reaction; that is, either as the undissociated HA *or* as the two ions H^+ and A^-. Similarly, if the reaction rate varies as the concentration of A^-, the transition state contains either A^- *or* HA and OH^-. This may be shown algebraically as follows.

Rearranging equation 5·2 for the ionization of an acid gives

$$[HA] = \frac{[A^-][H^+]}{K_a} \qquad (2·75)$$

The concentration of the acid is related to the product of the concentrations of its conjugate base and the proton. Because of this, it is not possible to tell whether a reaction that depends on the concentration of HA really involves the undissociated acid or whether it involves the combination of an H^+ with the conjugate base in the transition state. Similarly, since the concentration of the basic form is related to that of the hydroxide ion and the acid by

$$[A^-] = \frac{[HA][OH^-]K_a}{K_w} \qquad (2·76)$$

(where K_w is the ionic product of water), it is not possible to distinguish a reaction involving A^- from one involving a combination of HA and OH^- by examining the concentration dependence. This is the principle of kinetic equivalence. For example, mechanisms 2·77 and 2·78 for general-base catalysis follow the same rate law:

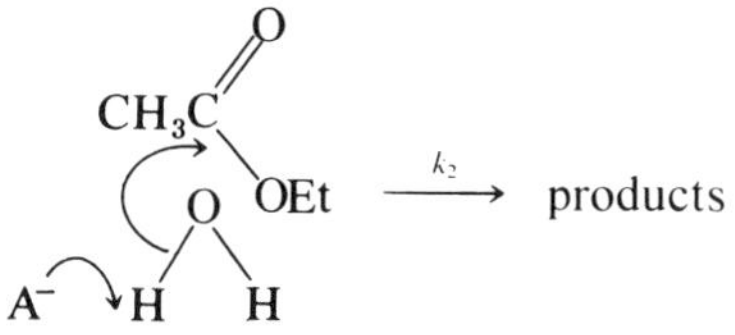

$$\text{products} \qquad (2\cdot77)$$

$$\text{products} \qquad (2\cdot78)$$

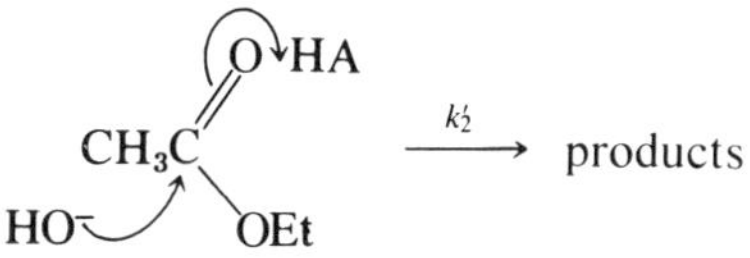

For mechanism 2·77,

$$v = k_2[\text{A}^-][\text{ester}] \qquad (2\cdot79)$$

For mechanism 2·78,

$$v = k_2'[\text{HA}][\text{OH}^-][\text{ester}] \qquad (2\cdot80)$$

$$= \frac{k_2'[\text{A}^-][\text{ester}]K_w}{K_a} \qquad (2\cdot81)$$

$$= k_2''[\text{A}^-][\text{ester}] \qquad (2\cdot82)$$

Because equations 2·79 and 2·82 are equivalent, the two mechanisms cannot be distinguished by the concentration or pH dependence. Nor can the two mechanisms be distinguished by the sign of the Brönsted β for the variation of the pK_a of the catalyst. One can see this intuitively from transition state theory, since both reactions involve a dispersion of negative charge in the transition state. Alternatively, this result can be derived mathematically. Mechanism 2·77 involves a positive value of β since the catalysis will be stronger for stronger bases. Mechanism 2·78 also involves a positive β value, because although the general-acid component involves a negative value of β, the term $1/K_a$ in equation 2·81 means that a component of $1 \times pK_a$ must be added to log v, which more than compensates for the fractionally negative β value for the chemical step.

The only time that kinetically equivalent mechanisms can be distinguished is when one of the mechanisms involves an "impossible" step and generates, say, a second-order rate constant that is faster than is feasible for a diffusion-controlled reaction, or, as in the case of mechanism 2·78 for aspirin hydrolysis, a negative energy of activation.[12]

G. Kinetic isotope effects[65]

Information about the extent and nature of the bond making and breaking steps in the transition state may sometimes be obtained by studying the effects of isotopic substitution on the reaction rates. The effects may be divided into two classes, depending on the position of the substitution.

1. Primary isotope effects

A primary isotope effect results from the cleavage of a bond to the substituted atom. For example, it is often found that the cleavage of a C—D bond is several times slower than that of a C—H bond. Smaller decreases in rate, up to a few percent, are sometimes found on the substitution of ^{15}N for ^{14}N, or of ^{18}O for ^{16}O. The magnitude of the change in rate gives some idea of the extent of the breaking of the bond in the transition state.

A simple way of analyzing isotope effects is to compare an enzymatic reaction with a simple chemical model whose chemistry has been established by other procedures. In this section we are interested primarily in the empirical results of the model experiments, but the following oversimplified account of the theoretical origins of the effects is helpful in understanding their nature. The plot of the energy of a carbon-hydrogen bond against interatomic distance gives the characteristic curve shown in Figure 2·10. The carbon-deuterium bond gives an identical plot, since the shape is determined by the electrons in the orbitals. According to quantum theory, the lowest energy level is at a value of $\frac{1}{2}h\nu$ above the bottom of the well, i.e., above the *zero-point energy*, where ν is the frequency of vibration. The value of ν may be found from the infrared stretching fre-

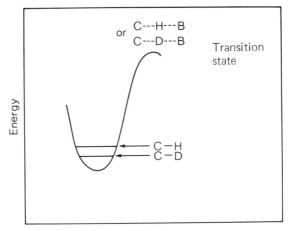

C—H or C—D interatomic distance

FIGURE 2·10. The energy changes during the transfer of hydrogen or deuterium from carbon. The energy of the transition state is the same for both (subject to the provisos in the text), but the hydrogen is at a higher energy in the starting materials because of its higher zero-point energy. The activation energy for the transfer of hydrogen is therefore less than that for deuterium.

quencies to give values of 17.4 and 12.5 kJ/mol (4.15 and 3 kcal/mol) for the zero-point energies of the C—H and C—D bonds, respectively. If during the transition state for the reaction the hydrogen or deuterium atom is at a potential energy maximum rather than in a well, there will be no zero-point energy. It is therefore easier to break the C—H bond by 17.4 to 12.5 kJ/mol (4.15 to 3.0 kcal/mol), a factor of 7 at 25°C. In practice, the kinetic isotope effect may be higher than this because of quantum mechanical tunneling, or lower because there are compensating bending motions in the transition state. As a general rule, values of 2 to 15 are good evidence that a C—H bond is being broken in the transition state. It will be seen in Chapter 15 that values of 3 to 5 are found for k_H/k_D in hydride transfer between the substrate and NAD^+ in the reactions of alcohol dehydrogenase.

Carbon-tritium bonds are broken even more slowly than C—D bonds because of the greater mass of tritium. A simple relationship between the deuterium and tritium isotope effects holds at the temperatures at which enzymatic reactions occur:[66]

$$\frac{k_H}{k_T} = \left(\frac{k_H}{k_D}\right)^{1.442}$$

(2·83)

Oxygen and nitrogen kinetic isotope effects have been used to probe the nature of the rate-determining step in the reactions of chymotrypsin and papain. In a model system, the alkaline- and general-base-catalyzed hydrolyses of methyl formate, $HC(=O)^{16,18}OCH_3$, are characterized by $k_{16O}/k_{18O} = 1.01$ (equation 2·84). Hydrazinolysis, in which the rate-determining step is breakdown of the tetrahedral intermediate (equation 2·85), has an isotope effect of 1.062. These values may be compared with

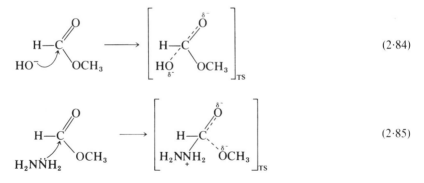

(2·84)

(2·85)

a value of 1.052 calculated for the complete loss of the zero-point energies of the bonds.[67] The kinetic isotope effects on the chymotrypsin-catalyzed hydrolysis of ester substrates are close to those found for the general-base-catalyzed hydrolysis of methyl formate.[68] That is, there is probably the rate-determining formation of a tetrahedral intermediate. The papain-catalyzed hydrolysis of benzoyl-L-arginine amide has a $^{14}N/^{15}N$ kinetic isotope effect of about 1.024, close to the upper limit of 1.01 to 1.025 found for C—N bond cleavage in model reactions.[69] This indicates a considerable degree of C—N bond cleavage in the transition state in the enzymatic reaction, and is almost certainly caused by the rate-determining breakdown of a tetrahedral intermediate (Chapter 15, section B2). A kinetic nitrogen isotope effect of 1.006 to 1.01 found for the chymotrypsin-catalyzed hydrolysis of acetyltryptophan amide has been interpreted as evidence for a tetrahedral intermediate in which both formation and breakdown contribute to the rate-determing step.[70]

$^{16}O/^{18}O$ kinetic isotope effects have been used to show that there is a change in the rate-determining step in the reaction of a series of substrates with β-galactosidase[71] (Chapter 7, section C3).

2. Secondary isotope effects

These result from bond cleavage between atoms *adjacent* to the isotopically substituted atom. Secondary isotope effects are caused by a change in the electronic hybridization of the bond linking the isotope, rather than by cleavage of the bond. Perhaps the best-known example of this in enzymatic reactions is the substitution of deuterium or tritium for hydrogen on the C-1 carbon of substrates for lysozyme (Chapter 15). A carbonium ion is formed in the reaction (equation 2·86), and the C-1 carbon changes from sp^3 to sp^2. Model reactions give a value of 1.14 for k_H/k_D, compared with the 1.11 found for the enzymatic reaction.[72,73]

$$(2·86)$$

3. Solvent isotope effects

These are found from comparing the rates of a reaction in H_2O and D_2O. They are usually a result of proton transfers between electronegative atoms *accompanying* the bond making and breaking steps between the heavier atoms in reactions such as the following:

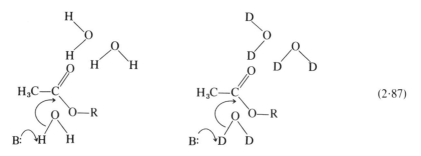

$$(2·87)$$

Solvent isotope effects differ in origin from those generated by the cleavage of a C—H bond, and their cause is not fully understood. Proton transfers between electronegative atoms are very rapid compared with normal bond making and breaking steps, and the proton probably sits at the bottom of a potential energy well during the reaction. There is no loss of zero-point energy such as occurs in the transfer of a proton from carbon, where the actual C—H bond breaking step is slow. Solvation of the reagents and secondary isotope effects caused by the exchange of deuterium for hydrogen in the reagents may make contributions to the solvent isotope effect.

Solvent isotope effects are a useful diagnostic tool in simple chemical reactions, although they can be variable. For example, it is found that general-base-catalyzed reactions such as 2·87 have a k_H/k_D of about 2, whereas the nucleophilic attack on an ester has a k_H/k_D of about 1. The isotope effects in enzymatic reactions are more difficult to analyze because the protein has so many protons that may exchange with deuterons from D_2O.[74] Also there may be slight changes in the structure of the protein on the change of solvent.

References

1 H. Pelzer and E. Wigner, *Z. Phys. Chem.* **B15**, 445 (1932).
2 H. Eyring, *Chem. Rev.* **17**, 65 (1935).
3 M. G. Evans and M. Polanyi, *Trans. Faraday Soc.* **31**, 875 (1935).
4 A. A. Frost and R. G. Pearson, *Kinetics and mechanism*, Wiley (1961).
5 G. S. Hammond, *J. Am. Chem. Soc.* **77**, 334 (1955).
6 J. N. Brönsted and K. Pedersen, *Z. Phys. Chem.* **A108**, 185 (1923).
7 W. P. Jencks and J. Carriuolo, *J. Am. Chem. Soc.* **83**, 1743 (1961).
8 T. H. Fife, *Accts. Chem. Res.* **5**, 264 (1972).
9 G. A. Craze and A. J. Kirby, *J. Chem. Soc. Perk. II* **61** (1974).
10 R. F. Atkinson and T. C. Bruice, *J. Am. Chem. Soc.* **96**, 819 (1974).
11 F. G. Bordwell and W. J. Boyle, Jr., *J. Am. Chem. Soc.* **94**, 3907 (1972).
12 A. R. Fersht and A. J. Kirby, *J. Am. Chem. Soc.* **89**, 4853, 4857 (1967); **90**, 5818, 5826 (1968).
13 T. St. Pierre and W. P. Jencks, *J. Am. Chem. Soc.* **90**, 3817 (1968).
14 Extrapolated from the data of E. Gaetjens and H. Morawetz, *J. Am. Chem. Soc.* **82**, 5328 (1960), and V. Gold, D. G. Oakenfull, and T. Riley, *J. Chem. Soc.* **1968B**, 515 (1968).
15 A. R. Fersht and A. J. Kirby, *J. Am. Chem. Soc.* **90**, 5833 (1968).
16 T. Higuchi, L. Eberson, and J. D. Macrae, *J. Am. Chem. Soc.* **89**, 3001 (1967).
17 W. P. Jencks, F. Barley, R. Barnett, and M. Gilchrist, *J. Am. Chem. Soc.* **88**, 4464 (1966).
18 M. I. Page and W. P. Jencks, *Proc. Natn. Acad. Sci. USA* **68**, 1678 (1971).
19 M. I. Page, *Chem. Soc. Revs.* **2**, 295 (1973).
20 D. E. Koshland, Jr., *J. Theoret. Biol.* **2**, 75 (1962).
21 D. R. Storm and D. E. Koshland, Jr., *Proc. Natn. Acad. Sci. USA* **66**, 445 (1970); *J. Am. Chem. Soc.* **94**, 5805 (1972).
22 T. C. Bruice, A. Brown, and D. C. Harris, *Proc. Natn. Acad. Sci. USA* **68**, 658 (1971).

23   W. P. Jencks and M. I. Page, *Biochem. Biophys. Res. Commun.* **57**, 887 (1974).

24   M. F. Perutz, *Proc. R. Soc.* **B167**, 448 (1967).

25   C. A. Vernon, *Proc. R. Soc.* **B167**, 389 (1967).

26   A. Warshel, *Proc. Natn. Acad. Sci. USA* **75**, 5250 (1978).

27   A. Warshel, *Biochemistry* **20**, 3167 (1981).

28   W. G. J. Hol, L. M. Halie, and C. Sander, *Nature, Lond.* **294**, 532 (1981).

29   R. J. P. Williams, *Pure and Appl. Chem.* **54**, 1889 (1982).

30   D. A. Buckingham, C. E. Davis, D. M. Foster, and A. M. Sargeson, *J. Am. Chem. Soc.* **92**, 5571 (1970).

31   D. A. Buckingham, J. MacB. Harrowfield, and A. M. Sargeson, *J. Am. Chem. Soc.* **96**, 1726 (1974).

32   E. Chaffee, T. P. Dasgupta, and G. M. Harris, *J. Am. Chem. Soc.* **95**, 4169 (1973).

33   D. A. Palmer and G. M. Harris, *Inorg. Chem.* **13**, 965 (1974).

34   D. A. Buckingham and L. M. Engelhardt, *J. Am. Chem. Soc.* **97**, 5915 (1975).

35   J. E. Coleman, *Prog. Bioorg. Chem.* **1**, 159 (1971), and references therein.

36   P. Woolley, *Nature, Lond.* **258**, 677 (1975).

37   K. K. Kannan, M. Petef, K. Fridborg, H. Cid-Dresdener, and S. Lovgren, *FEBS Lett.* **73**, 115 (1977).

38   D. A. Buckingham, F. R. Keene, and A. M. Sargeson, *J. Am. Chem. Soc.* **96**, 4981 (1974).

39   T. H. Finlay, J. Valinsky, K. Sato, and R. H. Abeles, *J. Biol. Chem.* **247**, 4197 (1974).

40   B. T. Golding and L. Radom, *J. Am. Chem. Soc.* **98**, 6331 (1976).

41   E. E. Snell and S. J. di Mari, *The Enzymes* **2**, 335 (1976).

42   G. A. Hamilton and F. H. Westheimer, *J. Am. Chem. Soc.* **81**, 6332 (1959).

43   S. G. Warren, B. Zerner, and F. H. Westheimer, *Biochemistry* **5**, 817 (1966).

44   E. Grazi, T. Cheng, and B. L. Horecker, *Biochem. Biophys. Res. Commun.* **7**, 250 (1962).

45   J. C. Speck, Jr., P. T. Rowley, and B. L. Horecker, *J. Am. Chem. Soc.* **85**, 1012 (1963).

46   B. L. Horecker, O. Tsolas, and C. Y. Lai, *The Enzymes* **7**, 213 (1972).

47   O. Tsolas and B. L. Horecker, *The Enzymes* **7**, 259 (1972).

48   L. Davis and D. E. Metzler, *The Enzymes* **7**, 33 (1972).

49   A. R. Fersht and W. P. Jencks, *J. Am. Chem. Soc.* **92**, 5442 (1970).

50   A. R. Fersht, *J. Am. Chem. Soc.* **93**, 3504 (1971).

51   R. G. Pearson, *J. Chem. Educ.* **45**, 103, 643 (1968).

52   W. P. Jencks and J. Carriuolo, *J. Am. Chem. Soc.* **82**, 1778 (1960).

53   W. P. Jencks and M. Gilchrist, *J. Am. Chem. Soc.* **90**, 2622 (1968).

54   E. G. Sander and W. P. Jencks, *J. Am. Chem. Soc.* **90**, 6154 (1968).

55   B. A. Connolly and F. Eckstein, *J. Biol. Chem.* **256**, 9450 (1981).

56   S. L. Huang and M. D. Tsai, *Biochemistry* **21**, 1530 (1982).

57   J. Fastrez and A. R. Fersht, *Biochemistry* **12**, 1067 (1973).

58   M. L. Bender and K. Nakamura, *J. Am. Chem. Soc.* **84**, 2577 (1962).

59   M. Caplow and W. P. Jencks, *Fedn. Proc.* **21**, 248 (1962).

60   G. Lowe and Y. Yuthavong, *Biochem. J.* **124**, 107 (1971).

61   J. Klinman, *Biochemistry* **15**, 2018 (1976).

62   J. Järv, T. Kesvatera, and A. Aaviksaar, *Eur. J. Biochem.* **67**, 315 (1976).

63   J. S. Thomsen, *Phys. Rev.* **91**, 1263 (1953).

64   R. M. Krupka, H. Kaplan, and K. J. Laidler, *Trans. Faraday Soc.* **62**, 2754 (1966).

65   D. B. Northrop, *Ann. Rev. Biochem.* **50**, 103 (1981).
66   C. G. Swain, E. C. Stivers, J. F. Reuwer, Jr., and L. J. Staod, *J. Am. Chem. Soc.* **80**, 5885 (1958).
67   C. B. Sawyer and J. F. Kirsch, *J. Am. Chem. Soc.* **95**, 7375 (1973).
68   C. B. Sawyer and J. F. Kirsch, *J. Am. Chem. Soc.* **97**, 1963 (1975).
69   M. H. O'Leary, M. Urberg, and A. P. Young, *Biochemistry* **13**, 2077 (1974).
70   M. H. O'Leary and M. D. Kluetz, *J. Am. Chem. Soc.* **94**, 3584 (1972).
71   S. Rosenberg and J. F. Kirsch, *Biochemistry* **20**, 3189 (1981).
72   F. W. Dahlquist, T. Rand-Meir, and M. A. Raftery, *Proc. Natn. Acad. USA* **61**, 1194 (1968).
73   L. E. H. Smith, L. H. Mohr, and M. A. Raftery, *J. Am. Chem. Soc.* **95**, 7497 (1973).
74   A. J. Kresge, *J. Am. Chem. Soc.* **95**, 3065 (1972).

Further reading

W. P. Jencks, *Catalysis in chemistry and enzymology*, McGraw-Hill (1969).
R. D. Gandour and R. L. Schowen, *Transition states of biochemical processes*, Plenum (1978).
T. C. Bruice and S. J. Benkovic, *Bioorganic mechanisms*, Benjamin (1966).
C. Walsh, *Enzymatic reaction mechanisms*, W. H. Freeman and Company (1979).
H. Dugas and C. Penney, *Bioorganic chemistry*, Springer-Verlag (1981).

3

The basic equations of enzyme kinetics

A. Steady state kinetics

The concept of the steady state is used widely in dynamic systems. It generally refers to the situation in which the value of a particular quantity is constant—is "in a steady state"—because its rate of formation is balanced by its rate of destruction. For example, the population of a country is in a steady state when the birth and immigration rates equal those of death and emigration. Similarly, the concentration of a metabolite in a cell is at a steady state level when it is being produced as rapidly as it is being degraded. In enzyme kinetics, the concept is applied to the concentrations of enzyme-bound intermediates. When an enzyme is mixed with a large excess of substrate, there is an initial period, known as the *pre-steady state*, during which the concentrations of these intermediates build up to their steady state levels. Once the intermediates reach their steady state concentrations, the reaction rate changes relatively slowly with time. It is during this steady state period that the rates of enzymatic reactions are traditionally measured. The steady state is an approximation, since the substrate is gradually depleted during the course of an experiment. But, provided that the rate measurements are restricted to a short time interval over which the concentration of the substrate does not greatly change, it is a very good approximation.

Although the use of pre–steady state kinetics is undoubtedly superior as a means of analyzing the chemical mechanisms of enzyme catalysis (Chapters 4 and 7), steady state kinetics is more important for the understanding of metabolism, since it measures the catalytic activity of an enzyme in the steady state conditions in the cell.

1. The experimental basis: The Michaelis-Menten equation[1]

In deriving the following kinetic expressions, we assume that the concentration of the enzyme is negligible compared with that of the substrate.

Except in the procedures for the rapid reaction measurements described in the next chapter, this is generally true in practice because of the high catalytic efficiency of enzymes. We also assume that what is being measured is the *initial rate* v of formation of products (or depletion of substrates)—that is, the rate of formation of the first few percent of the products—so that they have not significantly accumulated and the substrates have not been appreciably depleted. Under these conditions, changes in reagent concentrations are generally linear with time.

It is found experimentally in most cases that v is directly proportional to the concentration of enzyme, $[E]_0$. However, v generally follows *saturation kinetics* with respect to the concentration of substrate, $[S]$, in the following way (Figure 3·1). At sufficiently low $[S]$, v increases linearly with $[S]$. But as $[S]$ is increased, this relationship begins to break down and v increases less rapidly than $[S]$ until, at sufficiently high or *saturating* $[S]$, v tends toward a *limiting* value termed $V_{\max}$. This is expressed quantitatively in the Michaelis-Menten equation, the basic equation of enzyme kinetics:

$$v = \frac{[E]_0[S]k_{cat}}{K_M + [S]} \qquad (3·1)$$

where

$$k_{cat}[E]_0 = V_{\max} \qquad (3·2)$$

The concentration of substrate at which $v = \frac{1}{2}V_{\max}$ is termed K_M, the Michaelis constant. Note that at low $[S]$, where $[S] \ll K_M$,

$$v = \frac{k_{cat}}{K_M}[E]_0[S] \qquad (3·3)$$

2. Interpretation of the kinetic phenomena for single-substrate reactions: The Michaelis-Menten mechanism

In 1913, L. Michaelis and M. L. Menten developed the theories of earlier workers and proposed the following scheme:

$$E + S \xrightleftharpoons{K_S} ES \xrightarrow{k_{cat}} E + P \qquad (3·4)$$

The catalytic reaction is divided into two processes. The enzyme and the substrate first combine to give an enzyme–substrate complex, ES. This step is assumed to be rapid and reversible with no chemical changes taking place; the enzyme and the substrate are held together by physical forces. The chemical processes then occur in a second step with a first-order rate constant k_{cat} (the turnover number). The rate equations are solved in the following manner.

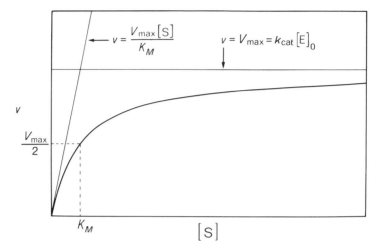

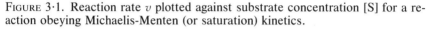

FIGURE 3·1. Reaction rate v plotted against substrate concentration [S] for a reaction obeying Michaelis-Menten (or saturation) kinetics.

From equation 3·4,

$$\frac{[E][S]}{[ES]} = K_S \tag{3·5}$$

and

$$v = k_{cat}[ES] \tag{3·6}$$

Also, the total enzyme concentration, $[E]_0$, and that of the free enzyme, $[E]$, are related by

$$[E] = [E]_0 - [ES] \tag{3·7}$$

Thus

$$[ES] = \frac{[E]_0[S]}{K_S + [S]} \tag{3·8}$$

and

$$v = \frac{[E]_0[S]k_{cat}}{K_S + [S]} \tag{3·9}$$

This is identical to equation 3·1, where K_M is equal to the dissociation constant of the enzyme–substrate complex, K_S.

The concept of the enzyme–substrate complex is the foundation stone of enzyme kinetics and our understanding of the mechanism of enzyme catalysis. In honor of its introducer, this noncovalently bound complex is often termed the *Michaelis* complex.

There is, of course, an enzyme–product complex, EP, through which the reverse reaction proceeds. We assume in these analyses that the dissociation of EP is fast and so can be ignored in the forward reaction. The initial-rate assumption allows us to ignore the accumulation of the EP complex and the reverse reaction, since [P] is always very low.

3. Extensions and modifications of the Michaelis-Menten mechanism

A distinction must be drawn between the equation and the mechanism proposed by Michaelis and Menten. Their equation holds for many mechanisms, but their mechanism is not always appropriate.

The Michaelis-Menten mechanism assumes that the enzyme–substrate complex is in thermodynamic equilibrium with free enzyme and substrate. This is true only if, in the following scheme, $k_2 \ll k_{-1}$:

$$E + S \underset{k_{-1}}{\overset{k_1}{\rightleftharpoons}} ES \overset{k_2}{\longrightarrow} E + P \tag{3·10}$$

The case in which k_2 is comparable to k_{-1} was first analyzed by G. E. Briggs and J. B. S. Haldane in 1925.

a. Briggs-Haldane kinetics:[2] $K_M > K_S$

The solution of scheme 3·10 is somewhat more complicated than the solution of the Michaelis-Menten scheme: the steady state approximation is applied to the concentration of ES. That is, if the reaction rate measured is approximately constant over the time interval concerned, then [ES] is also constant:

$$\frac{d[ES]}{dt} = 0 = k_1[E][S] - k_2[ES] - k_{-1}[ES] \tag{3·11}$$

Substituting equation 3·7 gives

$$[ES] = \frac{[E]_0[S]}{[S] + (k_2 + k_{-1})/k_1} \tag{3·12}$$

and, since $v = k_2[ES]$,

$$v = \frac{[E]_0[S]k_2}{[S] + (k_2 + k_{-1})/k_1} \tag{3·13}$$

This is identical to the Michaelis-Menten equation (3·1), where now

$$K_M = \frac{k_2 + k_{-1}}{k_1} \tag{3·14}$$

Since K_S for the dissociation of [ES] is equal to k_{-1}/k_1, we have

$$K_M = K_S + \frac{k_2}{k_1} \tag{3·15}$$

Of course, when $k_{-1} \gg k_2$, equation 3·14 simplifies to $K_M = K_S$.

b. Intermediates occurring after ES: $K_M < K_S$
The Michaelis-Menten scheme may be extended to cover a variety of cases in which additional intermediates, covalently or noncovalently bound, occur on the reaction pathway. It is found in all examples that the Michaelis-Menten equation still applies, although K_M and k_{cat} are now combinations of various rate and equilibrium constants. K_M is always less than or equal to K_S in these cases. Suppose that, as for example in the following scheme, there are several intermediates and the final catalytic step is slow:

$$\text{E} + \text{S} \underset{}{\overset{K_S}{\rightleftharpoons}} \text{ES} \underset{}{\overset{K}{\rightleftharpoons}} \text{ES}' \underset{}{\overset{K'}{\rightleftharpoons}} \text{ES}'' \xrightarrow[\text{slow}]{k_4} \text{E} + \text{P} \tag{3·16}$$

where $[\text{ES}'] = K[\text{ES}]$ and $[\text{ES}''] = K'[\text{ES}']$. Then

$$K_M = \frac{K_S}{1 + K + KK'} \tag{3·17}$$

and

$$k_{cat} = \frac{k_4 KK'}{1 + K + KK'} \tag{3·18}$$

The chymotrypsin-catalyzed hydrolysis of esters and amides proceeds through scheme 3·19:

$$\text{E} + \text{S} \overset{K_S}{\rightleftharpoons} \text{ES} \overset{k_2}{\underset{\text{P}_1}{\searrow}} \text{EAc} \xrightarrow{k_3} \text{E} \tag{3·19}$$

where EAc is an "acylenzyme." Applying the steady state assumption to [EAc], it may be shown that

$$v = [\text{E}]_0[\text{S}]\left\{ \frac{k_2 k_3/(k_2 + k_3)}{K_S k_3/(k_2 + k_3) + [\text{S}]} \right\} \tag{3·20}$$

This is a Michaelis-Menten equation in which

$$K_M = K_S \frac{k_3}{k_2 + k_3} \tag{3·21}$$

and

$$k_{cat} = \frac{k_2 k_3}{k_2 + k_3} \tag{3·22}$$

or, alternatively,

$$\frac{1}{k_{\text{cat}}} = \frac{1}{k_2} + \frac{1}{k_3} \tag{3.23}$$

(The physical basis of this reciprocal relationship is discussed in section K.) In more complex reactions, the generalization that $K_M < K_S$ can break down.

c. All three mechanisms occur in practice

In the Briggs-Haldane mechanism, when k_2 is much greater than k_{-1}, k_{cat}/k_M is equal to k_1, the rate constant for the association of enzyme and substrate. It is shown in Chapter 4 that association rate constants should be on the order of 10^8 s^{-1} M^{-1}. This leads to a diagnostic test for the Briggs-Haldane mechanism: the value of k_{cat}/K_M is about 10^7 to 10^8 s^{-1} M^{-1}. Catalase, acetylcholinesterase, carbonic anhydrase, crotonase, fumarase, and triosephosphate isomerase all exhibit Briggs-Haldane kinetics by this criterion (see Chapter 4, Table 4.4).

It is extremely common for intermediates to occur after the initial enzyme–substrate complex does, as in equation 3.19. However, it is often found for the physiological substrates that these intermediates do not accumulate and that the slow step in equation 3.19 is k_2. (A theoretical reason for this is discussed in Chapter 12, where examples are given.) Under these conditions, K_M is equal to K_S, the dissociation constant, and the original Michaelis-Menten mechanism is obeyed to all intents and purposes. The opposite occurs in many laboratory experiments. The enzyme kineticist often uses synthetic, highly reactive substrates to assay enzymes, and covalent intermediates frequently accumulate.

B. The significance of the Michaelis-Menten parameters

1. The meaning of k_{cat}: The catalytic constant

In the simple Michaelis-Menten mechanism in which there is only one enzyme–substrate complex and all binding steps are fast, k_{cat} is simply the first-order rate constant for the chemical conversion of the ES complex to the EP complex. For more complicated reactions, k_{cat} is a function of all the first-order rate constants, and it cannot be assigned to any particular process except when simplifying features occur. For example, in the Briggs-Haldane mechanism, when the dissociation of the EP complex is fast, k_{cat} is equal to k_2 (equation 3.10). But if dissociation of the EP complex is slow, the rate constant for this process contributes to k_{cat}, and, in the extreme case in which EP dissociation is far slower than the chemical steps, k_{cat} will be equal to the dissociation rate constant. In the example of equation 3·19, k_{cat} was seen to be a function of k_2 and k_3. But if one of these constants is much smaller than the other, it becomes equal to k_{cat}. For example, if $k_3 \ll k_2$, then, from equation 3·22, $k_{cat} = k_3$. An extension of this is that k_{cat} cannot be greater than any first-order rate constant on the forward reaction pathway.[3] It thus sets a *lower* limit on the chemical rate constants.

The constant k_{cat} is often called the *turnover number* of the enzyme because it represents the maximum number of substrate molecules converted to products per active site per unit time, or the number of times the enzyme "turns over" per unit time.

Rule: The k_{cat} is a first-order rate constant that refers to the properties and reactions of the enzyme–substrate, enzyme–intermediate, and enzyme–product complexes.

2. The meaning of K_M: Real and apparent equilibrium constants

Although it is only for the simple Michaelis-Menten mechanism or in similar cases that $K_M = K_S$, the true dissociation constant of the enzyme–substrate complex, K_M may be treated for *some* purposes as an *apparent* dissociation constant. For example, the concentration of free enzyme in solution may be calculated from the relationship

$$\frac{[E][S]}{\Sigma\,[ES]} = K_M \tag{3·24}$$

where $\Sigma\,[ES]$ is the sum of *all* the bound enzyme species.†

The concept of *apparent* values is very useful, and it appears in other phenomena, such as in pK_a values. Quite often, a pK_a value does not represent the microscopic ionization of a particular group but is a combination of this value and various equilibrium constants between different conformational states of the molecule. The result is an *apparent* pK_a which may be handled titrimetrically as a simple pK_a. This simple-minded approach must not be taken too far, and, when one is considering the effects of temperature, pH, etc. on an apparent K_M, one must realize that

---

† It is possible to devise kinetic curiosities that give Michaelis-Menten kinetics without the enzyme being saturated with the substrate. For example, in the following scheme—where the active form of the enzyme reacts with the substrate in a second-order reaction to give the products and an inactive form of the enzyme, E′, which slowly reverts to the active form—apparent saturation kinetics are followed with $k_{cat} = k_2$ and $K_M = k_2/k_1$. Equation 3·24 applies to this example if E′ is treated as a "bound" form of the enzyme:

$$E + S \xrightarrow{k_1} E' + P$$
$$\downarrow{\scriptstyle k_2}$$
$$E$$

the rate-constant components of this term are also affected. The same applies to k_{cat} values. The literature contains examples in which breaks in the temperature dependence of k_{cat} have been interpreted as indicative of conformational changes in the enzyme, when, in fact, they are due to a different temperature dependence of the individual rate constants in k_{cat}, e.g., k_2 and k_3 in equation 3·22.

An illustration of the way that K_M is a measure of the amount of enzyme that is bound in any form whatsoever to the substrate is given by the following mechanism (cf. the chymotrypsin mechanism, equation 3·19):

$$E + S \underset{}{\overset{K_S}{\rightleftharpoons}} ES \xrightarrow{k_2} ES' \xrightarrow{k_3} E + P \qquad (3·25)$$

Application of the steady state approximation to [ES'] gives

$$[ES'] = [ES] \frac{k_2}{k_3} \qquad (3·26)$$

When k_2 is much greater than k_3, [ES'] is much greater than [ES], so that ES' makes a more important contribution to K_M than does ES and is the predominant enzyme-bound species. Without solving the equations for the reaction, we can say intuitively that K_M must be smaller than K_S by a factor of about k_3/k_2; i.e.,

$$K_M \approx K_S \frac{k_3}{k_2} \qquad (3·27)$$

In all cases, K_M is the substrate concentration at which $v = V_{max}/2$.

Rule: The K_M is an apparent dissociation constant that may be treated as the overall dissociation constant of all enzyme-bound species.

3. The meaning of k_{cat}/K_M: The specificity constant

It was pointed out earlier that the reaction rate for low substrate concentrations is given by $v = (k_{cat}/K_M)[E]_0[S]$ (equation 3·3); that is, k_{cat}/K_M is an *apparent* second-order rate constant. It is not a true microscopic rate constant except in the extreme case in which the rate-determining step in the reaction is the encounter of enzyme and substrate.

The importance of k_{cat}/K_M is that it relates the reaction rate to the concentration of free, rather than total, enzyme. This is readily seen from equation 3·3, mentioned above, since at low substrate concentrations the enzyme is largely unbound and $[E] \approx [E]_0$. At such concentrations the reaction rate is thus given by

$$v = [E][S] \frac{k_{cat}}{K_M}$$

It will be shown later (equation 3·41) that this result holds at *any* substrate concentration. It will also be shown later (equation 3·44) that k_{cat}/K_M determines the specificity for competing substrates. For this reason, k_{cat}/K_M is sometimes referred to as the "specificity constant."

The value of k_{cat}/K_M cannot be greater than that of any second-order rate constant on the forward reaction pathway.[3] It thus sets a *lower* limit on the rate constant for the association of enzyme and substrate.

Rule: The k_{cat}/K_M is an apparent second-order rate constant that refers to the properties and the reactions of the free enzyme and free substrate.

C. Graphical representation of data

It is very useful to transform the Michaelis-Menten equation into a linear form for analyzing data graphically and detecting deviations from the ideal behavior. One of the best known methods is the double-reciprocal or Lineweaver-Burk plot. Inverting both sides of equation 3·1 and substituting equation 3·2 gives the Lineweaver-Burk plot:[4]

$$\frac{1}{v} = \frac{1}{V_{max}} + \frac{K_M}{V_{max}[S]} \tag{3·28}$$

Plotting $1/v$ against $1/[S]$ (Figure 3·2) gives an intercept of $1/V_{max}$ on the y axis as $1/[S]$ tends toward zero, and of $1/[S] = -1/K_M$ on the x axis. The slope of the line is K_M/V_{max}.

Another common plot is that of G. S. Eadie and B. H. J. Hofstee (equation 3·29).[5,6] Equations 3·1 and 3·2 may be rearranged to give

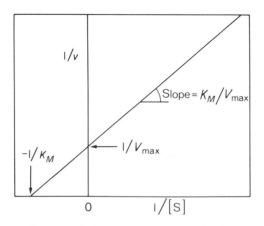

FIGURE 3·2. The Lineweaver-Burk plot.

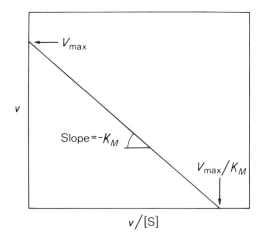

FIGURE 3·3. The Eadie-Hofstee plot.

$$v = V_{max} - \frac{K_M v}{[S]} \qquad (3·29)$$

Plotting v against $v/[S]$ (Figure 3·3) gives an intercept of V_{max} on the y axis as $v/[S]$ tends toward zero. The slope of the line is equal to $-K_M$. The intercept on the x axis is at $v/[S] = V_{max}/K_M$.

The Lineweaver-Burk plot has the disadvantage of compressing the data points at high substrate concentrations into a small region and emphasizing the points at lower concentrations. It does have the advantage that the values of v for a given value of [S] are easy to read from it.

The Eadie plot does not compress the higher values, but the values of v against [S] are more difficult to determine rapidly from it. The Eadie plot is considered more accurate and generally superior.[7,8]

D. Inhibition

As well as being irreversibly inactivated by heat or chemical reagents, enzymes may be *reversibly* inhibited by the noncovalent binding of inhibitors. There are four main types of inhibition.

1. Competitive inhibition

If an inhibitor I binds reversibly to the active site of the enzyme and prevents S binding and *vice versa*, I and S compete for the active site and I is said to be a *competitive* inhibitor. In the case of the simple Michaelis-

Menten mechanism (equation 3·4, where $K_M = K_S$), an additional equilibrium must be considered, i.e.,

$$\begin{array}{c} \text{E} \xrightleftharpoons{\text{S, } K_M} \text{ES} \xrightarrow{k_{cat}} \text{E} + \text{P} \\ \scriptstyle{\text{I, } K_I} \big\updownarrow \\ \text{EI} \end{array} \qquad (3\cdot30)$$

Solving the equilibrium and rate equations by using

$$[\text{E}]_0 = [\text{ES}] + [\text{EI}] + [\text{E}] \qquad (3\cdot31)$$

gives

$$v = \frac{[\text{E}]_0[\text{S}]k_{cat}}{[\text{S}] + K_M(1 + [\text{I}]/K_I)} \qquad (3\cdot32)$$

K_M is apparently increased by a factor of $(1 + [\text{I}]/K_I)$. This equation holds

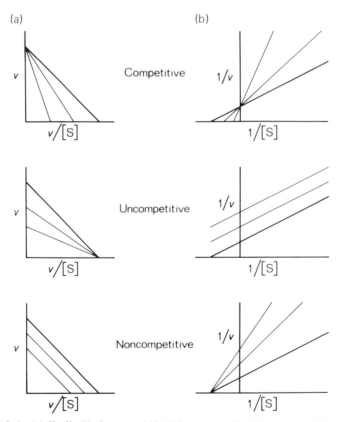

FIGURE 3·4. (a) Eadie-Hofstee and (b) Lineweaver-Burk plots of different types of inhibition. The heavier line in each plot shows the reaction in the absence of inhibitor; the lighter lines are for the reaction in the presence of inhibitor.

for all mechanisms obeying the Michaelis-Menten equation. Competitive inhibition affects K_M only and not V_{max}, since infinitely high concentrations of S displace I from the enzyme.

2. Noncompetitive, uncompetitive, and mixed inhibition

Different inhibition patterns occur if I and S bind simultaneously to the enzyme instead of competing for the same binding site (Figure 3·4):

$$
\begin{array}{ccc}
\text{E} & \underset{}{\overset{S,\ K_M}{\rightleftharpoons}} & \text{ES} \xrightarrow{k_{cat}} \\
{\scriptstyle I,\ K_I}\ \updownarrow & & \updownarrow\ {\scriptstyle I,\ K_I} \\
\text{EI} & \underset{}{\overset{S,\ K_M}{\rightleftharpoons}} & \text{ESI} \xrightarrow{k'}
\end{array}
\tag{3.33}
$$

It may be shown from the Michaelis-Menten mechanism—in the simplifying case in which the dissociation constant of S from EIS is the same as that from ES (i.e., $K_M = K'_M$) but in which ESI does not react (i.e., $k' = 0$)—that

$$
v = \frac{[E]_0[S]k_{cat}/(1 + [I]/K_I)}{[S] + K_M}
\tag{3.34}
$$

This is termed *noncompetitive inhibition*: K_M is unaffected, but k_{cat} is lowered by a factor of $(1 + [I]/K_I)$. More commonly, the dissociation constant of S from EIS is different than that from ES. In this case, both K_M and k_{cat} are altered and the inhibition is termed *mixed*. A further type of inhibition, *uncompetitive*, occurs when I binds to ES but not to E (Figure 3·5).

E. Nonproductive binding

In some reactions, a substrate binds in an alternative unreactive mode at the active site of the enzyme, in competition with the productive mode of binding:

$$\tag{3.35}$$

This is known as nonproductive binding. The effect of such binding on the Michaelis-Menten mechanism is to lower both the k_{cat} and the K_M. The k_{cat} is lowered since, at saturation, only a fraction of the substrate

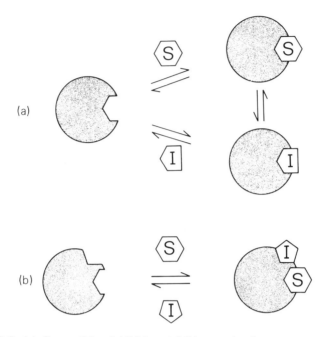

FIGURE 3·5. (a) Competitive inhibition: inhibitor and substrate compete for the same binding site. For example, indole, phenol, and benzene bind in the binding pocket of chymotrypsin and inhibit the hydrolysis of derivatives of tryptophan, tyrosine, and phenylalanine. (b) Noncompetitive inhibition: inhibitor and substrate bind simultaneously to the enzyme. An example is the inhibition of fructose 1,6-diphosphatase by AMP. This type of inhibition is very common with multi-substrate enzymes. A rare example of uncompetitive inhibition of a single-substrate enzyme is the inhibition of alkaline phosphatase by L-phenylalanine. This enzyme is composed of two identical subunits, so presumably the phenylalanine binds at one site and the substrate at the other. [From N. K. Ghosh and W. H. Fishman, *J. Biol. Chem.* **241**, 2516 (1966); see also M. Caswell and M. Caplow, *Biochemistry* **19**, 2907 (1980).]

is bound productively. The K_M is lower than the K_S because the existence of additional binding modes must lead to apparently tighter binding.

Solving equation 3·35 by the usual procedures gives

$$v = \frac{[E]_0 [S] k_2}{K_S + [S](1 + K_S/K_S')} \tag{3·36}$$

Comparison with the Michaelis-Menten equation (3·1) gives

$$k_{cat} = \frac{k_2}{1 + K_S/K_S'} \tag{3·37}$$

and

$$K_M = \frac{K_S}{1 + K_S/K_S'} \tag{3.38}$$

It should be noted that

$$\frac{k_{cat}}{K_M} = \frac{k_2}{K_S} \tag{3.39}$$

That is, k_{cat}/K_M is unaffected by the presence of the additional binding mode, since k_{cat} and K_M are altered in a compensating manner. For example, if the nonproductive site binds a thousand times more strongly than the productive site, K_M will be a thousand times lower than K_S, but since only one molecule in a thousand is productively bound, k_{cat} is a thousand times lower than k_2 (strictly speaking, the factor is 1001).

F. $k_{cat}/K_M = k_2/K_S$

The result of equation 3.39 for nonproductive binding is quite general. It applies to cases in which intermediates occur on the reaction pathway as well as in the nonproductive modes. For example, in equation 3.19 for the action of chymotrypsin on esters with accumulation of an acylenzyme, it is seen from the ratios of equations 3.21 and 3.22 that $k_{cat}/K_M = k_2/K_S$. This relationship clearly breaks down for the Briggs-Haldane mechanism in which the enzyme–substrate complex is not in thermodynamic equilibrium with the free enzyme and substrates. It should be borne in mind that K_M might be a complex function when there are several enzyme-bound intermediates in rapid equilibrium, as in equation 3.16. Here k_{cat}/K_M is a function of all the bound species.

G. Competing substrates

1. An alternative formulation of the Michaelis-Menten equation

Suppose that two substrates compete for the active site of the enzyme:

$$\tag{3.40}$$

The reaction rates may be calculated by the usual steady state or

Michaelis-Menten assumptions. However, there is an alternative approach for rapidly calculating the ratio of the reaction rates. Substitution of equation 3·24 into the Michaelis-Menten equation (3·1) gives

$$v = \frac{k_{cat}}{K_M}[E][S] \tag{3·41}$$

where [E] is the concentration of free or unbound enzyme. This is a useful equation since it is based on [E] rather than $[E]_0$. Several important relationships may be inferred directly from this equation without the need for a detailed mechanistic analysis, as shown below.

2. Specificity for competing substrates
If two substrates A and B compete for the enzyme, then

$$-\frac{d[A]}{dt} = v_A = \left(\frac{k_{cat}}{K_M}\right)_A [E][A] \tag{3·42}$$

and

$$-\frac{d[B]}{dt} = v_B = \left(\frac{k_{cat}}{K_M}\right)_B [E][B] \tag{3·43}$$

which give

$$\frac{v_A}{v_B} = \frac{(k_{cat}/K_M)_A[A]}{(k_{cat}/K_M)_B[B]} \tag{3·44}$$

The important conclusion is that specificity, in the sense of discrimination between two competing substrates, is determined by the ratios of k_{cat}/K_M and not by K_M alone. Since k_{cat}/K_M is unaffected by nonproductive binding (section E) and by the accumulation of intermediates (section F), these phenomena do not affect specificity (see Chapter 13). Note that equation 3·44 holds at all concentrations of substrates.

H. Reversibility: The Haldane equation

1. Equilibria in solution

$$S \underset{k_r}{\overset{k_f}{\rightleftharpoons}} P \tag{3·45}$$

$$K_{eq} = \frac{[P]}{[S]} = \frac{k_f}{k_r} \tag{3·46}$$

An enzyme cannot alter the equilibrium constant between the free-solution concentrations of S and P. This places constraints on the relative values of k_{cat}/K_M for the forward and reverse reactions. Specifically, since the rates of formation of P and S are equal at equilibrium, application of equation 3·44 gives

$$\left(\frac{k_{cat}}{K_M}\right)_S [E][S] = \left(\frac{k_{cat}}{K_M}\right)_P [E][P] \qquad (3\cdot47)$$

so that

$$\frac{(k_{cat}/K_M)_S}{(k_{cat}/K_M)_P} = K_{eq} \qquad (3\cdot48)$$

This relationship is known as the Haldane equation, after J. B. S. Haldane, who derived it in 1930.[9]

2. Equilibria on the enzyme surface

The Haldane equation does not relate the equilibrium constant between ES and EP to that between S and P in solution. The equilibrium constant for the enzyme-bound reagents is often very different from that in solution for several reasons:

1. *Strain.* The geometry of the active site may be such that, for example, P is bound more tightly than S. The equilibrium on the enzyme surface will favor P more than the equilibrium in solution.
2. *Nonproductive binding.* If the enzyme has binding modes for S other than the catalytically productive mode, these will favor [ES] in the equilibrium.
3. *Entropy.* In the case where the product is two separate molecules,

$$S \rightleftharpoons P + P' \qquad (3\cdot49)$$

the equilibrium constant in solution has a term reflecting the favorable gain in entropy on formation of two molecules from one. However, if both P and P' are bound on the enzyme surface—

$$ES \rightleftharpoons EPP' \qquad (3\cdot50)$$

—the relevant equilibrium constant will not have this entropic contribution.

One example in which an enzyme-bound equilibrium is vastly different from the equilibrium in solution is indicated by the hydrolysis constant for ATP in equilibrium with ADP and orthophosphate bound to myosin. The constant is only 9, compared with a value of 10^5 to 10^6 for the equilibrium in solution.[10] This phenomenon is discussed further in Chapter 12, section E2.

I. Breakdown of the Michaelis-Menten equation

Apart from essentially trivial reasons such as an experimental inability to measure initial rates, there are two main reasons for the failure of the Michaelis-Menten equation.

The first possibility is substrate inhibition. A second molecule of substrate binds to give an ES_2 complex that is catalytically inactive. If, in a simple Michaelis-Menten mechanism, the second dissociation constant is K'_S, then

$$v = \frac{[E]_0[S]k_{cat}}{K_S + [S] + [S]^2/K'_S} \tag{3.51}$$

At low concentrations of S, the rate is given by $v = [E]_0[S]k_{cat}/K_S$, as usual. But as [S] increases, there is first a maximum value of v followed by a decrease.

The second possibility is substrate activation: an ES_2 complex is formed that is more active than ES.

J. Multisubstrate systems

We have dealt so far with enzymes that react with a single substrate only. The majority of enzymes, however, involve two substrates. The dehydrogenases, for example, bind both NAD^+ and the substrate that is to be oxidized. Many of the principles developed for the single-substrate systems may be extended to multisubstrate systems. However, the general solution of the equations for such systems is complicated and well beyond the scope of this book. Many books devoted almost solely to the detailed analysis of the steady state kinetics of multisubstrate systems have been published, and the reader is referred to these for advanced study.[11-14] The excellent short accounts by W. W. Cleland[15] and K. Dalziel[16] are highly recommended.

From the point of view of this book, the most important experimental observation is that most reactions obey Michaelis-Menten kinetics when the concentration of one substrate is held constant and the other is varied. Furthermore, in practice, only a limited range of mechanisms is commonly observed. In this section we shall just list some common pathways and give a glossary of terms.

Reactions in which all the substrates bind to the enzyme before the first product is formed are called *sequential*. Reactions in which one or more products are released before all the substrates are added are called *ping-pong*. Sequential mechanisms are called *ordered* if the substrates combine with the enzyme and the products dissociate in an obligatory order. A *random* mechanism implies no obligatory order of combination or release. The term *rapid equilibrium* is applied when the chemical steps are slower than those for the binding of reagents. Some examples follow.

1. The random sequential mechanism

$$(3 \cdot 52)$$

The complex EAB is called a *ternary* or *central* complex.

2. The ordered mechanism

$$E \xrightleftharpoons{A} EA \xrightleftharpoons{B} EAB \longrightarrow \qquad (3 \cdot 53)$$

Ordered mechanisms often occur in the reactions of the NAD^+-linked dehydrogenases, with the coenzyme binding first. The molecular explanation for this is that the binding of the dinucleotide causes a conformational change that increases the affinity of the enzyme for the other substrate (see Chapter 15).

3. The Theorell-Chance mechanism

The Theorell-Chance mechanism is an ordered mechanism in which the ternary complex does not accumulate under the reaction conditions, as is found for horse liver alcohol dehydrogenase:

$$E \xrightleftharpoons{A} EA \xrightarrow{\;\;B\;\;P\;\;} EQ \qquad (3 \cdot 54)$$

(P is one product and Q the other—acetaldehyde and NADH, respectively, for the liver alcohol dehydrogenase.)

4. The ping-pong (or substituted-enzyme or double-displacement) mechanism

The following type of reaction, in which the enzyme reacts with one substrate to give a covalently modified enzyme and releases one product, and then reacts with the second substrate, gives rise to the characteristic family of Lineweaver-Burk plots illustrated in Figure 3·6:

$$E + A \rightleftharpoons E \cdot A \rightleftharpoons E{-}P + Q \qquad (3 \cdot 55)$$

$$E{-}P + B \rightleftharpoons E{-}P \cdot B \longrightarrow E + P{-}B \qquad (3 \cdot 56)$$

A ping-pong reaction occurs, for example, when a phosphate-transferring enzyme, such as phosphoglycerate mutase, is phosphorylated by one substrate to form a phosphorylenzyme (E—P in equation 3·55), which then transfers the phosphoryl group to a second substrate (equation 3·56). In the case of phosphoglycerate mutase (equation 3·57), N is the imidazole

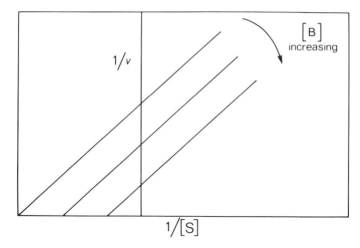

FIGURE 3·6. The characteristic parallel-line reciprocal plots of ping-pong kinetics. As the concentration of the second substrate in the sequence increases, V_{max} increases, as does the K_M for the first substrate. V_{max}/K_M, the reciprocal of the slope of the plot, remains constant (see equation 3·68).

side chain of a histidine residue. Another example occurs in the transfer

$$E{-}N + ROPO_3^{2-} \rightleftharpoons E{-}N{\cdot}ROPO_3^{2-} \xrightarrow[ROH]{} E{-}N{-}PO_3^{2-} \xrightarrow{R'OH}$$

$$E{-}N + R'OPO_3^{2-} \qquad (3\cdot57)$$

of an acyl group from acetyl–coenzyme A to sulfanilimide or to another amine in a reaction catalyzed by acetyl–coenzyme A : arylamine acetyl-transferase (equation 3·58). This reaction almost certainly involves the

$$E{-}SH + CH_3COS\text{-}CoA \rightleftharpoons E{-}SH{\cdot}CH_3COS\text{-}CoA \xrightarrow[HS\text{-}CoA]{} E{-}S{-}COCH_3 \xrightarrow{RNH_2}$$

$$E{-}SH + RNHCOCH_3 \qquad (3\cdot58)$$

formation of an acylthioenzyme in which the —SH of a cysteine residue is acylated.[17] The rate equation for reactions 3·57 and 3·58 is given later (3·66).

In many ways, ping-pong kinetics is the most mechanistically informative of all the types of steady state kinetics, since information is given about the occurrence of a covalent intermediate. The finding of ping-pong kinetics is often used as evidence for such an intermediate, but because other kinetic pathways can give rise to the characteristic parallel double-reciprocal plots of Figure 3·6, the evidence must always be treated with caution and confirming data should be sought.

Steady state kinetics may be used to distinguish between the various mechanisms mentioned above. Under the appropriate conditions, their application can determine the order of addition of substrates and the order of release of products from the enzyme during the reaction. For this reason, the term "mechanism" when used in steady state kinetics often refers just to the sequence of substrate addition and product release.

K. Useful kinetic short cuts

We end with two ideas that should provide insight into why many rate equations have their particular mathematical forms. These ideas point to useful short cuts for quickly noting the effects of additional intermediates on mechanisms, and even for solving certain complicated mechanisms by inspection instead of by analyzing the full steady state rate equations.

1. Calculation of net rate constants[18]

It is possible to reduce the rate constants for a series of reactions as in equation 3·59 to a single net rate constant, or to a series of single

$$A \underset{k_{-1}}{\overset{k_1}{\rightleftharpoons}} B \underset{k_{-2}}{\overset{k_2}{\rightleftharpoons}} C \underset{k_{-3}}{\overset{k_3}{\rightleftharpoons}} D \underset{k_{-4}}{\overset{k_4}{\rightleftharpoons}} E \overset{k_5}{\longrightarrow} F \qquad (3·59)$$

rate constants, by just considering a net rate constant for the flux going through each step. To illustrate this, we consider the simpler reaction

$$X \underset{k_{-1}}{\overset{k_1}{\rightleftharpoons}} Y \overset{k_2}{\longrightarrow} Z \qquad (3·60)$$

The rate of X going to Z via Y is given by the rate of X going to Y ($= k_1[X]$) times the probability of Y going to Z rather than reverting to X [i.e., $k_2/(k_{-1} + k_2)$]. The net rate constant for $X \rightarrow Y$, k'_1, is thus given by

$$k'_1 = \frac{k_1 k_2}{k_{-1} + k_2} \qquad (3·61)$$

The same treatment can be applied to equation 3·59, starting from the irreversible step on the right-hand side and progressively working to the left. For example, the net rate constant for $D \rightarrow E$, k'_4, equals $k_4 k_5/(k_{-4} + k_5)$, as in equation 3·61. The net rate constant for $C \rightarrow D$, k'_3, is calculated by analogy with equation 3·61 to be $k_3 k'_4/(k_{-3} + k'_4)$. This is continued sequentially to eventually give the net rate constant for $A \rightarrow B$, i.e., $k'_1 = k_1 k'_2/(k_{-1} + k'_2)$.

This procedure provides a very simple means of calculating partitioning ratios where there is a branch point in a pathway. For example, suppose that in equation 3·59, A may also give directly a product P with rate constant k_P as well as giving F via B, C, D, and E:

$$(3·62)$$

The partitioning is simply calculated by using the net rate constant k_1' in equation 3·62. The rate of formation of F relative to that of P is thus equal to k_1'/k_P. Cleland has shown how the net rate constant method may be applied to more complex problems.[18]

2. Use of transit times instead of rate constants

The value of V_{max}, k_{cat}, or any rate constant for a series of sequential reactions may be derived by considering the *time* taken for each step, as follows. The dimensions of v are moles per second. The dimensions of $1/v$ are seconds per mole, and $1/v$ is the time taken for 1 mol of reagents to give products. Similarly, the reciprocal of k_{cat} (i.e., $[E]_0/V_{max}$) has the dimensions of seconds and is the time taken for one molecule of reagent to travel the whole reaction pathway in the steady state at saturating $[S]$. The reciprocal rate constant may be considered a *transit time*.

For a series of reactions as in equation 3·63, the reciprocal of the rate constant for any individual step is its transit time. The total time for

$$E·P_1 \xrightarrow{k_1} E·P_2 \xrightarrow{k_2} E·P_3 \xrightarrow{k_3} E·P_4 \xrightarrow{k_4} \cdots \xrightarrow{k_{n-1}} E·P_n \qquad (3·63)$$

one molecule to be converted from P_1 to P_n, $1/k$, is given by the sum of the transit times for each step. That is,

$$\frac{1}{k} = \frac{1}{k_1} + \frac{1}{k_2} + \frac{1}{k_3} + \frac{1}{k_4} + \cdots + \frac{1}{k_{n-1}} \qquad (3·64)$$

where $1/k = 1/k_{cat} = [E]_0/V_{max}$ for saturating conditions. This is precisely the relationship derived earlier for k_{cat} in the acylenzyme mechanism for chymotrypsin (equation 3·23), and it gives the physical reason for the reciprocal relationship between k_{cat} and the first-order rate constants on the pathway: the reciprocals of the rate constants, i.e., the transit times, are additive, so that the time taken for a molecule to traverse the whole reaction pathway is the sum of the times taken for each step. For concentrations of S below saturating, the transit time for the whole reaction is $[E]_0/v$. (The transit times are analogous to resistances in electrical circuits, and may be summed in the same way: for a series of reactions, the

residence times are additive; for parallel reaction pathways, the reciprocals of the residence times are additive.) This is the physical reason
why the rate laws for steady state mechanisms are usually written in terms
of $[E]_0/v$.

As an example, we consider the Briggs-Haldane mechanism:

$$E + A \underset{k_{-1}}{\overset{k_1}{\rightleftharpoons}} EA \xrightarrow{k_2} E + P \tag{3.65}$$

The binding step is reduced to the net rate constant

$$\frac{k_1[A]k_2}{k_{-1} + k_2}$$

The total transit time is given by

$$\frac{[E]_0}{v} = \frac{k_{-1} + k_2}{k_1[A]k_2} + \frac{1}{k_2} \tag{3.66}$$

Equation 3·66 is, in fact, the Lineweaver-Burk double-reciprocal plot
(equation 3·28) in slight disguise: $[E]_0$ has been moved to the left-hand
side, and from equation 3·14, $K_M = (k_{-1} + k_2)/k_1$.

The ping-pong mechanisms of equations 3·57 and 3·58, for example,
may similarly be solved by inspection. Restating those equations as

$$E + A \underset{k_{-1}}{\overset{k_1}{\rightleftharpoons}} EA \xrightarrow{k_2} E—P \xrightarrow{k_3[B]} E + P—B \tag{3.67}$$

the transit times are summed to give

$$\frac{[E]_0}{v} = \frac{k_{-1} + k_2}{k_1[A]k_2} + \frac{1}{k_2} + \frac{1}{k_3[B]} \tag{3.68}$$

The total transit time is longer because of the third step. If the reciprocals
of equation 3·68 are taken, a Michaelis-Menten-type equation is generated.

This method is very useful for solving the kinetics of polymerization
reactions (e.g., DNA polymerization, Chapter 14) by inspection: the time
taken to synthesize a polymer is the sum of the transit times for the
addition of each monomer. For example, suppose that each step in equation 3·63 is a Michaelis-Menten process of

$$E·P_x \underset{k_{-y}}{\overset{k_y[Y]}{\rightleftharpoons}} E·P_x·Y \xrightarrow{k_x} E·P_{x+1} \tag{3.69}$$

Then, the step may be reduced to a single net rate constant k_x', as above,
given by

$$k_x' = \frac{k_y[Y]k_x}{k_{-y} + k_x} \tag{3.70}$$

and the transit times may be summed to give $[E]_0/v$.

References

1   L. Michaelis and M. L. Menten, *Biochem. Z.* **49**, 333 (1913).
2   G. E. Briggs and J. B. S. Haldane, *Biochem. J.* **19**, 338 (1925).
3   L. Peller and R. A. Alberty, *J. Am. Chem. Soc.* **81**, 5907 (1959).
4   H. Lineweaver and D. Burk, *J. Am. Chem. Soc.* **56**, 658 (1934).
5   G. S. Eadie, *J. Biol. Chem.* **146**, 85 (1942).
6   B. H. J. Hofstee, *Nature, Lond.* **184**, 1296 (1959).
7   J. E. Dowd and D. S. Riggs, *J. Biol. Chem.* **249**, 863 (1965).
8   G. L. Atkins and I. A. Nimmo, *Biochem. J.* **149**, 775 (1975).
9   J. B. S. Haldane, *Enzymes*, Longmans, Green and Co. (1930). M.I.T. Press (1965).
10.   C. R. Bagshaw and D. R. Trentham, *Biochem. J.* **133**, 323 (1973).
11   H. J. Fromm, *Initial rate enzyme kinetics*, Springer (1975).
12   J. T.-F. Wong, *Kinetics of enzyme mechanisms*, Academic Press (1975).
13   I. H. Segel, *Enzyme kinetics*, Wiley (1975).
14   A. Cornish-Bowden, *Principles of enzyme kinetics*, Butterworth (1975).
15   W. W. Cleland, *The Enzymes* **2**, 1 (1970).
16   K. Dalziel, *The Enzymes* **10**, 2 (1975).
17   W. P. Jencks, M. Gresser, M. S. Valenzuela, and F. C. Huneeus, *J. Biol. Chem.* **247**, 3756 (1972).
18   W. W. Cleland, *Biochemistry* **14**, 3220 (1974).

4

Measurement and magnitude of enzymatic rate constants

PART 1 Methods for measurement:
An introduction to pre–steady state kinetics

Steady state kinetic measurements on an enzyme usually give only two pieces of kinetic data, the K_M value, which may or may not be the dissociation constant of the enzyme–substrate complex, and the k_{cat} value, which may be a microscopic rate constant but may also be a combination of the rate constants for several steps. The kineticist does have a few tricks that may be used on occasion to detect intermediates and even measure individual rate constants, but these are not general and depend on mechanistic interpretations. (Some examples of these methods will be discussed in Chapter 7.) In order to measure the rate constants of the individual steps on the reaction pathway and detect transient intermediates, it is necessary to measure the rate of approach to the steady state. It is during the time period in which the steady state is set up that the individual rate constants may be observed.

Since values of k_{cat} lie between 1 and 10^7 s^{-1}, measurements must be made in a time range of 1 to 10^{-7} s. This requires either techniques for rapidly mixing and then observing the enzyme and substrate, or totally new methods. Also, since the events that are to be observed occur on the enzyme itself, the enzyme must be available in substrate quantities. The development of apparatus for measuring these rapid reactions and of techniques for isolating large quantities of pure proteins has revolutionized enzyme kinetics.

In sections A to C, we shall discuss four types of techniques. The first is *rapid mixing*. This is extremely useful since it is possible to mix two solutions in a fraction of a millisecond, and the majority of enzyme turnover numbers are less than 1000 s^{-1}. Rapid mixing techniques are now standard laboratory practice because of their ease and their wide range of application.

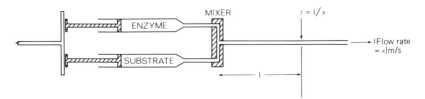

FIGURE 4·1. Continuous-flow apparatus.

A. Rapid mixing and sampling techniques

1. The continuous-flow method

In 1923, H. Hartridge and F. J. W. Roughton introduced the continuous-flow method to solution kinetics in order to study the combination of deoxyhemoglobin with ligands.[1] The principle of the method is illustrated in Figure 4·1. Two syringes are connected by a mixing chamber to a flow tube. One syringe is filled with enzyme, the other is filled with substrate, and the two are compressed at a constant rate. The two solutions mix thoroughly in the mixing chamber, pass down the flow tube, and "age." At a constant flow rate the age of the solution is linearly proportional to the distance down the flow tube and the flow rate; e.g., if the flow rate is 10 m s^{-1}, then 1 cm from the mixing chamber the solution is 1 ms old, 10 cm from the chamber it is 10 ms old, and so on. The flow rate of the liquid must be kept above a critical velocity in order to ensure "turbulent flow." Below this value, which is about 2 m s^{-1} for a tube with a 1-mm diameter, the flow may be laminar, the liquid at the center traveling faster than that near the wall. This places an upper time limit on the apparatus for a particular length of tube.

2. The stopped-flow method

The stopped-flow method was introduced by Roughton in 1934[2] and greatly improved by B. Chance some six years later.[3] The principle is illustrated in Figure 4·2.[4] In contrast to the setup in the continuous-flow system, the two driving syringes are compressed to express about 50 to 200 μL from each, and then they are mechanically stopped. Suppose that there is an observation point 1 cm from the mixing chamber. If the flow rate is 10 m s^{-1} during the period of compression, during this *continuous-flow* period the detector sees a solution that is 1 ms old. When the flow is stopped, the solution ages normally with time and the detector sees the events occurring after 1 ms. The age of the solution at the initial observation is known as the "dead time" of the apparatus.

The stopped-flow method is a routine laboratory tool, whereas the

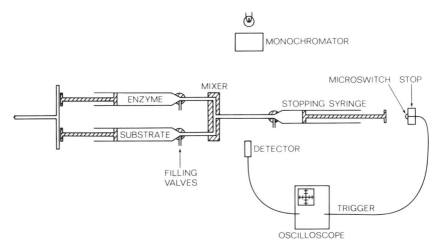

FIGURE 4·2. Stopped-flow apparatus.

continuous-flow apparatus is used in a few specialized cases only. The stopped-flow technique requires only 100 to 400 μL of solution or less for the complete time course of a reaction; the dead time is as low as 0.5 ms or so; and observations may be extended to several minutes. Stopped flow does, however, require a rapid detection and recording system. The continuous-flow system requires very large reaction volumes, and readings may be taken only up to about 100 ms due to the impracticability of using a longer observation tube. Another difficulty is that the whole length of the flow tube must be scanned by the detector. Besides mechanical problems, this may lead to systematic errors if the tube is not uniform in dimensions, thermostatting, etc. Continuous flow is, however, a slightly faster technique, with dead times as low as 100 to 200 μs, since there are no mechanical problems of stopping the flow: the stopping takes a fraction of a millisecond and can set up shock waves if it is too vigorous. A second advantage is that slowly responding detectors may be used, because at a particular point on the flow tube the age of the solution is constant. This was particularly important in the original experiments of Hartridge and Roughton, who, before the introduction of photomultipliers, used a hand (reversion) spectroscope as a detector!

3. Rapid quenching techniques

Instead of using a photomultiplier or other detector in the flow systems, the solutions may be quenched by, say, the addition of an acid, such as trichloroacetic acid, and the reaction products directly analyzed by chromatographic or other techniques.

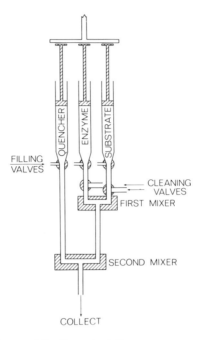

FIGURE 4·3. Quenched-flow apparatus.

a. The quenched-flow technique

The simplest form of the method is to submerge the end of the observation tube of a continuous-flow apparatus in a beaker of acid. A somewhat more sophisticated version is illustrated in Figure 4·3. A third syringe mixes the quenching acid with the reagent solutions via a second mixing chamber. Such an apparatus may have a dead time of only 4 or 5 ms. However, the maximum practical reaction time that may be measured using small volumes of reagents is about 100 to 150 ms; otherwise, excessively long reaction tubes are required.

b. The pulsed quenched-flow technique

The time range of the quenched-flow technique may be extended by a procedure similar to that of stopped flow.[5] As illustrated in Figure 4·4, the enzyme and the substrate (25 μL of each) are first mixed and driven into an incubation tube by a plunger actuated by compressed air. After the desired time interval—15 ms or longer—a second plunger is actuated to drive the incubated mixture with a pulse of distilled water into a second mixer, where it is quenched.

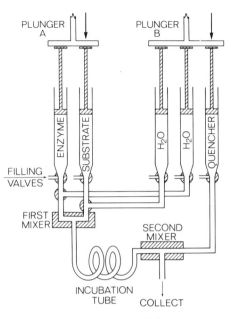

FIGURE 4·4. Pulsed quenched-flow apparatus.

B. Flash photolysis

The time involved in mixing places a limit on the dead time of flow techniques. The only way to increase the time resolution is to cut out the mixing by using a premixed solution of reagents that can be perturbed in some way to allow a measurable reaction to occur. A classic method from physical chemistry is flash photolysis, in which a particular bond in a reagent is cleaved by a pulse of light so that reactive intermediates are formed. This method was introduced in 1959 by Q. H. Gibson[6] for studying the dissociation of CO from carbonmonoxy-hemoglobin and carbonmonoxy-myoglobin, and the subsequent recombination with CO and other ligands. The procedure has been refined for these compounds by using mode-locked dye lasers to study dissociation on the femtosecond time scale, and recombination on the picosecond time scale.[7]

Unlike rapid mixing and sampling methods, flash photolysis cannot be generally applied to reactions because a suitable target for the flash is rarely available. However, flash photolysis has been adapted as a general procedure for initiating reactions that require ATP.[8] A derivative of ATP termed caged ATP (equation 4·1) may be photodissociated by a flash at 347 nm with a half-time of some 7 ms to generate ATP.[9] As caged ATP is generally unreactive, this provides a general means of studying ATP-utilizing reactions, e.g., muscle contraction,[9] on the millisecond time scale.

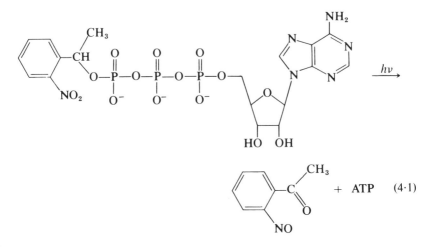

C. Relaxation methods

1. Temperature jump

An alternative method of overcoming the time delay of mixing is to use a relaxation method. An equilibrium mixture of reagents is preincubated and the equilibrium is perturbed by an external influence. The rate of return, or relaxation, to equilibrium is then measured. The most common procedure for this is temperature jump (Figure 4·5).[10] A solution is incubated in an absorbance or fluorescence cell and its temperature is raised through 5 to 10°C in less than a microsecond by the discharge of a capacitor (or, in more recent developments, in 10 to 100 ns by the discharge of an infrared laser). If the equilibrium involves an enthalpy change, the equilibrium position will change. The system will proceed to its new equilibrium position via a series of *relaxation times*, τ ($\equiv$ the reciprocal of the rate constant).

Clearly, this method cannot be applied to systems in which there are irreversible chemical processes. It is most suitable for situations involving simple ligand binding (such as NAD^+ with a dehydrogenase), inhibitor binding, or conformational changes in the protein. There have been some attempts to combine the temperature-jump with the stopped-flow method.

2. Nuclear magnetic resonance

The details of nuclear magnetic resonance are beyond the scope of this book; the reader is referred elsewhere for an introduction and further details.[11,12] It is a technique that does not perturb the chemistry of a system but that can measure the rate constants in an equilibrium mixture.

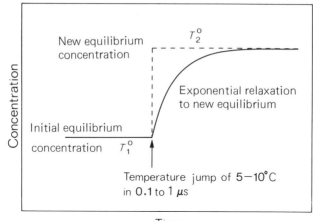

FIGURE 4·5. Illustration of temperature jump.

The accessible time is usually limited; depending upon both the isotope involved and various physical processes, the time range can be slower than that for stopped flow, or one or two orders of magnitude faster.

NMR has been used for measuring the *dissociation* rate constants of enzyme–inhibitor complexes from *exchange broadening*. This phenomenon was discussed in Chapter 1, section E2d, where its role in quantifying the rotation of amino acid side chains between two states was described. The method may be used in an analogous way for measuring the rate of exchange between enzyme-bound and free inhibitor.[13–16] The line-width analysis method is limited to a narrow range of rate constants, and, in practice, it cannot be applied to complex reaction pathways.

Recently, more sophisticated technology has generated two new, promising methods of applying NMR to kinetic measurements.

The first, which uses magnetization transfer,[17] has been applied to the carbonic anhydrase reaction at equilibrium in the presence of ^{13}C-enriched substrates:[18]

$$CO_2 + H_2O \rightleftharpoons HCO_3^- + H^+ \tag{4·2}$$

The ^{13}C signals are in slow exchange. One signal (e.g., that from CO_2) is selectively inverted by using a saturating pulse sequence, and the time course for recovery of both is determined. This is followed by the complementary experiment whereby the other signal (that from HCO_3^-) is inverted. Then, provided that the exchange rates between the two substrates are greater than $T_1/7$ ($T_1 =$ the spin-lattice relaxation time), the rate constants for the chemical interconversion may be calculated. Reactions with half-times of 60 ms to 40 s may be measured by using either ^1H or (enriched) ^{13}C NMR.

The second method uses two-dimensional Fourier-transform NMR with nuclear Overhauser enhancement (called 2D NOESY). This avoids the difficulties that may arise in applying a selective irradiating pulse in the above technique when the spectrum is crowded. The isomerization of glucose 6-phosphate and fructose 6-phosphate in the steady state catalyzed by glucose 6-phosphate isomerase has been studied by using ^{31}P NMR and this method.[19] Studies using this isotope should also be feasible with intact biological samples.

D. Analysis of pre–steady state and relaxation kinetics

Some of the fundamental differences between steady state and pre–steady state enzyme kinetics will become apparent later. Pre–steady state kinetics is concerned with the detection and analysis of transient enzyme-bound species as they arise and decay during the early phase of reaction. The processes generally observed are just the first-order rates of change from one enzyme species to another. Their time courses are generally exponential curves, unlike the linear plots of steady state kinetics. During these processes, the enzyme undergoes a single turnover in contrast to the multiple recycling of the steady state.

It is relatively straightforward to solve the differential equations for the time dependence of the transients in simple cases. However, it is important to understand the physical meaning of why a particular case gives rise to a particular form of solution. In this section we will concentrate on an intuitive approach to this understanding. Once a feel for the subject has been developed, algebraic mistakes will not be made and some complex kinetic schemes may be solved by inspection.

1. Simple exponentials

a. Irreversible reactions

Suppose that a compound A transforms into B with a first-order rate constant k_f, and the reaction proceeds to completion:

$$A \xrightarrow{k_f} B \qquad (4\cdot3)$$

Then

$$\frac{d[A]}{dt} = -k_f[A] \qquad (4\cdot4)$$

This is solved by integration to give

$$[A]_t = [A]_0 \exp(-k_f t) \qquad (4\cdot5)$$

where $[A]_0$ is the initial concentration of A. Since

$$[A]_t + [B]_t = [A]_0 \qquad (4\cdot6)$$

it is also true that

$$[B]_t = [A]_0 \{1 - \exp(-k_f t)\} \qquad (4\cdot7)$$

Both [A] and [B] follow simple exponentials.

It should be noted that the half-life of the reaction, $t_{1/2}$, where $[A] = [B] = [A]_0/2$, is given by

$$\exp(-k_f t) = \tfrac{1}{2} \qquad (4\cdot8)$$

i.e.,

$$t_{1/2} = \frac{0.6931}{k_f} = 0.6931\tau \qquad (4\cdot9)$$

b. The method of initial rates

When discussing more complex examples we shall use an adaptation of the *method of initial rates* to illustrate the physical meaning of some of the expressions. This method is often used in experiments that are too slow to follow over a complete time course, or in which there are complicating side reactions.

The initial rate v_0 of equation 4·3 is given by

$$v_0 = k_f [A]_0 \qquad (4\cdot10)$$

The value of k_f is determined by dividing v_0 by the expected change in reagent concentrations:

$$k_f = \frac{v_0}{\Delta[A]_0} \qquad (4\cdot11)$$

(Note that for an irreversible reaction, $\Delta[A]_0 = [A]_0$.)

c. Reversible reactions

In this case, A does not completely transform to B, but there is an equilibrium concentration of A:

$$A \underset{k_r}{\overset{k_f}{\rightleftharpoons}} B \qquad (4\cdot12)$$

and

$$K_{eq} = \frac{[B]}{[A]} = \frac{k_f}{k_r} \tag{4.13}$$

Here,

$$\frac{d[A]}{dt} = -k_f[A] + k_r[B] \tag{4.14}$$

Substitution of equation 4·6 gives

$$\frac{d[A]}{dt} = -k_f[A] + k_r([A]_0 - [A]) \tag{4.15}$$

This equation may be integrated by separating the variables and multiplying each side by an exponential factor:

$$\frac{d[A]}{dt} + [A](k_f + k_r) = k_r[A]_0 \tag{4.16}$$

$$\frac{d[A]}{dt} \exp (k_f + k_r)t + [A](k_f + k_r) \exp (k_f + k_r)t$$
$$= k_r[A]_0 \exp (k_f + k_r)t \tag{4.17}$$

$$\therefore \frac{d}{dt} \{[A] \exp (k_f + k_r)t\} = k_r[A]_0 \exp (k_f + k_r)t$$

$$\therefore [A] \exp (k_f + k_r)t = \frac{k_r}{k_f + k_r} [A]_0 \exp (k_f + k_r)t + \text{constant} \tag{4.18}$$

Using the boundary conditions that at $t = 0$, $[A] = [A]_0$, and at $t = \infty$, $[A]_{eq}$, the equilibrium concentration of $[A]$, is given by

$$[A]_{eq} = \frac{[A]_0 k_r}{k_f + k_r} \tag{4.19}$$

(from equations 4·6 and 4·13), the solution is

$$[A]_t = \frac{[A]_0 k_f}{k_f + k_r} \{\exp [-(k_f + k_r)t] + k_r\} \tag{4.20}$$

The expression for the time course (equation 4·20) may be divided into various factors. There is first the exponential term with the rate constant, or in terms of relaxation kinetics, the reciprocal relaxation time $1/\tau$ given by

$$\frac{1}{\tau} = k_f + k_r \tag{4.21}$$

Second, there is an *amplitude* factor given by

$$\frac{k_f}{k_f + k_r} \tag{4.22}$$

Now suppose that the reaction is started from the other direction, with B initially present, but not A. Then

$$[B]_t = \frac{[B]_0 k_r}{k_f + k_r} \{\exp\left[-(k_f + k_r)t\right] + k_f\} \tag{4.23}$$

This expression has the same relaxation time as equation 4·20, but a different amplitude factor.

The first important point to be noted is that the rate constant for the approach to equilibrium is greater than either of the individual first-order rate constants, k_f and k_r, and is equal to their sum. The reason why this is so is readily understood by applying the principle of initial rates. The initial velocity in the reversible reaction (4·12) is the same as in the irreversible one (4·3), but the former reaction does not have to proceed so far. For example, before B accumulates, only A is present, so again

$$v_0 = k_f[A]_0$$

—however, the total change in [A] is given by

$$\Delta[A]_0 = [A]_0 - [A]_{eq} \tag{4.24}$$

Substitution of equation 4·19 gives

$$\Delta[A]_0 = \frac{[A]_0 \, k_f}{k_f + k_r} \tag{4.25}$$

$$\therefore \frac{1}{\tau} = \frac{v_0}{\Delta[A]_0} = k_f + k_r \tag{4.26}$$

The second point to be noted is that k_f and k_r cannot be assigned without a knowledge of the amplitude factor. This basic symmetry in the relaxation times occurs in many cases, and, in general, the rate constants for unimolecular reactions cannot be assigned unless the concentrations of A and B at equilibrium may be determined. It will be seen later that when the reactions are not unimolecular but pseudo-unimolecular because of the presence of a second reagent, the relaxation time will have a concentration dependence that removes this ambiguity.

2. Association of enzyme and substrate

$$[E] + [S] \rightleftharpoons [ES] \tag{4.27}$$

If $[S] \geqslant [E]$, the reaction is effectively first-order since the concentration of S is hardly affected by the reaction. If the second-order rate constant for the association is k_{on} and that for dissociation is k_{off}, then the system reduces to

$$[E] \underset{k_{off}}{\overset{k_{on}[S]}{\rightleftharpoons}} [ES] \qquad (4\cdot28)$$

The relaxation time for this reaction is, from equation 4·21,

$$\frac{1}{\tau} = k_{off} + k_{on}[S] \qquad (4\cdot29)$$

Two points should be noted: (1) Because the rate constants are *pseudo*-unimolecular, there is a concentration dependence, so k_{on} and k_{off} may be resolved without the amplitude factor. (2) There is a lower limit to $1/\tau$; that is, $1/\tau$ cannot be less than k_{off}. This sets a limit on the measurement of these rate constants. A good stopped-flow spectrophotometer can cope only with rate constants of 1000 s^{-1} or less, and many enzyme–substrate dissociation constants are faster than this.

A favorable example is given in Figure 4·6. The dissociation rate constant of tyrosine from its complex with tyrosyl-tRNA synthetase is low, so the association and dissociation rate constants can be measured

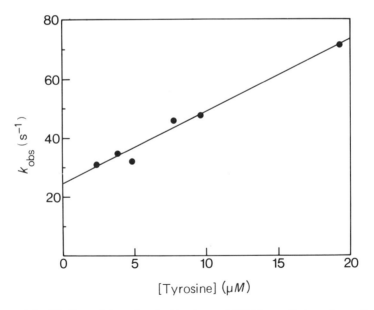

FIGURE 4·6. Binding of tyrosine to the tyrosyl-tRNA synthetase from *Bacillus stearothermophilus*. [From A. R. Fersht, R. S. Mulvey, and G. L. E. Koch, *Biochemistry* **14**, 13 (1975).]

by stopped flow. (Note that sometimes a two-step process may appear to be a single-step reaction: see section 6.)

Where there is no subsequent turnover of a substrate, such as occurs on the omission of a cosubstrate in a multisubstrate reaction, or on inhibitor binding, the temperature-jump technique is generally the most useful tool for the determination of these constants.

3. Consecutive reactions

a. Irreversible reactions

The simplest case of consecutive reactions is

$$A \xrightarrow{k_1} B \xrightarrow{k_2} C \tag{4·30}$$

This is solved by simply using the conservation equation and the integration procedures above to give equations 4·31.

$$[A] = [A]_0 \exp(-k_1 t)$$

$$[B] = \frac{[A]_0 k_1}{k_2 - k_1} [\exp(-k_1 t) - \exp(-k_2 t)] \tag{4·31}$$

$$[C] = [A]_0 \left\{ 1 + \frac{1}{k_1 - k_2} [k_2 \exp(-k_1 t) - k_1 \exp(-k_2 t)] \right\}$$

B is a transient intermediate that appears and then disappears (Figure 4·7). If $k_1 \gg k_2$, it is formed with rate constant k_1 and then slowly de-

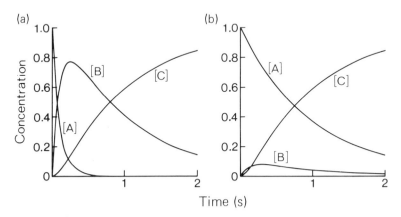

FIGURE 4·7. Plots of the concentrations of A, B, and C in the reaction $A \rightarrow B \rightarrow C$ (equation 4·30). (a) $k_1 = 10\ s^{-1}$, $k_2 = 1\ s^{-1}$. (b) $k_1 = 1\ s^{-1}$, $k_2 = 10\ s^{-1}$. Note that: (1) The progress curves for C are identical; (2) the *shapes* of the two curves for B are identical—they differ only in amplitude; and (3) [A] decreases 10 times more rapidly in (a). Thus, unless [A] is monitored, the two examples cannot be distinguished on the basis of measured rate constants only.

composes with rate constant k_2. However, if $k_2 \gg k_1$, [B] reaches a steady state level with rate constant k_2 and decays slowly with rate constant k_1. The apparently paradoxical situation is that the intermediate appears to be formed with its decomposition rate constant and to decompose with its formation rate constant! This is readily understood on the initial-rate treatment. When $k_1 \ll k_2$, [B] reaches a steady state level given by

$$\frac{d[B]}{dt} = 0 = k_1[A] - k_2[B] \tag{4.32}$$

The initial concentration of B when the steady state is set up is given by

$$[B]_{SS} \approx \frac{k_1}{k_2}[A]_0 \qquad (\text{i.e., } \ll [A]_0) \tag{4.33}$$

and

$$v_0 = k_1[A]_0 \tag{4.34}$$

$$\therefore \frac{1}{\tau} = \frac{v_0}{[B]_{SS}} = k_2 \tag{4.35}$$

In this latter case, where $k_2 > k_1$, B is at a very low concentration; in the former case, where $k_1 > k_2$, it accumulates. The two cases can be resolved if [B] can be monitored in absolute terms.

This type of kinetic situation sometimes occurs in protein renaturation experiments, in which the kinetics are often monitored by fluorescence changes. The biphasic traces cannot be resolved in such circumstances unless the quantum yield of the transient intermediate is known, so that its absolute concentration can be determined.

An example of the application of these equations is found in Chapter 7, section D. The aminoacyl-tRNA synthetase that specifically esterifies the tRNA molecule that accepts valine, $tRNA^{Val}$, "corrects" the error when it mistakenly forms an aminoacyl adenylate with threonine by the following scheme:

$$E \cdot Thr\!-\!AMP \cdot tRNA \xrightarrow{\text{transfer}} E \cdot Thr\text{-}tRNA \xrightarrow{\text{hydrolysis}} E + Thr + tRNA$$
$$\searrow$$
$$AMP \tag{4.36}$$

It will be seen that the rate of disappearance of $E \cdot Thr\text{-}AMP \cdot tRNA$ was directly measured by the formation of AMP, that the intermediate $E \cdot Thr\text{-}tRNA$ was directly measured, and that the second step (hydrolysis) was measured directly and independently by isolating the mischarged tRNA and adding it to the enzyme.

b. Quasireversible reactions: Steady state
A more common situation in enzyme kinetics is the following:

$$E \xrightarrow{k_1[S]} ES' \xrightarrow{k_2} E + P_2 \tag{4.37}$$
$$+ P_1$$

For example, chymotrypsin reacts with p-nitrophenyl acetate (AcONp) according to the above scheme (when $[\text{AcONp}] \ll K_S$ for the first step) to give an intermediate acylenzyme, EAc:

$$\text{E} \xrightarrow[\text{HONp}]{k_1[\text{AcONp}]} \text{EAc} \xrightarrow[\text{AcOH}]{\text{H}_2\text{O}} \text{E} \tag{4·38}$$

(where $k_1 = k_{\text{cat}}/K_M$ for this step).

Since the acylenzyme is continuously being formed and turning over, its concentration is in the steady state (provided that $[\text{AcONp}] \gg [\text{E}]$). The steady state concentration of the acylenzyme is given by

$$\frac{d}{dt}[\text{EAc}] = 0 = k_1[\text{AcONp}][\text{E}] - k_2[\text{EAc}]_{\text{ss}} \tag{4·39}$$

Now, since

$$[\text{E}] + [\text{EAc}] = [\text{E}]_0 \tag{4·40}$$

equation 4·39 can be written as

$$0 = k_1[\text{AcONp}]([\text{E}]_0 - [\text{EAc}]) - k_2[\text{EAc}]_{\text{ss}} \tag{4·41}$$

or

$$[\text{EAc}]_{\text{ss}} = \frac{k_1[\text{AcONp}][\text{E}]_0}{k_2 + k_1[\text{AcONp}]} \tag{4·42}$$

Applying the initial-rate treatment, we have

$$v_0 = k_1[\text{E}]_0[\text{AcONp}] \tag{4·43}$$

$$\frac{1}{\tau} = \frac{v_0}{[\text{EAc}]_{\text{ss}}} \tag{4·44}$$

$$= k_2 + k_1[\text{AcONp}] \tag{4·45}$$

Just as in the case of reversible reactions, the intermediate is formed with a rate constant that is greater than the rate constant for the transformation of the preceding intermediate.

The analytical solution for the rate constant is

$$[\text{HONp}] = [\text{E}]_0 \left(\frac{k_1'}{k_1' + k_2} \right)$$

$$\times \left(\frac{k_1'}{k_1' + k_2} \{1 - \exp[-(k_1' + k_2)t]\} + k_2 t \right) \tag{4·46}$$

where $k_1' = k_1[\text{AcONp}]$. If the rules of saturation kinetics are observed for the acylation step, k_1' is of the form $k_{\text{cat}}[\text{S}]/(K_M + [\text{S}])$. There are an initial exponential phase, which dies out after t is about 5 times greater than τ, and a linear term that eventually predominates.

c. Consecutive reversible reactions

The general solution for these reactions is given in section 6. We shall deal here with cases in which one step is fast compared with the other. Under these circumstances the relaxation times are on different time scales and do not "mix" with each other.

1. *First step fast* (*pre-equilibrium*).

$$\text{E} \underset{k_{-1}}{\overset{k_1[\text{S}]}{\rightleftharpoons}} \text{ES} \underset{k_{-2}}{\overset{k_2}{\rightleftharpoons}} \text{ES}' \tag{4.47}$$

This example may be readily solved by inspection if two simple rules are applied: (1) There must be two sets of relaxation times, since there are two sets of reactions involved. (2) Because the reactions occur on different time scales, they may be dealt with separately.

The first relaxation time is for the binding step. By analogy with equation 4·29, this is given by

$$\frac{1}{\tau_1} = k_{-1} + k_1[\text{S}] \tag{4.48}$$

The second relaxation time is for the slow step. This is an example of a reversible reaction, and, by analogy with equation 4·21, the reciprocal relaxation time is given by the sum of the forward and the reverse rate constants for the step. However, the effective forward rate constant for this step is given by k_2 multiplied by the fraction of the enzyme that is in the form of ES; i.e.,

$$\frac{1}{\tau_2} = k_{-2} + \frac{k_2[\text{S}]}{[\text{S}] + K_\text{S}} \tag{4.49}$$

where

$$K_\text{S} = \frac{k_{-1}}{k_1} \tag{4.50}$$

2. *Second step fast.* Reaction 4·47 could involve a substrate-induced conformational change in the enzyme, where ES' is just a different conformational state, or, alternatively, it could involve the accumulation of an intermediate on the pathway. The following reaction illustrates the displacement of an equilibrium between two conformational states of an enzyme, caused by the binding of a substrate to one form only:

$$\text{E}' \underset{k_{-1}}{\overset{k_1}{\rightleftharpoons}} \text{E} \underset{\text{fast}}{\overset{\text{S}, K_\text{S}}{\rightleftharpoons}} \text{ES} \tag{4.51}$$

This situation is found for the binding of ligands to chymotrypsin, which exists in two conformational states. Only one of these states binds aromatic substrates. It may be shown from the formal analysis to be given in section 6 that

$$\frac{1}{\tau_2} = k_1 + k_{-1}\left(\frac{K_S}{[S] + K_S}\right) \qquad (4 \cdot 52)$$

Case 2 may be distinguished from case 1 in that $1/\tau_2$ *decreases* with increasing [S]. This may be understood by analogy with the examples of irreversible and reversible reactions (equations 4·3 and 4·12). Clearly, when [S] is very high the reaction is essentially irreversible, since E' is transformed completely to ES and so $1/\tau_2$ tends toward k_1. Similarly, as [S] tends toward zero there is very little ES, and $1/\tau_2$ tends toward $k_1 + k_{-1}$. Hence the concentration dependence.

4. Parallel reactions

Parallel reactions are said to arise when a compound undergoes two or more reactions simultaneously. Enzymatic reactions often occur in parallel, when an activated intermediate may react with several competing acceptors:

$$\begin{array}{c} B \\ k_B \nearrow \\ A \\ k_C \searrow \\ C \end{array} \qquad (4 \cdot 53)$$

The kinetic equations are easily solved by integration, but it is instructive to solve them intuitively. It is obvious that [A] decreases with a rate constant that is the sum of k_B and k_C, and also that B and C are formed in the ratio of the rate constants. Since the rates of formation of B and C depend on [A], B and C must each be formed with a rate constant that is the same as for the disappearance of A. Therefore, we have

$$[A] = [A]_0 \exp\left[-(k_B + k_C)t\right] \qquad (4 \cdot 54)$$

$$[B] = \frac{[A]_0 k_B}{k_B + k_C}\left\{1 - \exp\left[-(k_B + k_C)t\right]\right\} \qquad (4 \cdot 55)$$

$$[C] = \frac{[A]_0 k_C}{k_B + k_C}\left\{1 - \exp\left[-(k_B + k_C)t\right]\right\} \qquad (4 \cdot 56)$$

The situation is similar to that for reversible reactions (equation 4·12), in that the relaxation time is composed of the sum of those for two reactions.

Chapter 7 includes examples of parallel reactions, e.g., the attack of various nucleophiles on acylchymotrypsins, measured by steady state and pre–steady state kinetics.

5. Derivation of equations for temperature jump

As an illustration, consider the association of an enzyme and a substrate in a one-step reaction:

$$E + S \underset{k_{-1}}{\overset{k_1}{\rightleftharpoons}} ES \tag{4.57}$$

Suppose that because of the change in temperature, the equilibrium moves to a new position, so that

$$[E] = [E]_{eq} + e \tag{4.58}$$

$$[S] = [S]_{eq} + s \tag{4.59}$$

and

$$[ES] = [ES]_{eq} + es \tag{4.60}$$

where $[E]_{eq}$, $[S]_{eq}$, and $[ES]_{eq}$ are the equilibrium concentrations at the new temperature. Then

$$\frac{d[ES]}{dt} = k_1([E]_{eq} + e)([S]_{eq} + s) - k_{-1}([ES]_{eq} + es) \tag{4.61}$$

$$= k_1[E]_{eq}[S]_{eq} - k_{-1}[ES]_{eq}$$
$$+ k_1([E]_{eq}s + [S]_{eq}e + e \cdot s) - k_{-1}es \tag{4.62}$$

Equation 4.62 may be simplified, since the first two terms on the right-hand side cancel out (they are equal at equilibrium). Also, because the reagents are conserved, $e = s = -es$. And since $[ES]_{eq}$ is a constant, $d[ES]/dt = des/dt$. Therefore,

$$-\frac{des}{dt} = k_1([E]es + [S]es + e \cdot s) + k_{-1}es \tag{4.63}$$

Now, if the perturbation from equilibrium is small, the second-order term $e \cdot s$ may be ignored. Equation 4.63 may then be integrated to give the relaxation time:

$$\frac{1}{\tau} = k_1([E] + [S]) + k_{-1} \tag{4.64}$$

If the equilibrium is perturbed only slightly, the return to equilibrium is always a first-order process, even though the reagents may be present at similar concentrations.

6. A general solution of two-step consecutive reversible reactions

The solution of the following equation involves simultaneous linear differential equations:

$$A \underset{k_{-1}}{\overset{k_1}{\rightleftharpoons}} B \underset{k_{-2}}{\overset{k_2}{\rightleftharpoons}} C \tag{4.65}$$

Two relaxation times are obtained:

$$\frac{1}{\tau_1} = \frac{p + q}{2} \tag{4.66}$$

$$\frac{1}{\tau_2} = \frac{p - q}{2} \tag{4.67}$$

where

$$p = k_1 + k_{-1} + k_2 + k_{-2}$$

and   (4.68)

$$q = [p^2 - 4(k_1 k_2 + k_{-1} k_{-2} + k_1 k_{-2})]^{1/2}$$

These basic equations may be manipulated to cover many cases. A useful trick is to express the rate constants as the sums and products of the relaxation times:

$$\frac{1}{\tau_1} + \frac{1}{\tau_2} = k_1 + k_{-1} + k_2 + k_{-2} \tag{4.69}$$

$$\frac{1}{\tau_1 \tau_2} = k_1 k_2 + k_{-1} k_{-2} + k_1 k_{-2} \tag{4.70}$$

The equations are easy to solve if a concentration dependence is involved. For example, if the sequence is the pseudo-first-order series

$$E \underset{k_{-1}}{\overset{k_1'[S]}{\rightleftharpoons}} ES \underset{k_{-2}}{\overset{k_2}{\rightleftharpoons}} ES' \tag{4.71}$$

then $k_1'[S]$ may be substituted for k_1 in equations 4·69 and 4·70. Also, by analogy with equation 4·64 for temperature jump, $k_1'([S] + [E])$ may be substituted for k_1 in a relaxation experiment.

The equations may be simplified if one of the relaxation times is much faster than the other. For example, if in equation 4·71 the first step is fast, $1/\tau_2$ and $k_2 + k_{-2}$ may be ignored in 4·69. The value of τ_2 may then be obtained by substituting 4·69 into 4·70. In the case of a temperature-jump experiment, this gives

$$\frac{1}{\tau_1} = k_1'([E] + [S]) + k_{-1} \tag{4.72}$$

$$\frac{1}{\tau_2} = k_{-2} + \frac{k_2([E] + [S])}{k_{-1}/k_1' + [E] + [S]} \tag{4.73}$$

The same manipulations may be performed for

$$E \underset{k_{-1}}{\overset{k_1}{\rightleftharpoons}} E' \underset{k_{-2}}{\overset{k_2'[S]}{\rightleftharpoons}} E'S \tag{4.74}$$

where the first step is slow, to give

$$\frac{1}{\tau_1} = k_{-2} + k_2'([E] + [S]) \tag{4.75}$$

$$\frac{1}{\tau_2} = k_1 + \frac{k_{-1}\{(k_{-2}/k_2') + [E']\}}{k_{-2}/k_2' + [E'] + [S]} \tag{4.76}$$

Two practical points should be noted. The kinetic mechanisms in equations 4.71 and 4.74 may be distinguished by the concentration dependence of $1/\tau_2$. For 4.71 this increases with increasing [S]; for 4.74 it decreases. But there are situations that are difficult to resolve. For example, in equation 4.74, if $[E'] \gg [E]$ there will be a burst of formation of ES' with relaxation time τ_1, followed by a small increase at relaxation time τ_2 as E converts to E'. The concentration dependence of τ_2 will be small, since $k_1 \gg k_{-1}$ for $[E'] \gg [E]$ (Figure 4.8). This can be mistaken for the scheme in equation 4.71, where only a little ES' is formed. In this case also, the concentration dependence of $1/\tau_2$ is small, because $k_{-2} \gg k_2$. In both cases the amplitudes of the changes will often be small and the rate constants difficult to measure precisely.

A more common situation that leads to difficulties is the two-step combination of an enzyme and substrate, as in equation 4.71, where the dissociation constant for the first step, k_{-1}/k_1', is high. If measurements are made only in the region where $k_{-1}/k_1' > [E] + [S]$, equation 4.73 reduces to†

$$\frac{1}{\tau_2} = k_{-2} + k_1' \frac{k_2}{k_{-1}} ([E] + [S]) \tag{4.77}$$

This has the form of a simple one-step association of an enzyme and a substrate, as in equation 4.64, and may mistakenly be interpreted as this.

---

† Equation 4.73 was derived on the assumption that $k_{-1} \gg k_2$. If k_2 is appreciable, equation 4.77 should be modified to

$$\frac{1}{\tau_2} = k_{-2} + \frac{k_1' k_2}{k_{-1} + k_2} ([E] + [S])$$

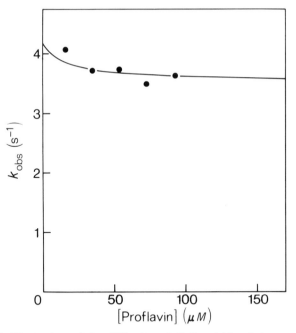

FIGURE 4·8. Illustration of the difficulty of distinguishing between mechanisms 4·71 and 4·74 by equations 4·73 and 4·76. The rate constant for the binding of proflavin to α-chymotrypsin at pH 6.84 and 25°C is plotted against the concentration of proflavin. The reaction scheme is

$$\text{E} \underset{0.6\,s^{-1}}{\overset{3.1\,s^{-1}}{\rightleftharpoons}} \text{E}' \underset{\text{PF}}{\overset{\text{fast}}{\rightleftharpoons}} \text{E}'\text{PF}$$

The rate constant decreases with increasing proflavin as predicted by equation 4·21, but the decrease is small; the maximum possible change is only 19%. [From A. R. Fersht and Y. Requena, *J. Molec. Biol.* **60**, 279 (1971).]

In such a case, the association rate constant would appear to be $k_1'(k_2/k_{-1})$, a value lower than the true rate constant of k_1'. Some of the low values to be shown in Table 4·3 are undoubtedly caused by this. Measurements should always be extended to high substrate concentrations to search for a leveling off of rate as predicted by equation 4·73 (Figure 4·9).

7. Experimental application of pre–steady state kinetics

Later, in Chapter 7, we shall discuss several examples of the successful application of transient kinetics to the solution of enzyme mechanisms. Here, we briefly describe some of the strategies and tactics used by the kineticist to initiate a transient kinetic study. On many occasions, steady state kinetics and other studies have set kineticists a well-defined and

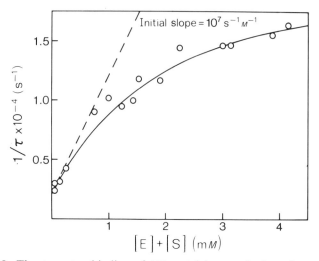

FIGURE 4·9. The two-step binding of (N-acetylglucosamine)$_2$ to lysozyme. Solutions of enzyme (0.03 to 0.2 mM) and ligand (0.02 to 4.1 mM) were temperature-jumped from 29 to 38°C at pH 6. Experiments at low concentration are in the linear region of the curve and give an apparent second-order rate constant of 10^7 s^{-1} M^{-1} for binding. However, measurements at higher concentrations reveal that the rate reaches a plateau, indicating a two-step process. [From E. Holler, J. A. Rupley, and G. P. Hess, *Biochem. Biophys. Res. Commun.* **37**, 423 (1969).]

specific question to answer. At other times, they just wish to study a particular system to gather information. In both cases there is no substitute for imagination and insight in designing the incisive experiment. But there is a systematic approach that can be used.

The analysis of a relaxation time may be divided into two basic steps:

1. The relaxation time must be assigned to a specific physical event.
2. This physical event must be fitted into the overall kinetic mechanism of the process under study.

Let us now consider two examples. The first is the binding of a ligand to a protein under noncatalytic conditions. Two likely physical events are the initial binding step and a ligand-induced conformational change. The first experiment is the measurement of the concentration dependence and the number of relaxation times in order to determine the number of intermediate states and the rate constants for their interconversion. Under ideal circumstances, the number of relaxation times will be equal to the number of steps in the reaction. Even if only one relaxation time is found, its concentration dependence may indicate a two-step process by being nonlinear (e.g., equations 4·73 and 4·76). Additional information may be obtained by using more than one physical probe, for example fluorescence and absorbance, and by studying both the ligand and the protein, since some steps may show up in one of these and not in the other. Further physical processes may occur, such as proton release or uptake during the reaction, or a change in the state of aggregation of the protein. The former provides an additional convenient probe since changes of pH may be measured by a chromophoric pH indicator. Aggregation complicates the kinetics, but may be detected and measured to provide additional information. Relaxation techniques are often more powerful than flow methods for these simple reactions because they can measure faster processes. However, there are times when stopped flow is more useful: For example, processes that are too slow for detection by temperature jump may be measured by stopped flow; and also, certain experiments, such as those to determine the effect of a large change in pH, can be performed only by mixing (although small changes in pH can be made in a temperature-jump experiment by using a buffer whose pK_a is temperature-dependent).

The second example is the reaction mechanism of an enzyme under catalytic conditions. In addition to determination of the binding steps and conformational changes described above, it is even more important to measure the bond making and breaking steps and to detect the chemical intermediates in the reaction. The chemical steps are usually best studied by the methods that directly measure the concentration of chemical species—for example, stopped-flow spectrophotometry and quenched flow. Indirect measurements of chemical steps, such as a change in protein fluorescence, must somehow be assigned. The ideal situation for study is the accumulation of an intermediate that may be detected and measured. In general, as many different probes as possible should be used in order to confirm existing information and add further details. Good examples of the application of such methods are to be found in Chapter 7.

E. The absolute concentration of enzymes

1. Active-site titration and the magnitudes of "bursts"

The calculation of rate constants from steady state kinetics and the determination of binding stoichiometries requires a knowledge of the concentration of active sites in the enzyme. It is not sufficient to calculate this specific concentration value from the relative molecular mass of the protein and its concentration, since isolated enzymes are not always 100% pure. This problem has been overcome by the introduction of the technique of active-site titration, a combination of steady state and pre–steady state kinetics whereby the concentration of active enzyme is related to an initial burst of product formation. This type of situation occurs when an enzyme-bound intermediate accumulates during the reaction. The first mole of substrate rapidly reacts with the enzyme to form stoichiometric amounts of the enzyme-bound intermediate and product, but then the subsequent reaction is slow since it depends on the slow breakdown of the intermediate to release free enzyme.

$$E + S \xrightarrow{k_1'} EI \xrightarrow{k_2} E + P_2 \atop + P_1 \tag{4.78}$$

Clearly, if in equation 4·78 k_1' is very fast and k_2 is negligibly slow, the release of P_1 is easily measured and related to the concentration of enzyme. However, in practice, k_2 is generally not negligible, so that there is an initial burst of formation of P_1 followed by a progressive increase as the intermediate turns over. The mathematics of this situation was described previously (equations 4·37 to 4·46). It was shown that the overall release of products is linear with time after an initial transient. From equation 4·46 it can be seen that the linear portion extrapolates back to a burst, π, given by

$$\pi = [E]_0 \left(\frac{k_1'}{k_1' + k_2} \right)^2 \tag{4.79}$$

(Figures 4·10 and 4·11). It should be noted that the burst depends on a "squared" relationship with the rate constants. If the ratio $k_1':k_2$ is high, the squared term is close to 1, so that the burst is equal to the enzyme concentration. If this condition does not hold, the concentration will be underestimated unless both rate constants are measured and substituted into equation 4·79.

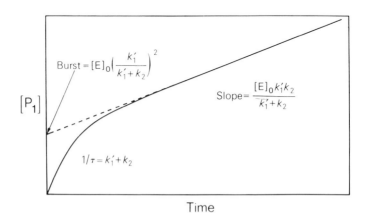

FIGURE 4·10. The principle of active-site titration.

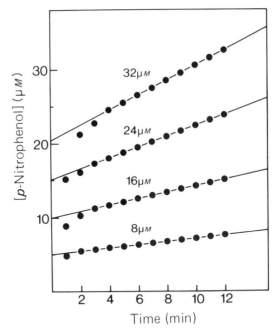

FIGURE 4·11. The "original" active-site titration experiment. The indicated concentrations of chymotrypsin were mixed with p-nitrophenyl ethyl carbonate,

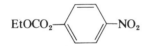

The acylenzyme E—O—COOEt is rapidly formed but hydrolyzes slowly. Note that about 0.63 mol of p-nitrophenol is released per mole of enzyme in the "burst." Either the enzyme is only 63% pure (active), or the rate constant for the formation of the acylenzyme is not sufficiently greater than that for deacylation for the acylenzyme to accumulate fully. [From B. S. Hartley and B. A. Kilby, *Biochem. J.* **56**, 288 (1954).]

2. The dependence of the burst on substrate concentration

In equation 4·78, the term k_1' is the apparent first-order rate constant for the formation of the intermediate under the particular reaction conditions. In general, this will follow the Michaelis-Menten equation, that is,

$$k_1' = \frac{k_{cat}[S]}{[S] + K_M} \tag{4·80}$$

where k_{cat} and k_M refer to the first step. At sufficiently low concentrations of S, there will be no burst, but, provided that k_{cat} is greater than k_2, one will occur at higher concentrations. Substituting equation 4·80 into 4·79 gives

$$\frac{1}{\sqrt{\pi}} = \frac{1}{\sqrt{[E]_0}}\left(1 + \frac{k_2}{k_{cat}} + \frac{K_M k_2}{[S] k_{cat}}\right) \tag{4·81}$$

If k_{cat} is much greater than k_2, equation 4·81 may be used to extrapolate the burst from measurements at various substrate concentrations. It is obvious that when k_2 is not negligible, care must be taken not to underestimate the concentration of the enzyme.

3. Active-site titration vs. rate assay

Active-site titration is not always applicable, since it requires the accumulation of an intermediate in the reaction. The more usual procedure is to determine the concentration of an enzyme from a rate assay. This has the disadvantage that it does not give the absolute concentration of the enzyme unless it has been calibrated against an active-site titration. Further, rate measurements are sensitive to the reaction conditions. Whereas these may be controlled with some precision in a particular laboratory, they often vary from laboratory to laboratory. Active-site titration suffers from the disadvantage that several milligrams of enzyme are required for a spectrophotometric assay, but this quantity may be reduced a thousand times by using radioactive methods (Figure 4·12).[20] With its relative in-

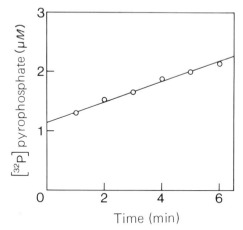

FIGURE 4·12. Active-site titration of isoleucyl-tRNA synthetase by using only 10 μg (100 pmol) of enzyme (0.1 mL of 1-μM enzyme). The reaction sequence is

$$E + Ile + [\gamma\text{-}^{32}P]ATP \xrightarrow{\text{fast}} E\cdot Ile\text{-}AMP \xrightarrow[H_2O]{\text{slow}} E + Ile + AMP$$
$$\searrow$$
$$[^{32}P]PP$$

[From A. R. Fersht and M. M. Kaethner, *Biochemistry* **15**, 818 (1976).]

sensitivity to precise reaction conditions and its yield of absolute values for the concentrations of enzyme solutions, active-site titration has been a most important factor in providing highly reproducible data and making possible the comparison of rate constants from steady state and pre–steady state kinetics. (The rate constants of pre–steady state kinetics generally involve exponential processes that do not depend on the concentration of enzyme, whereas steady state rates are typically directly proportional to the concentration.)

PART 2 The magnitude of rate constants for enzymatic processes

A. Upper limits on rate constants[21]

1. Association and dissociation

A simple way of analyzing the rate constants of chemical reactions is the *collision theory* of reaction kinetics. The rate constant for a bimolecular reaction is considered to be composed of the product of three terms: the frequency of collisions, Z; a steric factor, p, to allow for the fraction of the molecules that are in the correct orientation; and an activation energy term to allow for the fraction of the molecules that are sufficiently thermally activated to react. That is,

$$k_2 = Zp \exp\left(-\frac{E_{act}}{RT}\right) \tag{4.82}$$

The maximum value for the bimolecular rate constant occurs when the activation energy E_{act} is zero and the steric factor is 1. The rate is then said to be *diffusion-controlled*, and it is equal to the encounter frequency of the molecules. Assuming that the reacting molecules are uncharged spheres of radius r_A and r_B, the encounter frequency may be calculated as

$$Z = \left(\frac{2RT}{3000\eta}\right)\frac{(r_A + r_B)^2}{r_A r_B} \tag{4.83}$$

(where η is the viscosity). For two molecules of the same radius in water at 25°C, the encounter frequency is equal to $7 \times 10^9 \text{ s}^{-1} M^{-1}$. It should be noted that two large molecules collide at exactly the same rate as two small ones. This is because the increase of target area exactly compensates for the slower diffusion of the larger molecules. However, the rate of encounter of a small molecule with a large molecule is higher than this value because of the combination of the large target area of the latter and the high mobility of the former. More sophisticated calculations, allowing for the possibility of favorable electrostatic interactions (at one extreme) and an unfavorable geometry for a small molecule hitting a particular target area of a larger one (at the other extreme), give a range of 10^9 to 10^{11} s^{-1} M^{-1} for the encounter frequency. A similar treatment gives a range of 10^9 to 10^{12} s^{-1} for the upper limit on the *dissociation* rate constants of bimolecular complexes. Many of the second-order rate constants that do not involve the proton or hydroxide ion are found experimentally to be about 10^9 s^{-1} M^{-1}.

2. Chemical processes

The upper limit on the rate constant of any unimolecular or intramolecular reaction is the frequency of a molecular vibration, about 10^{12} to 10^{13} s^{-1}.

3. Proton transfers

Favorable proton transfers between electronegative atoms such as O, N, and S are extremely fast. The bimolecular rate constants are generally diffusion-controlled, being 10^{10} to 10^{11} s^{-1} M^{-1} (Table 4·1). For example, the rate constant for the transfer of a proton from H_3O^+ to imidazole, a favorable transfer since imidazole is a stronger base than H_2O, is 1.5×10^{10} s^{-1} M^{-1} (Table 4·2). The rate constant for the reverse reaction, the transfer of a proton from the imidazolium ion to water, may be calculated from the difference in their pK_a's by using the following equations:

$$\frac{[B][H^+]}{[BH^+]} = K_a \qquad \frac{[A^-][H^+]}{[HA]} = K_a' \tag{4·84}$$

$$\frac{[B][HA]}{[BH^+][A^-]} = K_a/K_a' \tag{4·85}$$

$$[B] + [HA] \underset{k_{-1}}{\overset{k_1}{\rightleftharpoons}} [BH^+] + [A^-] \tag{4·86}$$

so that

$$\frac{k_{-1}}{k_1} = \frac{K_a}{K_a'} \tag{4·87}$$

The rate constant for the transfer of a proton from the imidazolium ion ($pK_a = 6.95$) to water ($[H_2O] = 55$ M, $pK_a = -1.74$) is calculated from equation 4·87 to be 1.7×10^3 s^{-1}.

Proton transfers from carbon acids and to carbon bases are generally much slower. This is because the lower electronegativity of carbon requires that the negative charge on a carbon base be stabilized by electron delocalization. The consequent reorganization of structure and solvent may slow down the overall transfer rate.

It was once thought that the rate of equilibrium of the catalytic acid and basic groups on an enzyme with the solvent limited the rates of acid-

TABLE 4·1. *Proton-transfer rate constants (25°C, $s^{-1} M^{-1}$)[a]*

H$^+$ and	k	OH$^-$ and	k
OH$^-$	1.4×10^{11}		
Inorganic acid anions	10^{10}–10^{11}	Inorganic acids	~10^{10}
Carboxylates	~5×10^{10}	Carboxylic acids	~10^{10}
Phenolates	~5×10^{10}	Phenols	
Enolates	~5×10^{10}	Enols	~10^{10}
Amines	10^{10}	Ammonium	~10^{10}
		ions	~3×10^{10}
Carbanions	<1–~10^{10}	Carbon acids	<1–~10^9
		Phosphoric acids	10^8–10^{10}

[a] From M. Eigen, *Nobel Symp.* **5**, 245 (1967).

and base-catalyzed reactions to turnover numbers of 10^3 s^{-1} or less. This is because the rate constants for the transfer of a proton from the imidazolium ion to water and from water to imidazole are about 2×10^3 s^{-1}. However, protons are transferred between imidazole or imidazolium ion and buffer species in solution with rate constants that are many times higher than this. For example, the rate constants with ATP, which has a pK_a similar to imidazole's, are about 10^9 s^{-1} M^{-1}, and the ATP concentration is about 2 mM in the cell. Similarly, several other metabolites that are present at millimolar concentrations have acidic and basic groups that allow catalytic groups on an enzyme to equilibrate with the solvent at 10^7 to 10^8 s^{-1} or faster. Enzyme turnover numbers are usually considerably lower than this, in the range of 10 to 10^3 s^{-1}, although carbonic anhydrase and catalase have turnover numbers of 10^6 and 4×10^7 s^{-1}, respectively. The problem of carbonic anhydrase is discussed in Chapter 15.

TABLE 4·2. *Proton-transfer rates involving imidazole (pK_a = 6.95)[a]*

Donor (DH$^+$)	pK_a	k (DH$^+ \rightarrow$ Im) (s$^{-1} M^{-1}$)	k (ImH$^+ \rightarrow$ D) (s$^{-1} M^{-1}$)
H$_3$O$^+$	-1.74	1.5×10^{10}	31
H$_2$O	15.74	45	2.3×10^{10}
CH$_3$CO$_2$H	4.76	1.2×10^9	7.7×10^6
HATP^{3-}	6.7	2×10^9	1×10^9
p-Nitrophenol	7.14	4.5×10^8	7.0×10^8
HP$_2$O$_7^{3-}$	8.45	1.1×10^8	3.6×10^9
Phenol	9.95	1×10^7	1×10^{10}
CO$_3^{2-}$	10.33	1.9×10^7	2×10^{10}
Glucose	12.3	1.6×10^5	2×10^{10}

[a] At 25°C, ionic strength = 0. The H$_2$O rates are calculated on [H$_2$O] = 55 M. [From M. Eigen and G. G. Hammes, *Adv. Enzymol.* **25**, 1 (1963).]

B. Enzymatic rate constants and rate-determining processes

1. Association of enzymes and substrates

Calculations suggest that the diffusion-controlled encounter frequency of an enzyme and a substrate should be about 10^9 s^{-1} M^{-1}. The observed values in Table 4·3 tend to fall in the range of 10^6 to 10^8 s^{-1} M^{-1}. The faster ones are close to diffusion-controlled, but the slower ones are significantly lower than the limit. This may be partly due to desolvation

TABLE 4·3. *Association and dissociation rate constants for enzyme–substrate interactions*

Enzyme	Substrate	k_1 (s^{-1} M^{-1})	k_{-1} (s^{-1})	Ref.
Protein-small ligands				
Catalase	H_2O_2	5×10^6		1
Catalase–H_2O_2	H_2O_2	1.5×10^7		1
Chymotrypsin	Proflavin	1.2×10^8	8.3×10^3	2
	Acetyl-L-tryptophan p-nitrophenyl ester	6×10^7	6×10^4	3
	Furylacryloyl-L-tryptophanamide	6.2×10^6		4
	Trifluorylacetyl-D-tryptophan	1.5×10^7		5
	Indole	1.9×10^7	5.8×10^3	6
Creatine kinase	ADP	2.2×10^7	1.8×10^4	7
	MgADP	5.3×10^6	5.1×10^3	
Glyceraldehyde 3-phosphate dehydrogenase	NAD^+	1.9×10^7 1.4×10^6	1×10^3 210	8
Lactate dehydrogenase (rabbit muscle)	NADH	$\sim 10^9$	$\sim 10^4$	9
Lactate dehydrogenase (pig heart)	NADH Oxamate	5.5×10^7 8.1×10^6	39 17	10 10
Liver alcohol dehydrogenase	NADH	2.5×10^7	9	11
Lysozyme	$(NAG)_2$	4×10^7	1×10^5	12, 13
Malate dehydrogenase	NADH	5×10^8	50	14
Pyruvate carboxylase–Mn^{2+}	Pyruvate	4.5×10^6	2.1×10^4	15
Ribonuclease	Uridine 3'-phosphate	7.8×10^7	1.1×10^4	16
	Uridine 2', 3'-cyclic phosphate	1×10^7	2×10^4	17

TABLE 4·3. *Association and dissociation rate constants for enzyme–substrate interactions* (continued)

Enzyme	Substrate	k_1 $(s^{-1} M^{-1})$	k_{-1} (s^{-1})	Ref.
Tyrosyl-tRNA synthetase	Tyrosine	2.4×10^6	24	18
Protein–nucleic acids				
Phenylalanyl-tRNA synthetase	tRNA^{Phe}	1.6×10^8	27	19
Seryl-tRNA synthetase	tRNA^{Ser}	2.1×10^8	11	20
Tyrosyl-tRNA synthetase	tRNA^{Tyr}	2.2×10^8 1.4×10^8	1.5 53	21
Protein-protein				
Trypsin	Basic pancreatic	1.1×10^6	6.6×10^{-8}	22
Anhydrotrypsin	trypsin inhibitor	7.7×10^5	8.5×10^{-8}	22
Trypsin	Pancreatic	6.8×10^6	2.2×10^{-4}	22
Anhydrotrypsin	secretory trypsin inhibitor	4×10^6	1.4×10^{-3}	22
Insulin	Insulin	1.2×10^8	1.5×10^4	23
β-Lactoglobulin	β-Lactoglobulin	4.7×10^4	2.1	23
α-Chymotrypsin	α-Chymotrypsin	3.7×10^3	0.68	23

1 B. Chance, in *Currents in biochemical research* (D. E. Green, Ed.), Wiley, p. 308 (1956).
2 U. Quast, J. Engel, H. Heumann, G. Krause, and E. Steffen, *Biochemistry* **13**, 2512 (1974).
3 M. Renard and A. R. Fersht, *Biochemistry* **12**, 4713 (1973).
4 G. P. Hess, J. McConn, E. Ku, and G. McConkey, *Phil. Trans. R. Soc.* **B257**, 89 (1970).
5 S. H. Smallcombe, B. Ault, and J. H. Richards, *J. Am. Chem. Soc.* **94**, 4585 (1972).
6 R. Maehler and J. R. Whitaker, *Biochemistry* **21**, 4621 (1982).
7 G. G. Hammes and J. K. Hurst, *Biochemistry* **8**, 1083 (1969).
8 K. Kirschner, M. Eigen, R. Bittman, and B. Voigt, *Proc. Natn. Acad. Sci. USA* **56**, 1661 (1966).
9 G. H. Czerlinski and G. Schreck, *J. Biol. Chem.* **239**, 913 (1964).
10 H. d'A. Heck, *J. Biol. Chem.* **244**, 4375 (1969).
11 J. D. Shore and H. Gutfreund, *Biochemistry* **9**, 4655 (1970).
12 E. Holler, J. A. Rupley, and G. P. Hess, *Biochem. Biophys. Res. Commun.* **37**, 423 (1969).
13 J. H. Baldo, S. E. Halford, S. L. Patt, and B. D. Sykes, *Biochemistry* **14**, 1893 (1975).
14 G. Czerlinski and G. Schreck, *Biochemistry* **3**, 89 (1964).
15 A. S. Mildvan and M. C. Scrutton, *Biochemistry* **6**, 2978 (1967).
16 G. G. Hammes and F. G. Walz, Jr., *J. Am. Chem. Soc.* **91**, 7179 (1969).
17 E. J. del Rosario and G. G. Hammes, *J. Am. Chem. Soc.* **92**, 1750 (1970).
18 A. R. Fersht, R. S. Mulvey, and G. L. E. Koch, *Biochemistry* **14**, 13 (1975).
19 G. Krauss, R. Römer, D. Riesner, and G. Maass, *FEBS Letts.* **30**, 6 (1973).
20 A. Pingoud, D. Riesner, D. Boehme, and G. Maass, *FEBS Letts.* **30**, 1 (1973).
21 A. Pingoud, D. Boehme, D. Riesner, R. Kownatski, and G. Maass, *Eur. J. Biochem.* **56**, 617 (1975).
22 J.-P. Vincent, M. Peron-Renner, J. Pudles, and M. Lazdunski, *Biochemistry* **13**, 4205 (1974).
23 R. Koren and G. G. Hammes, *Biochemistry* **15**, 1165 (1976).

requirements in some cases, or, as is more likely in others, to a two-step process that appears as a single step. For example, at low concentrations the binding of $(NAG)_2$ to lysozyme appears to occur at about 5×10^6 $s^{-1} M^{-1}$. But extension of the measurements to higher concentrations shows that the binding is a two-step process with an association rate constant of $4 \times 10^7 s^{-1} M^{-1}$ at pH 4.4 and 31°C (see section D6 and Figure 4·9):[16,22]

$$E + (NAG)_2 \underset{1.2 \times 10^5 \, s^{-1}}{\overset{4 \times 10^7 \, s^{-1} M^{-1}}{\rightleftharpoons}} E \cdot (NAG)_2 \underset{1.3 \times 10^3 \, s^{-1}}{\overset{1.7 \times 10^4 \, s^{-1}}{\rightleftharpoons}} E' \cdot (NAG)_2 \qquad (4 \cdot 88)$$

2. Association can be rate-determining for k_{cat}/K_M

Table 4·4 shows that for some efficient enzymes, k_{cat}/K_M may be as high as $3 \times 10^8 s^{-1} M^{-1}$. In these cases, the rate-determining step for this parameter, which is the apparent second-order rate constant for the reaction of free enzyme with free substrate, is close to the diffusion-controlled encounter of the enzyme and the substrate. Briggs-Haldane kinetics holds for these enzymes (Chapter 3, section B3).

TABLE 4·4. *Enzymes for which k_{cat}/K_M is close to the diffusion-controlled association rate*

Enzyme	Substrate	k_{cat} (s^{-1})	K_M (M)	k_{cat}/K_M $(s^{-1} M^{-1})$	Ref.
Acetylcholin-esterase	Acetylcholine	1.4×10^4	9×10^{-5}	1.6×10^8	1
Carbonic anhydrase	CO_2	1×10^6	0.012	8.3×10^7	2
	HCO_3^-	4×10^5	0.026	1.5×10^7	3
Catalase	H_2O_2	4×10^7	1.1	4×10^7	4
Crotonase	Crotonyl-CoA	5.7×10^3	2×10^{-5}	2.8×10^8	5
Fumarase	Fumarate	800	5×10^{-6}	1.6×10^8	6
	Malate	900	2.5×10^{-5}	3.6×10^7	6
Triosephosphate isomerase	Glyceraldehyde 3-phosphate	4.3×10^3	4.7×10^{-4}	2.4×10^{8a}	7
β-Lactamase	Benzylpenicillin	2.0×10^3	2×10^{-5}	1×10^8	8

[a] The observed value is $9.1 \times 10^6 s^{-1} M^{-1}$. The tabulated value is calculated on the basis of only 3.8% of the substrate being reactive, since 96.2% is hydrated under the conditions of the experiment.

1   T. I. Rosenberry, *Adv. Enzymol.* **43**, 103 (1975).
2   J. C. Kernohan, *Biochim. Biophys. Acta* **81**, 346 (1964).
3   J. C. Kernohan, *Biochim. Biophys. Acta* **96**, 304 (1965).
4   Y. Ogura, *Archs. Biochem. Biophys.* **57**, 288 (1955).
5   R. M. Waterson and R. L. Hill, *Fedn. Proc.* **30**, 1114 (1971).
6   J. W. Teipel, G. M. Hass, and R. L. Hill, *J. Biol. Chem.* **243**, 5684 (1968).
7   S. J. Putman, A. F. W. Coulson, I. R. T. Farley, B. Riddleston, and J. R. Knowles, *Biochem. J.* **129**, 301 (1972).
8   J. Fisher, J. G. Belasco, S. Khosla, and J. R. Knowles, *Biochemistry* **19**, 2985 (1980).

3. Dissociation of enzyme–substrate and enzyme–product complexes

Dissociation rate constants are much lower than the diffusion-controlled limit, since the forces responsible for the binding must be overcome in the dissociation step. In some cases, enzyme–substrate dissociation is slower than the subsequent chemical steps, and this gives rise to Briggs-Haldane kinetics.

4. Enzyme–product release can be rate-determining for k_{cat}

Product release is sometimes rate-determining at saturating substrate concentrations with some dehydrogenases. Examples of this are the dissociation of NADH from glyceraldehyde 3-phosphate dehydrogenase at high pH,[23] of NADH from horse liver alcohol dehydrogenase at low salt,[24,25] and of NADPH from glutamate dehydrogenase.[26] Note that the phrase product release is used, and not product dissociation. This is because the overall release of products can involve steps in addition to dissociation, such as conformational changes, and these may be the rate-determining steps rather than the dissociation itself (see the next section).

5. Conformational changes

There are many documented cases of substrate-induced conformational changes with rate constants in the range of 10 to 10^4 s^{-1}, and also instances in which discrepancies in rate constants indicate rate-determining protein isomerizations.[26] Isomerizations are often associated with slow steps; for example, the dissociation of NADH from some dehydrogenases involves a concomitant conformational change. However, there are few, if any, direct demonstrations that a conformational change is, by itself, rate-determining. There is evidence that the rate-determining step in the formation of glyceraldehyde 3-phosphate catalyzed by triosephosphate isomerase is a conformational change after product release.[27]

It should be noted that the rate-determining step of a reaction changes with substrate concentration, since the rate is proportional to k_{cat} at saturating concentrations of substrate, and to k_{cat}/K_M at low concentrations. When a step is said to be rate-determining and the reaction conditions are not stated, the reaction is usually at saturating substrate concentrations.

References

1   H. Hartridge and F. J. W. Roughton, *Proc. R. Soc.* **A104**, 376 (1923).
2   F. J. W. Roughton, *Proc. R. Soc.* **B115**, 475 (1934).
3   B. Chance, *J. Franklin Inst.* **229**, 455, 613, 637 (1940).
4   Q. H. Gibson, *J. Physiol.* **117**, 49P (1952).

5   A. R. Fersht and R. Jakes, *Biochemistry* **14**, 3350 (1975).

6   Q. H. Gibson, *Progr. Biophys. Biophys. Chem.* **2**, 1 (1959).

7   J. L. Martin, A. Migus, C. Poyart, Y. Lecarpentier, A. Antonetti, and A. Orszag, *Biochem. Biophys. Res. Commun.* **107**, 803 (1982).

8   J. H. Kaplan, B. Forbush III, and J. F. Hoffman, *Biochemistry* **17**, 1929 (1978).

9   Y. E. Goldman, M. G. Hibberd, J. A. McRay, and D. R. Trentham, *Nature, Lond.* **300**, 701 (1982).

10 G. Czerlinski and M. Eigen, *Z. Electrochem.* **63**, 652 (1959).

11 O. Jardetzky and G. C. K. Roberts, *NMR in molecular biology*, Academic Press (1981).

12 C. R. Cantor and P. R. Schimmel, *Biophysical chemistry*, W. H. Freeman and Company (1980).

13 A. S. Mildvan and M. C. Scrutton, *Biochemistry* **6**, 2978 (1967).

14 B. D. Sykes, *J. Am. Chem. Soc.* **91**, 949 (1969).

15 S. H. Smallcombe, B. Ault, and J. H. Richards, *J. Am. Chem. Soc.* **94**, 4585 (1972).

16 J. H. Baldo, S. E. Halford, S. L. Patt, and B. D. Sykes, *Biochemistry* **14**, 1893 (1975).

17 J. J. Led and H. Gesmar, *J. Mag. Res.* **49**, 444 (1982).

18 J. J. Led, E. Neesgard, and J. T. Johansen, *FEBS Lett.* **147**, 74 (1982).

19 R. S. Balaban and J. A. Ferretti, *Proc. Natn. Acad. Sci. USA* **80**, 1241 (1983).

20 A. R. Fersht, J. A. Ashford, C. J. Bruton, R. Jakes, G. L. E. Koch, and B. S. Hartley, *Biochemistry* **14**, 1 (1975).

21 M. Eigen and G. G. Hammes, *Adv. Enzymol.* **25**, 1 (1963).

22 E. Holler, J. A. Rupley, and G. P. Hess, *Biochem. Biophys. Res. Commun.* **37**, 423 (1969).

23 D. R. Trentham, *Biochem. J.* **122**, 71 (1971).

24 H. Theorell and B. Chance, *Acta Chem. Scand.* **5**, 1127 (1951).

25 J. D. Shore and H. Gutfreund, *Biochemistry* **9**, 4655 (1970).

26 A. di Franco, *Eur. J. Biochem.* **45**, 407 (1974).

27 I. A. Rose and R. Iyengar, *Biochemistry* **21**, 1591 (1982).

Further reading

K. Hiromi, *Kinetics of fast enzyme reactions*, Halsted Press (Wiley) (1979).

5

The pH dependence of enzyme catalysis

The activities of many enzymes vary with pH in the same way that simple acids and bases ionize. This is not surprising, since, as we saw in Chapter 1, the active sites generally contain important acidic or basic groups (Table 5·1). It is to be expected that if only one protonic form of the acid or base is catalytically active, the catalysis will somehow depend on the concentration of the active form. In this chapter we shall see that k_{cat}, K_M, and k_{cat}/K_M are affected in different ways by the ionizations of the enzyme and enzyme–substrate complex.

A. Ionization of simple acids and bases: The basic equations

It is usual to discuss the ionization of a base B in terms of its conjugate acid BH^+ in order to use the same set of equations for both acids and bases. The ionization constant K_a is defined by

$$K_a = \frac{[B][H^+]}{[BH^+]} \tag{5·1}$$

Or for an acid HA and its conjugate base A^-,

$$K_a = \frac{[A^-][H^+]}{[HA]} \tag{5·2}$$

The pK_a is defined by

$$pK_a = -\log K_a \tag{5·3}$$

Equations 5·1 and 5·3 (or 5·2) may be rearranged to give the Henderson-Hasselbalch equation:

$$pH = pK_a + \log \frac{[B]}{[BH^+]} \tag{5·4}$$

TABLE 5·1. pK_a's of ionizing groups[a]

	pK_a	
Group	Model compounds (small peptides)	Usual range in proteins
Amino acid α-CO_2H	3.6 ⎫	
Asp (CO_2H)	4.0 ⎬	2–5.5
Glu (CO_2H)	4.5 ⎭	
His (imidazole)	6.4	5–8
Amino acid α-NH_2	7.8	~8
Lys (ϵ-NH_2)	10.4	~10
Arg (guanidine)	~12	—
Tyr (OH)	9.7	9–12
Cys (SH)	9.1	8–11
Phosphates	1.3, 6.5	—

[a] Data mainly from C. Tanford, *Adv. Protein Chem.* **17**. 69 (1962); C. Tanford and R. Roxby, *Biochemistry* **11**, 2192 (1972); Z. Shaked, R. P. Szajewski, and G. M. Whitesides, *Biochemistry* **19**, 4156 (1980).

It is readily seen from this equation that the pK_a of an acid or a base is the pH of half neutralization when the concentrations of B and BH^+ are equal.

The variation of the concentrations of HA and A^- with the proton concentration is found from rearranging equation 5·2 as

$$[HA] = \frac{[A]_0[H^+]}{K_a + [H^+]} \tag{5·5}$$

and

$$[A^-] = \frac{[A]_0 K_a}{K_a + [H^+]} \tag{5·6}$$

where $[A]_0 = [HA] + [A^-]$

Suppose that there is some quantity L (an absorption coefficient, a rate constant, etc.) such that the corresponding property of a solution (the absorbance, the reaction rate, etc.) is the product of this quantity and the concentration. If the value of L for the molecule HA is L_{HA}, and the value for the molecule A^- is L_{A^-}, then the observed value of the property at a particular pH, $L_H[A]_0$, is given by $L_H[A]_0 = L_{HA}[HA] + L_{A^-}[A^-]$. That is:

$$L_H[A]_0 = \frac{L_{HA}[A]_0[H^+]}{K_a + [H^+]} + \frac{L_{A^-}[A]_0 K_a}{K_a + [H^+]} \tag{5·7}$$

so that

$$L_H = \frac{L_{HA}[H^+] + L_{A^-}K_a}{K_a + [H^+]}$$ (5·8)

An example of equation 5·8 in which L is an equilibrium constant K is plotted in Figure 5·1. A further example is shown in Figure 5·2, where L is a rate constant k for a reaction that depends on the basic form of an acid (i.e., $L_{HA} = 0$). This is plotted in two ways. Note in the logarithmic plot that there is a linear decrease with decreasing pH at low pH. The point of intersection of the two linear regions is the pK_a.

A *doubly ionizing* system such as

$$H_2A \xrightleftharpoons{K_1} HA^- + H^+ \xrightleftharpoons{K_2} A^{2-} + 2H^+$$ (5·9)

may be analyzed to give

$$L_H = \frac{[H^+]^2 L_{H_2A} + [H^+]K_1 L_{HA^-} + K_1 K_2 L_{A^{2-}}}{K_1 K_2 + [H^+]K_1 + [H^+]^2}$$ (5·10)

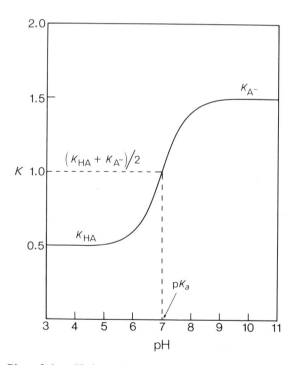

FIGURE 5·1. Plot of the pH dependence of an arbitrary constant K that has the value of K_{HA} for the acidic form and K_{A^-} for the basic form of an acid of pK_a 7.

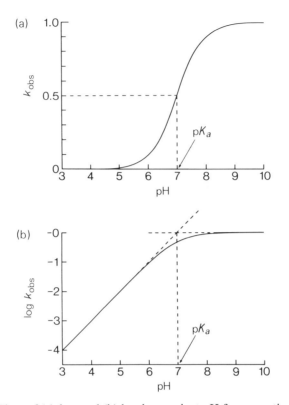

FIGURE 5·2. Plots of (a) k_{obs} and (b) log k_{obs} against pH for a reaction whose rate depends on the basic form of an acid of pK_a 7.

It is frequently found in enzyme reactions that activity depends on two groups, one in the acidic form and one in the basic. This is equivalent to the situation in equation 5·9, where rate varies as [HA$^-$], leading to equation 5·10, in which $L_{H_2A} = L_{A^{2-}} = 0$. Equation 5·10 is then the equation of a bell-shaped curve with its maximum at pH = (pK_{a1} + pK_{a2})/2.

1. Extraction of pK_a's by inspection of equations

The crucial point in equations 5·8 and 5·10 is that the points of inflection in the plots of L_H against pH—i.e., the pK_a's—are determined by the *denominator* of the fraction. The numerator determines the amplitudes of the functions. The apparent pK_a's of a complex kinetic equation may be found by rearranging the equation to the form of equation 5·8 (for a singly ionizing system) or to that of equation 5·10 (for a doubly ionizing system), and then comparing the denominators. This may be illustrated by considering the pH dependence of $1/L_H$ instead of L_H in equation 5·8; i.e.,

$$\frac{1}{L_H} = \frac{K_a + [H^+]}{L_{HA}[H^+] + L_{A^-}K_a} \tag{5·11}$$

Rearranging equation 5·11 to be of the same form as 5·8 gives

$$\frac{1}{L_H} = \frac{K_a/L_{HA} + [H^+]/L_{HA}}{K_aL_{A^-}/L_{HA} + [H^+]} \tag{5·12}$$

We can say directly from examining the denominator of equation 5·12 that the plot of $1/L_H$ against pH will have an apparent ionization constant, K_{app}, given by

$$K_{app} = K_a \frac{L_{A^-}}{L_{HA}} \tag{5·13}$$

B. The effect of ionizations of groups in enzymes on kinetics

Although enzymes contain a multitude of ionizing groups, it is usually found that plots of rate against pH take the form of simple single or double ionization curves. This is because the only ionizations that are of importance are those of groups that are directly involved in catalysis at the active site, or those of groups elsewhere that are responsible for maintaining the active conformation of the enzyme.

We shall now analyze some of the simple examples, beginning with the Michaelis-Menten mechanism. We can make four simplifying assumptions that may break down in some circumstances but that often hold in practice:

1. The groups act as perfectly titrating acids or bases (this is generally a fair approximation).
2. Only one ionic form of the enzyme is active (this is usually true).
3. All intermediates are in protonic equilibrium; i.e., proton transfers are faster than chemical steps (this is generally true—see Chapter 4).
4. The rate-determining step does not change with pH (this may break down, with interesting consequences).

1. The simple theory: The Michaelis-Menten mechanism

$$\begin{array}{ccc}
\text{E} & \underset{}{\overset{K_S}{\rightleftharpoons}} & \text{ES} & \xrightarrow[\text{slow}]{k_{cat}} \\
K_E \updownarrow & & \updownarrow K_{ES} & \\
\text{HE} & \underset{K_S'}{\rightleftharpoons} & \text{HES} &
\end{array} \tag{5·14}$$

In equation 5·14, the ionization constant of the free enzyme is K_E; that of the enzyme–substrate complex is K_{ES}; and the dissociation constants of HES and ES are K'_S and K_S, respectively.

All four equilibrium constants cannot vary independently, due to the cyclic nature of the equilibria. Once three are fixed, the fourth is defined by

$$K_E K'_S = K_{ES} K_S \tag{5.15}$$

Equation 5·15 may be derived by multiplying the various dissociation and ionization constants or simply by inspection, since the process HES → ES → E must give the same energy change as HES → HE → E.

The two important conclusions from equation 5·15 are:

1. If $K_E = K_{ES}$, then $K_S = K'_S$, and there is no pH dependence for binding.
2. If $K_E \neq K_{ES}$, i.e., if the pK_a is perturbed on binding, then

$$K_S = K'_S \frac{K_E}{K_{ES}} \tag{5.16}$$

and the binding of S must of necessity be pH-dependent.

2. The pH dependence of k_{cat}, k_{cat}/K_M, K_M, and $1/K_M$[1,2]

The pH dependence of v is obtained by expressing the concentration of ES in terms of $[E]_0$ to give, after the necessary algebra,

$$v_H = \frac{k_{cat}[E]_0[S]}{K_S + [S](1 + [H^+]/K_{ES}) + K_S[H^+]/K_E} \tag{5.17}$$

The pH dependence of k_{cat} is derived from equation 5·17 when [S] is much greater than K_S; i.e.,

$$(V_{max})_H = [E]_0(k_{cat})_H = \frac{k_{cat}[E]_0 K_{ES}}{K_{ES} + [H^+]} \tag{5.18}$$

Comparison of the denominator of equation 5·18 with that of 5·8 shows that the pH dependence of V_{max} or k_{cat} follows the ionization constant of the enzyme–substrate complex, K_{ES}.

The apparent value of K_M at each pH may be found by rearranging equation 5·17 to the form of the basic Michaelis-Menten equation (equation 3·1):

$$(K_M)_H = \frac{K_S K_{ES} + [H^+]K_S K_{ES}/K_E}{K_{ES} + [H^+]} \tag{5.19}$$

K_M also follows the ionization of the enzyme–substrate complex.

The pH dependence of k_{cat}/K_M is given by the variation of v at low values of [S] (or, alternatively, by the ratios of equations 5·18 and 5·19). Simplifying equation 5·17 by putting [S] close to zero and rearranging gives

$$\left(\frac{k_{cat}}{K_M}\right)_H = \frac{(k_{cat}/K_S)K_E}{K_E + [H^+]} \tag{5·20}$$

The value of k_{cat}/K_M (and v for [S] much less than K_M) follows the ionization of the free enzyme.

3. A simple rule for the prediction and assignment of pK_a's

The above results are part of the general rule that "the plot of the equilibrium constant K or the rate constant k for the process X $\rightarrow$ Y as a function of pH follows the ionization constants of X." Let us now apply this rule to scheme 5·14, bearing in mind that the substrate may also have ionizing groups:

1. *The pH dependence of k_{cat}.* The process concerned is

$$\text{ES} \xrightarrow{k_{cat}} \text{E} + \text{P} \tag{5·21}$$

The value of k_{cat} follows the pK_a of the enzyme–substrate complex.
2. *The pH dependence of K_M.* The process concerned is

$$\text{ES} \xrightarrow{K_M} \text{E} + \text{S} \tag{5·22}$$

The pH dependence of K_M follows the ionization of the enzyme–substrate complex.
3. *The pH dependence of $1/K_M$.*† The process concerned is

$$\text{E} + \text{S} \xrightarrow{1/K_M} \text{ES} \tag{5·23}$$

The pH dependence of $1/K_M$ follows the ionizations in the free enzyme and the free substrate.
4. *The pH dependence of k_{cat}/K_M.* The process concerned is

$$\text{E} + \text{S} \xrightarrow{k_{cat}/K_M} \text{E} + \text{P} \tag{5·24}$$

The pH dependence of k_{cat}/K_M follows the ionizations in the free enzyme and the free substrate.

---

† Some people find it confusing that plotting the inverse of a function, e.g., $1/K_M$ instead of K_M, gives a different pK_a. The mathematical reason for this was given in equations 5·11 to 5·13. The pH dependence of K_M contains the information for determining the pK_a's of E, S, and ES, but they are manifested in different ways in different plots. An alternative procedure for analyzing a plot of log K_M vs. pH has been given by M. Dixon.[1]

C. Modifications and breakdown of the simple theory

The simple theory outlined above has to be modified to account for the pH dependence of the catalytic parameters in mechanisms more complicated than the basic Michaelis-Menten.

1. Modifications due to additional intermediates

a. Intermediates on the reaction pathway[3]

$$\text{E} + \text{S} \overset{K_S}{\rightleftharpoons} \text{ES} \overset{k_2}{\longrightarrow} \text{EA} \overset{k_3}{\longrightarrow} \text{E} + \text{P} \tag{5.25}$$

The presence of additional intermediates does not affect the pH dependence of k_{cat}/K_M or $1/K_M$, since they represent changes from the free enzyme and free substrate only. The pH dependence of these still gives the pK_a's of the free enzyme and free substrate. But the pH dependence of k_{cat} and K_M now concerns changes from the intermediate (EA) as well as from the enzyme–substrate complex. If the intermediate EA in equation 5.25 is the major enzyme-bound species, the pH dependence of k_{cat} and K_M will give the pK_a's of EA. If both ES and EA accumulate, the pH profiles give pK_a's that are the weighted means of those of ES and EA.[4]

b. Nonproductive binding modes[5]

When a substrate binds in a nonproductive mode as well as in the productive mode, the pH dependence of k_{cat} and K_M may give an apparent pK_a for the catalytically important group at the active site; such a pK_a is far from the real value in the productive complex. This happens if the ratio of productive to nonproductive binding changes on ionization of the group. Suppose that the activity of the enzyme is dependent on a group being in the basic form at high pH. If, say, the substrate is bound with more being in the productive mode at low pH than at high pH, then as the pH decreases through the pK_a of the catalytic group, the decrease in rate as the group becomes protonated will be partially compensated for by an increase in productive binding. This has the effect of lowering the apparent pK_a controlling the pH dependence of k_{cat} from the value in the productive complex. The pH dependence of K_M is affected in an identical manner.

The algebraic solution of such a situation is shown in scheme 5.26. HES and ES are the productively bound complexes, and K' and K are

$$
\begin{array}{ccc}
\text{HE} & \overset{K_a}{\rightleftharpoons} & \text{E} \\
{\scriptstyle K'_S} \big\updownarrow & & \big\updownarrow {\scriptstyle K_S} \\
\text{HES} & \overset{K'_a}{\rightleftharpoons} & \text{ES} \overset{k_2}{\underset{\text{slow}}{\longrightarrow}} \text{E} + \text{P} \\
{\scriptstyle K'} \big\updownarrow & & \big\updownarrow {\scriptstyle K} \\
\text{HES}' & \overset{K''_a}{\rightleftharpoons} & \text{ES}'
\end{array}
\tag{5.26}
$$

the equilibrium constants between these and the nonproductively bound complexes HES' and ES' as defined in the scheme. Solving the rate equation by the usual means gives

$$k_{cat} = \frac{k_2}{K(1 + [H^+]/K_a'') + 1 + [H^+]/K_a'} \qquad (5 \cdot 27)$$

$$K_M = \frac{K_S(1 + [H^+]/K_a)}{K(1 + [H^+]/K_a'') + 1 + [H^+]/K_a'} \qquad (5 \cdot 28)$$

Through rearranging equations 5·27 and 5·28 to the form of equation 5·8 and inspecting the denominator, we can obtain an observed pK_a given by

$$pK_{a(obs)} = pK_a' - \log \frac{1 + K}{1 + K'} \qquad (5 \cdot 29)$$

Nonproductive binding modes do not affect the pH dependence of k_{cat}/K_M or $1/K_M$, for the reasons discussed in the previous case.

2. Breakdown of the simple rules: Briggs-Haldane kinetics and change of rate-determining step with pH: Kinetic pK_a's[4,6-8]

$$E + S \underset{k_{-1}}{\overset{k_1}{\rightleftharpoons}} ES \overset{k_2}{\longrightarrow} E + P \qquad (5 \cdot 30)$$

In the extreme case of Briggs-Haldane kinetics (equation 5·30), the rate constant for the chemical step is larger than that for the dissociation of the enzyme–substrate complex. In this case, k_{cat}/K_M is equal to k_1, the association constant of the enzyme and substrate. Suppose again that the catalytic activity of the enzyme depends on a group being in its basic form. At high pH, k_2 is faster than k_{-1}. But as the pH is lowered, k_2 decreases while the base becomes protonated, until k_2 is slower than k_{-1}. The apparent pK_a to which this leads in the pH profile of k_{cat}/K_M is lower than the pK_a of the important base (Figure 5·3).[7] It may be shown that if the ionization constant of the group on the enzyme is K_a, the apparent ionization constant is given by

$$K_{app} = \frac{K_a(k_2 + k_{-1})}{k_{-1}} \qquad (5 \cdot 31)$$

K_{app} is termed a *kinetic* pK_a, since it does not represent a real ionization but is composed of the ratios of rate constants that are not for proton transfers. Kinetic pK_a's occur whenever there is a change of rate-determining step with pH.

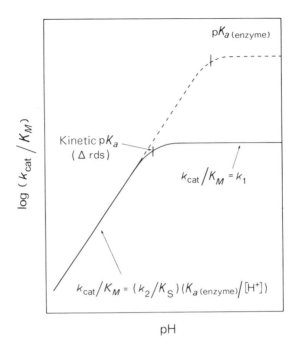

FIGURE 5·3. Illustration of the "kinetic pK_a" caused by a change in rate-determining step with pH. At low pH, k_{cat}/K_M falls off as

$$\left(\frac{k_2}{k_{-1}/k_1}\right)\left(\frac{K_a}{K_a + [H^+]}\right)$$

—which can be simplified because $K_S = k_{-1}/k_1$ and $[H^+] \gg K_a$. At high pH, k_{cat}/K_M levels off at k_1, the second-order rate constant for the association of enzyme and substrate.

3. An experimental distinction between kinetic and equilibrium pK_a's[7,8]

A pK_a that is composed of the combination of equilibrium constants and the individual pK_a's of titrating groups, such as that for the nonproductive binding scheme (equation 5·29), titrates as a real pK_a. If one were to add acid or base to the enzyme–substrate complexes in scheme 5·26, the catalytic group would titrate according to the pK_a given by equation 5·29. Similarly, if one were to measure the fraction of the base in the ionized form by some spectroscopic technique, it would be found to ionize with the pK_a of equation 5·29. This does not happen with kinetic pK_a's. In the case of the change of rate-determining step with pH for k_{cat}/K_M in the Briggs-Haldane mechanism, although the pK_a from the kinetics will be that given by equation 5·31, direct measurement of the titration of the catalytic group will give its true ionization constant K_a.

4. Microscopic and macroscopic pK_a's

When an acid exists in different forms, such as HES and HES' in scheme 5·26, the pK_a in each of the forms is termed a *microscopic* pK_a. The pK_a with which the system is found to titrate, e.g., $pK_{a(obs)}$ in equation 5·29, is called variously a *macroscopic,* an *apparent,* or a *group* pK_a. To all intents and purposes it is a real pK_a, unlike a kinetic pK_a.

D. The influence of surface charge on pK_a's of groups in enzymes

The surface of an enzyme contains many polar groups. Chymotrypsin-ogen, for example, has on its surface 4 arginine and 14 lysine residues, which are positively charged, and 7 aspartate and 5 glutamate residues, which are negatively charged. These provide an ionic atmosphere, or electrostatic field, which may stabilize or destabilize buried or partly buried ionic groups. Calculations have been made to show that the perturbation of the pK_a of a buried group is a complex function of the size and shape of the protein and also of the ionic strength of the solution.[9–11] The magnitude of these effects is nicely illustrated by some studies in which the surface carboxylates or ammonium ions are chemically modified.

The active site of chymotrypsin bears a negative charge at high pH and a net zero charge at low pH. It is expected that the basic form of the active site at high pH will be stabilized by a positively charged surface and destabilized by a negatively charged one. Table 5·2 shows that this is borne out experimentally. The conversion of the 14 positively charged lysines to negatively charged carboxylates in the reaction of chymotrypsin with succinic anhydride causes the pK_a controlling the pH dependence

TABLE 5·2. *Influence of surface charge on pK_a's*

Enzyme	Modification	pK_a	k_{cat} (s^{-1})
Chymotrypsin[a]	—	7.0^b	47
Succinyl-chymotrypsin[a]	Lys ($-NH_3^+$) → $-NHOCCH_2CH_2\ CO_2^-$	8.0^b	74
Ethylenediamine-chymotrypsin	Asp, Glu ($-CO_2^-$) → $-CONH\ (CH_2)_2NH_3^+$	6.1^b	50
Trypsin[c]	—	7.0^d	0.7
Acetyl-trypsin[c]	Lys ($-NH_3^+$) → $-NHCOCH_3$	7.2^d	1.2

[a] δ-Chymotrypsin, 25°C, ionic strength 0.1. [From P. Valenzuela and M. L. Bender, *Biochim. Biophys. Acta* **256**, 538 (1971).]
[b] The pK_a for k_{cat} for the hydrolysis of acetyl-L-tryptophan methyl ester.
[c] From W. E. Spomer and J. F. Wootton, *Biochim. Biophys. Acta* **235**, 164 (1971).
[d] The pK_a for k_{cat}/K_M for the hydrolysis of benzoyl-L-arginine amide.

of k_{cat} for the hydrolysis of acetyltryptophan methyl ester to increase from 7.0 to 8.0. On the other hand, the conversion of 13 negatively charged carboxyls to positively charged amines lowers the pK_a to 6.1. Similarly, the acylation of the surface amino groups of trypsin by acetic anhydride causes an increase of 0.2 units in the pK_a of the free enzyme.

These effects are mimicked by changes of pH. Below pH 5 the carboxylates become protonated, and above pH 9 the ammonium groups of lysine start to lose their positive charge as they deprotonate. This causes perturbations in the titration curves of the enzyme at extremes of pH. These effects are most marked at low ionic strengths of 0.2 M or less, but they may be depressed by high ionic strengths to become negligible at 1 M. Fortunately, many titration curves are measured between pH 5 and pH 9, where few surface groups titrate, and excellent results are obtained at ionic strengths of 0.1 M.[8,12]

As well as perturbing the pK_a of the catalytic acid or base, the large change of surface charge may alter the rate constants for the chemical steps. The change of surface charge is analogous to the change of ionic strength in nonenzymatic ionic reactions, and the two effects are analogous to secondary and primary salt effects in physical organic chemistry.

E. Graphical representation of data

Suppose that one of the kinetic quantities, such as k_{cat}, depends on the enzyme being in the acidic form. Then the rate will depend upon the ionization constant K_a and the pH according to

$$(k_{cat})_H = \frac{k_{cat}[H^+]}{K_a + [H^+]} \tag{5·32}$$

where $(k_{cat})_H$ is the observed value at the particular $[H^+]$ (equation 5·8). Equation 5·32 is in the same form as the Michaelis-Menten equation, so K_a may be found by plots analogous to the Lineweaver-Burk or Eadie plots (equations 3·28 and 3·29). For example, because

$$(k_{cat})_H = k_{cat} - \frac{K_a(k_{cat})_H}{[H^+]} \tag{5·33}$$

k_{cat} and K_a may be obtained by plotting $(k_{cat})_H$ against $(k_{cat})_H/[H^+]$. Similarly, if the rate depends on the enzyme being in the basic form, it may be shown that

$$(k_{cat})_H = k_{cat} - \frac{(k_{cat})_H[H^+]}{K_a} \tag{5·34}$$

Here $(k_{cat})_H$ must be plotted against $(k_{cat})_H[H^+]$ (see Figure 5·4).

For more complicated examples, in which the kinetic quantity does not fall to zero at either high or low pH (as is often found in plots of association or dissociation constants against pH), the procedure is modified by plotting the difference between the observed value at the particular pH and one of the limiting values at the extremes of pH.

F. Illustrative examples and experimental evidence

The most-studied enzyme in this context is chymotrypsin. Besides being well characterized in both its structure and its catalytic mechanism, it has

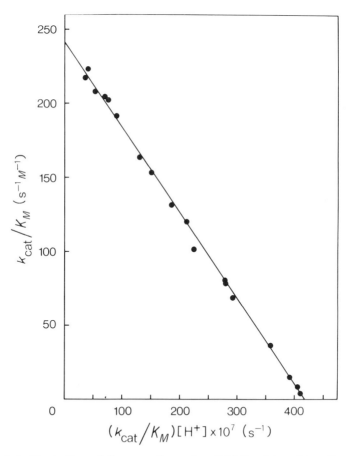

FIGURE 5·4. Illustration of the use of equation 5·34 for determining the pK_a of the active site of an enzyme. This example is the hydrolysis of acetyl-L-tyrosine *p*-acetylanilide by α-chymotrypsin at 25°C. Note that k_{cat}/K_M rather than k_{cat} is used here. [From A. R. Fersht and M. Renard, *Biochemistry* **13**, 1416 (1974).]

the advantage of a very broad specificity. Substrates may be chosen to obey the simple Michaelis-Menten mechanism, to accumulate intermediates, to show nonproductive binding, and to exhibit Briggs-Haldane kinetics with a change of rate-determining step with pH.

The pH dependence of k_{cat}/K_M for the hydrolysis of substrates follows a bell-shaped curve with a maximum at pH 7.8 and pK_a's of 6.8 and 8.8 for α-chymotrypsin, and a maximum at pH 7.9 and pK_a's of 6.8 and 9.1 for δ-chymotrypsin. The pK_a of 6.8 represents the ionization of the catalytically important base at the active site, whereas the high-pH ionization is due to the α-amino group of Ile-16, which holds the enzyme in a catalytically active conformation. This conformationally important ionization does not affect k_{cat}—which usually gives a sigmoid curve of pK_a 6 to 7—but it does cause an increase of K_M at high pH.

1. The pK_a of the active site of chymotrypsin

a. The free enzyme
The theory predicts that unless there is a change of rate-determining step with pH, the pH dependence of k_{cat}/K_M for all non-ionizing substrates should give the same pK_a: that for the free enzyme. With one exception, this is found (Table 5·3). At 25°C and ionic strength 0.1 M, the pK_a of the active site is 6.80 ± 0.03. The most accurate data available fit very precisely the theoretical ionization curves between pH 5 and 8, after allowance has been made for the fraction of the enzyme in the inactive conformation. The relationship holds for amides with which no intermediate accumulates and the Michaelis-Menten mechanism holds, and also for esters with which the acylenzyme accumulates.

TABLE 5·3. *Some pK_a's for the active site of chymotrypsin*

	pK_a		
Substrate	k_{cat}/K_M	k_{cat}	Ref.
Acetyl-L-phenylalanine alaninamide[a]	6.80	6.6	1
Formyl-L-phenylalanine semicarbazide[a]	6.84	6.32	1
Acetyl-L-tyrosine p-acetylanilide[a]	6.77	—	1
p-Nitrophenyl acetate[a]	6.85	—	2
Acetyl-L-phenylalanine ethyl ester[a]	6.8	6.85	3
Acetyl-L-tyrosine p-acetylanilide[b]	6.83	—	4
Acetyl-L-tryptophan p-nitrophenyl ester[b]	6.50	6.9	4

[a] α-Chymotrypsin, 25°C, ionic strength 0.1.
[b] δ-Chymotrypsin, 25°C, ionic strength 0.95.

1 A. R. Fersht and M. Renard, *Biochemistry* **13**, 1416 (1974).
2 M. L. Bender, G. E. Clement, F. J. Kézdy, and H. d'A. Heck, *J. Am. Chem. Soc.* **86**, 3680 (1964).
3 B. R. Hammond and H. Gutfreund, *Biochem. J.* **61**, 187 (1955).
4 M. Renard and A. R. Fersht, *Biochemistry* **12**, 4713 (1973).

b. Acetyl-L-tryptophan p-nitrophenyl ester: Briggs-Haldane kinetics with change of rate-determining step with pH

The one exception is the pH dependence of k_{cat}/K_M for the hydrolysis of acetyl-L-tryptophan p-nitrophenyl ester. This gives an apparent pK_a of 6.50 for the free enzyme. The reason for the low value is explained in section C2. At high pH, the association of enzyme and substrate is partly rate-determining for k_{cat}/K_M. The limiting value of k_{cat}/K_M at high pH is 3×10^7 s^{-1} M^{-1}, a value close to that for the diffusion-controlled encounter of enzyme and substrate. However, at low pH, the chemical steps slow down and become rate-determining as the enzyme becomes protonated and less active. The rate constant for the association step may be calculated from equation 5·31 to be 6×10^7 s^{-1} M^{-1}. The data are consistent with scheme 5·35 and a pK_a of 6.8 for the active site.[7,13] This is a rare example in which steady state kinetics may be analyzed to give rate constants for several specific steps on the reaction pathway.

$$\text{Ac-Trp-ON-Ph} + \text{CT} \underset{6 \times 10^4 \text{s}^{-1}}{\overset{6 \times 10^7 \text{s}^{-1} M^{-1}}{\rightleftharpoons}} \text{Ac-Trp-ON-Ph·CT}$$

$$\downarrow {\scriptstyle 7 \times 10^4 \text{s}^{-1}}$$

$$\text{Ac-Trp-CT} \qquad (5·35)$$

$$\downarrow {\scriptstyle 65 \text{s}^{-1}}$$

$$\text{Ac-Trp} + \text{CT}$$

(CT = chymotrypsin)

c. The enzyme–substrate complex

The pK_a of the enzyme–substrate complex is not constant like that of the free enzyme, since the binding of substrates perturbs the pK_a of the active site. The pK_a values of k_{cat} for the hydrolysis of amides range from 6 to 7. In consequence, the value of K_M increases at low pH according to equation 5·16.

The precise determination of the pK_a for the enzyme–substrate complex from the pH dependence of K_M is difficult for two reasons. First, the variation of K_M is relatively small, and highly accurate data are required. Secondly, two plateau regions must be determined, one at low pH as well as one at high pH. This means that measurements are required over a wider range of pH than for the determination of k_{cat} or k_{cat}/K_M, and are thus more susceptible to perturbations from the ionizations of other groups.

d. The ionization at high pH[14]

The pK_a at high pH is caused by a substrate-independent conformational change in the enzyme, a change that may be monitored directly by physical techniques such as optical rotation and fluorescence yield. Kinetic measurements give the same pK_a as that found by these methods.

G. Direct titration of groups in enzymes

Several methods have been developed for the direct titration of some ionizing side chains in proteins. One of the major problems is in identifying the desired group among all the similar ones.

1. The effect of D_2O on pH/pD and pK_a's

Many of the spectroscopic methods use D_2O as the solvent rather than H_2O, in order to separate the required signals from those coming from the solvent. It has been found experimentally that the glass electrode gives a lower reading in D_2O by 0.4 units:

$$pH = pD + 0.4 \qquad\qquad (5\cdot36)$$

In mixtures of H_2O and D_2O, the glass electrode reading is related to pH by

$$pH = pD + 0.3139\alpha + 0.0854\alpha^2 \qquad\qquad (5\cdot37)$$

where α is the atom fraction of deuterium, $[D]/([D] + [H])$.[15] However, pK_a's are higher in D_2O and in D_2O/H_2O mixtures than in H_2O. The increase in pK_a sometimes balances the decreased reading on the glass electrode, so that the measured pK_a in D_2O is often assumed to be the true pK_a in water. The effect of the solvent on ionization constants is somewhat variable, though, and the simplification can lead to errors of a few tenths of a pH unit.

2. Methods

a. Nuclear magnetic resonance
NMR is a most powerful method for determining pK_a values of residues in proteins because it can detect individual signals and their titration with pH.

^1H NMR has been particularly useful in determining the pK_a's of histidines in proteins. The signals from the C-2 and C-4 protons are further downfield than the bulk of the resonances of the protein, and are resolvable in D_2O solutions. The chemical shift changes on protonation, so the histidines may be readily titrated. Where there is more than one histidine it is difficult to assign the individual pK_a's. In the case of ribonuclease, chemical modification, selective deuteration, and a knowledge of the crystal structure have allowed the pK_a's of all four histidines to be assigned.[16,17]

The resonances of protons in hydrogen bonds may be shifted downfield to such an extent that they may be observed in H_2O solutions. The proton between Asp-102 and His-57 in chymotrypsinogen, chymotrypsin, and other serine proteases has been located and its resonance found to titrate with a pK_a of 7.5[18] (although the pK_a is for the dissociation of the proton on the other nitrogen of the imidazole ring).

^{15}N NMR has also been used to measure the pK_a value of the active-site histidine of the serine proteases,[19] confirming that it is the histidine and not the buried aspartate that ionizes with a pK_a of about 7.

^{13}C NMR may be used to measure the pK_a's of lysine and aspartate residues. ^{1}H NMR has been used to titrate the ionization of tyrosine residues.[20]

^{31}P NMR is a most valuable technique. Not only can it be used to titrate the pK_a values of phosphates, but it can also be used to measure the ionization of known phosphate residues *in vivo*, such as those of 2,3-diphosphoglycerate, and thus to determine intracellular pH.[21]

b. Infrared spectroscopy[22]

Carboxyl groups absorb at about 1710 cm^{-1}, and carboxylates at about 1570 cm^{-1}. Infrared difference spectra in D_2O solutions have been used to measure the pK_a's of abnormal carboxyl groups in α-lactoglobulin (7.5) and lysozyme (2.0, 6.5).

c. Ultraviolet difference spectroscopy[23]

Ultraviolet difference spectra have frequently been used to measure the ionization of the phenolic hydroxyl of tyrosines. The sulfydryls of cysteines and the imidazoles of histidines are also amenable to difference spectroscopy.

d. Fluorescence[23,24]

Fluorescence is useful when the ionization of a group perturbs the spectrum of a neighboring tryptophan, the major fluorescent species in proteins, or causes a conformational change that perturbs the fluorescence of the protein as a whole. Tyrosines may be titrated in the absence of tryptophans (which fluoresce more strongly).

e. Difference titration[25]

The direct titration of a protein with acid or base usually gives an uninterpretable ionization curve because of the overlapping titrations of the many groups. However, one or two ionizations may be isolated by comparing the titration curve with that of the enzyme after a particular group has been blocked. For example, Asp-52 of lysozyme may be specifically esterified with triethoxonium fluoroborate. The difference titration between this modified form and the native protein gives not only the pK_a of Asp-52 but also the effect of this ionization on the other important acid at the active site, Glu-35.

f. Denaturation difference titration[26]

Practically speaking, it is not possible to determine by titration whether or not a buried group in a protein, such as Asp-102 in chymotrypsin and chymotrypsinogen, is ionized. Accurate titration is possible only between pH 3 and 11 because of the high background concentrations of hydroxyl ions and protons outside this range. For this reason it is not possible to detect a missing ionization in the overall titration curve if the protein has any other abnormally titrating groups. Chymotrypsin, for example, has three abnormally low titrating carboxyls which are still ionized below pH 3. However, these groups titrate normally when the protein is denatured, and the number of carboxyls titrating in the denatured protein is easily determined. It was shown that Asp-102 is ionized in the high-pH forms of chymotrypsin and chymotrypsinogen by measuring the proton uptake on denaturation and adding this to the number of carboxyls known to be ionized in the denatured protein. In order to improve the accuracy, the majority of the surface carboxyls were converted to amides to lower the number of ionizing groups.

g. Chemical modification

The rate of inhibition of enzymes by irreversible inhibitors has been used in the same way as normal kinetics to give the pK_a's of the free enzyme and the enzyme–inhibitor complex. The pK_a's of other residues have been measured from the pH dependence of their reaction with chemical reagents. For example, since the basic forms of amines react with acetic anhydride[27] or dinitrofluorobenzene,[28-30] their extent of ionization is given from the relative rates of reaction as a function of pH. This has been used to measure the pK_a's of amino groups in proteins by modifying them with radioactive reagents, digesting the protein, separating the peptides, and measuring their specific radioactivities. In this way the pK_a's of several groups may be determined simultaneously, as has been done for elastase and chymotrypsin.

A useful variation of this technique is to measure the rate of exchange of tritium from tritiated water with the C-2 proton of the imidazole ring of a histidine.[31,32] The rate constant for the exchange depends on the state of ionization of the imidazole, and is faster for the unprotonated form. This procedure is a useful adjunct to NMR experiments, which also measure the pK_a's of histidine residues.

These experiments are tedious to perform, but the assignments of pK_a are unambiguous.

H. The effect of temperature, polarity of solvent, and ionic strength on pK_a's of groups in enzymes and in solution

Ions are stabilized by a polar solvent. The electrostatic dipoles of the solvent directly interact with the electrical charges of the ions, and the dielectric constant decreases the tendency of the ions to reassociate. The ionization of a neutral acid, as in equation 5·38, is depressed by the addition of a solvent of low polarity to an aqueous solution (Table 5·4).

$$HA = A^- + H^+ \tag{5·38}$$

On the other hand, the ionization of a cationic acid (equation 5·39) is insensitive to solvent polarity, since there is no change of charge in the equilibrium.

$$BH^+ = B + H^+ \tag{5·39}$$

The procedure of using the effect of solvent polarity on the pK_a of a group in an enzyme to tell whether it is a cationic or a neutral acid could be unreliable. A partly buried acid is shielded from the full effects of the solvent, and the electrostatic interactions with the protein may be more important. For example, the pK_a of the acylenzyme of benzoylarginine and trypsin is almost invariant in 0 to 50% dioxane/water and increases only slightly in 88% dioxane/water.[33] This suggests that the ionizing group is cationic, which is consistent with it being the imidazole moiety of His-57. (The pK_a of acetic acid increases by some 6 units under those conditions.) On the other hand, although increasing ionic strength decreases the pK_a's of carboxylic acids in solution, the pK_a's of Asp-52 and Glu-35 in lysozyme are increased by increasing ionic strength because of an effect on the surface charge of the protein.[25]

Imidazole groups in solution have enthalpies of ionization of about 30 kJ/mol (7 kcal/mol), whereas carboxylic acids have negligible enthalpies of ionization. But the changes in the enthalpy of the solvating water molecules also make important contributions to these values, so the solution values cannot be extrapolated to partly buried groups in proteins.

TABLE 5·4. *Effect of organic solvents on pK_a's at 25°C*[a]

Wt %	pK_a				
				Glycine	
dioxan	Acetic acid	Tris[b]	Benzoylarginine	—CO_2H	—NH_3^+
0	4.76	8.0	3.34	2.35	9.78
20	5.29	8.0	—	2.63	9.29
45	6.31	8.0	—	3.11	8.49
50	—	8.0	4.59	—	—
70	8.34	8.0	4.60	3.96	7.42

[a] From H. S. Harned and B. B. Owen, *The physical chemistry of electrolytic solutions*, Reinhold, pp. 755–56 (1958); T. Inagami and J. M. Sturtevant, *Biochim. Biophys. Acta* **38**, 64 (1960).
[b] $(HOCH_2)_3CNH_2$.

TABLE 5·5. *Some highly perturbed pK_a's of groups in proteins*[a]

Enzyme	Residue	pK_a
Lysozyme	Glu-35	6.5
Lysozyme–glycolchitin complex	Glu-35	~8.2
Acetoacetate decarboxylase	Lys (ϵ-NH$_2$)	5.9
Chymotrypsin	Ile-16 (α-NH$_2$)	10.0
α-Lactoglobulin	CO$_2$H	7.5
Papain	His-159	3.4

[a] From text references 25, 25, 6, 11, 20, and 34, respectively.

I. Highly perturbed pK_a's in enzymes

Many amine bases and carboxylic acids in proteins titrate with anomalously high or low pK_a's (Table 5·5). The reasons are quite straightforward, and depend on the microenvironment. If a carboxyl group is in a region of relatively low polarity, its pK_a will be raised, since the anionic form is destabilized. Alternatively, if the carboxylate ion forms a salt bridge with an ammonium ion, it will be stabilized by the positive charge so it will be more acidic. Conversely, if an amino group is buried in a nonpolar region, like the lysine in the active site of acetoacetate decarboxylase, protonation is inhibited and the pK_a is lowered. An ammonium ion in a salt bridge, such as Ile-16 in chymotrypsin, is stabilized by the negative charge on the carboxylate ion. Deprotonation is inhibited and the pK_a is raised.

References

1 M. Dixon, *Biochem. J.* **55**, 161 (1953).
2 R. A. Alberty and V. Massey, *Biochim. Biophys. Acta* **13**, 347 (1954).
3 M. L. Bender, G. E. Clement, F. J. Kézdy, and H. d'A. Heck, *J. Am. Chem. Soc.* **86**, 3680 (1964).
4 A. R. Fersht and Y. Requena, *J. Am. Chem. Soc.* **93**, 7079 (1971).
5 J. Fastrez and A. R. Fersht, *Biochemistry* **12**, 1067 (1973).
6 D. E. Schmidt, Jr., and F. H. Westheimer, *Biochemistry* **10**, 1249 (1971).
7 M. Renard and A. R. Fersht, *Biochemistry* **12**, 4713 (1973).
8 A. R. Fersht and M. Renard, *Biochemistry* **13**, 1416 (1974).
9 J. T. Edsall and J. Wyman, *Biophysical chemistry*, Academic Press, p. 510 (1958).
10 J. B. Matthews, S. H. Friend, and F. R. N. Gurd, *Biochemistry* **20**, 571 (1981).
11 J. G. Voet, J. Coe, J. Epstein, V. Matossian, and T. Shipley, *Biochemistry* **20**, 7182 (1981).
12 F. J. Kézdy, G. E. Clement, and M. L. Bender, *J. Am. Chem. Soc.* **86**, 3690 (1964).
13 A. C. Brower and J. F. Kirsch, *Biochemistry* **21**, 1302 (1982).
14 A. R. Fersht, *J. Molec. Biol.* **64**, 497 (1972), and references therein.

15 L. Pentz and E. R. Thornton, *J. Am. Chem. Soc.* **89**, 6931 (1967).
16 J. L. Markley, *Biochemistry* **14**, 3546 (1975).
17 S. M. Dudkin, M. Ya. Karpeisky, V. G. Sakharovskii, and G. I. Yakovlev, *Dokl. Akad. Nauk SSSR* **221**, 740 (1975).
18 G. Robillard and R. G. Shulman, *J. Molec. Biol.* **86**, 519 (1974).
19 W. W. Bachovchin, R. Kaiser, J. H. Richards, and J. D. Roberts, *Proc. Natn. Acad. Sci. USA* **78**, 7323 (1981).
20 S. Karplus, G. H. Snyder, and B. D. Sykes, *Biochemistry* **12**, 1323 (1973).
21 R. B. Moon and J. H. Richards, *J. Biol. Chem.* **248**, 7276 (1973).
22 S. N. Timasheff and J. A. Rupley, *Archs. Biochem. Biophys.* **150**, 318 (1972).
23 S. N. Timasheff, *The Enzymes* **2**, 371 (1970).
24 R. W. Cowgill, *Biochim. Biophys. Acta* **94**, 81 (1965).
25 S. M. Parsons and M. A. Raftery, *Biochemistry* **11**, 1623, 1630, 1633 (1972).
26 A. R. Fersht and J. Sperling, *J. Molec. Biol.* **74**, 137 (1973).
27 H. Kaplan, K. J. Stephenson, and B. S. Hartley, *Biochem. J.* **124**, 289 (1971).
28 A. L. Murdock, K. L. Grist, and C. H. W. Hirs, *Archs. Biochem. Biophys.* **114**, 375 (1966).
29 R. J. Hill and R. W. Davis, *J. Biol. Chem.* **242**, 2005 (1967).
30 W. H. Cruickshank and H. Kaplan, *Biochem. Biophys. Res. Commun.* **46**, 2134 (1972).
31 H. Matsuo, M. Ohe, F. Sakiyama, and K. Narita, *J. Biochem., Tokyo* **72**, 1057 (1972).
32 M. Ohe, H. Matsuo, F. Sakiyama, and K. Narita, *J. Biochem., Tokyo* **75**, 1197 (1974).
33 T. Inagami and J. M. Sturtevant, *Biochim. Biophys. Acta* **38**, 64 (1960).
34 F. A. Johnson, S. D. Lewis, and J. A. Shafer, *Biochemistry* **20**, 44 (1981).

6

Practical kinetics

An essential element in any kinetic study is the availability of convenient assays for measuring either the rate of formation of products or the rate of depletion of substrates. There exist a variety of methods for doing this, from the classic procedures of manometry, viscometry, and polarimetry to modern techniques such as NMR and EPR. However, the most commonly employed procedures are spectrophotometry, spectrofluorimetry, automatic titration, and the use of radioactively labeled substrates. In this chapter we shall discuss the more common methods, with an emphasis on their practical uses and limitations, rather than undertaking a general theoretical survey of all the techniques available.

A. Kinetic methods

1. Spectrophotometry

Many substances absorb light in the ultraviolet or visible regions of the spectrum. If the intensity of the light shining onto a solution of the compound is I_0 and that transmitted through is I, the absorbance A of the solution is defined by

$$A = \log \frac{I_0}{I} \tag{6·1}$$

The absorbance usually follows *Beer's law*,

$$A = \epsilon c l \tag{6·2}$$

where ϵ is the *absorption* or *extinction coefficient* of the compound, c is its concentration (usually in units of molarity), and l is the pathlength (in cm) of the light through the solution. A compound may be assayed by measuring A if ϵ is known.

Spectrophotometry is particularly useful with naturally occurring chromophores. For example, the rates of many dehydrogenases may be measured from the rate of appearance of NADH at 340 nm ($\epsilon = 6.23 \times 10^3 \, M^{-1} \, cm^{-1}$), because NAD^+ does not absorb at this wavelength. Otherwise artificial substrates may be used, such as p-nitrophenyl esters with esterases, since the p-nitrophenolate ion absorbs at 400 nm ($\epsilon = 1.8 \times 10^4 \, M^{-1} \, cm^{-1}$).

The sensitivity of the method depends on the extinction coefficient involved; for $\epsilon = 10^4$, the lower limit of detectability is about 0.5 nmol using a conventional spectrophotometer requiring at least 0.5 mL of solution and an absorbance of 0.01.

a. Some possible errors
The usual source of error is the breakdown of Beer's law. This may happen in several ways: The chromophore aggregates or forms micelles at high concentrations with a change in absorption coefficient; at high background absorbances, so much light of the correct wavelength is absorbed that light of other wavelengths leaking through the monochromator becomes significant ("stray light"); the bandwidth of the monochromator is set so wide that wavelengths other than those absorbed by the chromophore are transmitted to the sample; the solutions are turbid and scatter light.

2. Spectrofluorimetry

Some compounds absorb light and then re-emit it at a longer wavelength. This is known as fluorescence. The efficiency of the process is termed the quantum yield, q, which is equal to the number of quanta emitted per number absorbed (the ratio is always less than 1). Natural fluorophores include NADH, which absorbs at 340 nm and emits at about 460 nm. The major fluorophore in proteins is tryptophan, absorbing at 275 to 295 nm and re-emitting at 330 to 340 nm. Tyrosine fluoresces weakly in the same region. Many synthetic substrates for esterases, phosphatases, sulfatases, and glycosidases are based on 4-methylumbelliferone, a highly fluorescent derivative of phenol.

Fluorimetry is theoretically about 100 times more sensitive than spectrophotometry for the detection of low concentrations. Fluorescence is measured at right angles to the exciting beam against a dark background. A drift of 5% in the excitation intensity leads to only a 5% change in the signal. Absorbance measurements involve detecting a small change in the transmitted light; i.e., a small change against a high background. Any fluctuation in intensity is magnified: a 5% change is equivalent to an absorbance of 0.02.

Although the intensity of the fluoresced light is proportional to the intensity of the exciting beam, it is not possible to compensate indefinitely for decreasing concentration by increasing the excitation intensity. This is because the solvent scatters light—via Rayleigh scattering at the same wavelength as the excitation, and Raman scattering at a longer wavelength—and this eventually swamps the emitted light. Also, the compound that is being excited will be destroyed by photolysis at sufficiently high intensities.

a. Some properties of fluorescence
Fluorescence occurs by a photon being absorbed by a compound to give an excited state that decays by re-emission of a photon. (The lifetime of the excited state is about 10 ns.) Decay may also take place by a collision with another molecule, such as an iodide ion, or by a transfer of energy to another group in the molecule. The fluorescence is then said to be *quenched*. The quenching of the fluorescence of tryptophans of a protein on the binding of ligands is a useful way of measuring the extent of binding.

Fluorescence may also be *enhanced*. Sometimes a compound has a low quantum yield in aqueous solution but a higher one in nonpolar media. The dyes toluidinyl- and anilinyl-naphthalene sulfonic acid fluoresce very weakly in water, but strongly when they are bound in the hydrophobic pockets of proteins. Interestingly enough, if they are bound next to a tryptophan residue, they may be excited by light that is absorbed by the tryptophan at 275 to 295 nm and whose energy is transferred to them. Tryptophan and NADH fluoresce relatively weakly in water, and their fluorescence may be enhanced in the nonpolar regions of proteins.

b. Some possible errors
Errors usually arise through some form of quenching, such as the inadvertent addition of a substance such as potassium iodide, or more commonly by *concentration quenching*. This occurs when the fluorophore or ligand absorbs significantly through a Beer's law effect and reduces the intensity of the exciting or emitted light. For example, if the solution has an absorbance of A, then the average intensity of the exciting light in the cell is derived from equation 6·1:

$$I_{ave} = I_0 \times 10^{-\frac{1}{2}A} \tag{6·3}$$

The true fluorescence may be calculated from the above equation, from more sophisticated correction formulae, or from standard curves.[1]

3. Automated spectrophotometric and spectrofluorimetric procedures

The spectroscopic assays are simplest and most accurate when the products have a spectrum that is different from that of the reagents, so that the products may be observed directly. Assays in which aliquots of the reaction mixture are developed with a reagent to give a characteristic

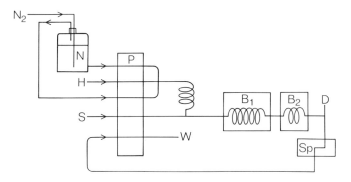

FIGURE 6·1. Flow diagram for the automated ninhydrin assay for the production of an amino acid during a reaction. Samples of the reaction mixture are continuously drawn off and the amino acid assayed as follows. The peristaltic pump (P) mixes ninhydrin (N) with hydrazine sulfate (H) and nitrogen (N_2). (Nitrogen bubbles form in the flow tube and separate the liquid into successive segments that do not mix with one another.) The hydrazine and ninhydrin mix thoroughly in the first coil to give the active assay solution. They then meet the reaction mixture (S) that is incubating and mix with it in the second coil. The color develops during the 20-min travel time at 95°C through coil B_1. After the mixture is cooled to 25°C in coil B_2, the bubbles are removed by the "debubbler" (D) and the absorbance is monitored by passing the mixture through a flow cell in the spectrophotometer (Sp). The amount of each solution mixed depends on the bore of the tubing used. As well as pumping all the solutions into the spectrophotometer, the peristaltic pump sucks out the solution and pumps it to waste: the bore of the sucking tube is sufficiently smaller than the sum of the tubes pumping in so that some of the solution and all of the nitrogen bubbles are forced out through the debubbler (D). [From A. R. Fersht and M. Renard, *Biochemistry* **13**, 1416 (1974), modified from J. Lenard, S. L. Johnson, R. W. Hyman, and G. P. Hess, *Analyt. Biochem.* **11**, 30 (1965).]

color, such as amino acids with ninhydrin, may be automated by using a proportionating pump that mixes the reagents in the desired ratios and pumps them through a flow cell in a spectrophotometer or fluorimeter (Figure 6·1). Besides being less tedious for routine work, these procedures are much more reproducible and accurate than the conventional approach.

4. Coupled assays

Some reactions that do not give chromophoric or fluorescent products may be coupled with another enzymatic reaction that does. Many of these coupled reactions are based on the formation or the disappearance of NADH. The formation of pyruvate may be linked with the conversion of NADH to NAD^+ by lactate dehydrogenase:

$$\text{Pyruvate} + \text{NADH} \xrightleftharpoons{\text{lactate dehydrogenase}} \text{lactate} + NAD^+ \qquad (6·4)$$

The formation of ATP may be coupled to the above reaction by the use of pyruvate kinase or another enzyme. For example, phosphofructokinase may be assayed by the following scheme:

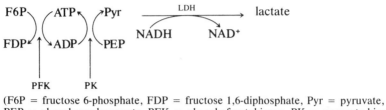

(F6P = fructose 6-phosphate, FDP = fructose 1,6-diphosphate, Pyr = pyruvate, PEP = phosphoenolpyruvate, PFK = phosphofructokinase, PK = pyruvate kinase, LDH = lactate dehydrogenase)

Another assay for phosphofructokinase involves converting the fructose 1,6-diphosphate to dihydroxyacetone phosphate and glyceraldehyde 3-phosphate with aldolase, equilibrating the triosephosphates with triosephosphate isomerase, and then measuring the production of NADH on the oxidation of the glyceraldehyde phosphate by glyceraldehyde 3-phosphate dehydrogenase.

5. Automatic titration of acid or base

A hydrolytic reaction that releases acid may be followed by titration with base. This is best done automatically by use of the pH-stat. A glass electrode registers the pH of the solution, which is kept constant by the automatic addition of base from a syringe controlled by an electronic circuit. Reaction volumes as low as 1 mL may be used, and the limit of detectability is about 50 nmol (5 to 10 μL of base at 5×10^{-3} to 1×10^{-2} M). The usual source of error with this apparatus is the buffering effect of dissolved CO_2.

6. Radioactive procedures

The most sensitive assay methods available involve the use of radioactively labeled substrates and reaction volumes of 20 to 100 μL.

Radioactivity is measured either in curies (Ci) (1 Ci is 2.22×10^{12} decompositions per minute) or in becquerels (Bq) (1 Bq is 1 decomposition per second). In practice, radioactive decay is measured in terms of counts per minute by using a scintillation counter with the common isotopes ^{3}H, ^{14}C, ^{35}S, and ^{32}P. These isotopes emit β radiation (electrons) when they decay. The radiation may be monitored by using a scintillant (see the Appendix to this chapter), which converts the radiation into light quanta that are registered as "counts" by a photomultiplier. The low energy emission of ^{3}H is counted with an efficiency of between 15 and 40%, and the higher energy emissions from ^{14}C, ^{35}S, and ^{32}P with an efficiency of about 80%. Because the energy of the ^{3}H emission is so different from the others, it may be counted in their presence in a "double labeling" experiment by monitoring different regions of the energy spectrum separately (Table 6·1).

The energy of the emission from ^{32}P is so high that it may be monitored in the absence of scintillant due to the Čerenkov effect. On passing through water or a polyethylene scintillation vial, the electrons move so rapidly that they spontaneously emit photons, which may be detected at about 40% efficiency by using the counter at an open-window setting. This technique has been used to count ^{32}P and ^{14}C when both are present, by first monitoring the Čerenkov radiation and then counting again after the addition of scintillant.

The sensitivity of detection depends on the specific activity of the compound. Some examples are given in Table 6·2. Detection is possible for as low as 10^{-18} to 10^{-17} mol, which is some 6 or 7 orders of magnitude below the lower limit of spectrophotometry.

a. Some possible errors

1. *Quenching.* The emission from tritium is so weak that it is readily absorbed by a filter paper or chromatographic strip on which it is adsorbed, or by precipitates. Fewer problems occur with ^{14}C, but the position of the energy maximum may be lowered on rare occasions. Colored materials or charcoal may absorb the light emitted from the scintillant.

2. *^{3}H transfer.* Tritium is sometimes transferred from substrates to solvents or proteins. For this reason and more importantly for that above, the use of ^{14}C is preferable to ^{3}H, although the higher specific activities and relative cheapness of ^{3}H-labeled compounds often more than compensate for the disadvantages.

TABLE 6·1. *Some common radioactive isotopes*

Isotope	Half-life	Specific activity of 100% isotopic abundance (Ci/mol)	Type of emission	Maximum energy of emission (MeV)
^{14}C	5730 yr	62.4	β	0.156
^{3}H	12.35 yr	2.9×0^4	β	0.0186
^{35}S	87.4 days	1.49×10^6	β	0.167
^{32}P	14.3 days	9.13×10^6	β	1.709
^{125}I	60 days	2.18×10^6	γ	—
^{131}I	8.06 days	1.62×10^7	β, γ	0.247–0.806
^{75}Se	120 days	1.09×10^6	γ	—

TABLE 6·2. *Some radioactively labeled substrates available from Amersham International*

Substrate	Specific activity (Ci/mol)	Limit of detectability[a] (pmol)
[^{14}C]methionine (90% enriched)	280	0.1
[CH_3-^{3}H]methionine	8×10^4	1×10^{-3}
[^{35}S]methionine	1×10^6	3×10^{-5}
[γ-^{32}P]ATP	1×10^7	3×10^{-6}

[a] Based on 50 counts per minute as the minimum measurable.

3. *Miscellaneous.* The addition of water to scintillants often lowers the efficiency of counting, so it should always be added in constant amounts. Sometimes the addition of base causes artifacts. Strip lighting can cause some scintillants to phosphoresce, and the effect can last several minutes.

b. Separation of reaction products
Clearly, in order to assay a reaction, it is necessary to separate the products from the starting materials to measure the radioactivity. Chromatography and high-voltage electrophoresis are very useful since the supporting paper may be cut into strips and the *relative* activities of the various regions accurately measured. Much more convenient, though, is the use of a filter pad to adsorb or trap the desired compound selectively. For example, labeled ATP may be adsorbed onto charcoal and collected on a glass fiber disk, or adsorbed onto a disk impregnated with charcoal, and the remaining reagents washed away. Filter disks of diethylammoniumethyl-cellulose have been used to adsorb anionic reagents selectively, and disks of carboxymethyl-cellulose to trap cations. Proteins and any strongly bound ligand often adsorb to nitrocellulose filters. Covalently labeled proteins or polynucleotides may be precipitated with acid and collected on glass fiber filters if a heavy precipitate is formed, or on nitrocellulose if the precipitate is light. (These disks do not lower the efficiency of counting of ^{14}C or ^{32}P.) Double-stranded DNA adsorbs to glass fiber filters; single-stranded DNA adsorbs to nitrocellulose.

B. Plotting kinetic data

1. Exponentials
a. Single

$$A \xrightarrow{k} B \tag{6·5}$$

In the first-order reaction of equation 6·5, the concentration of B at time t, $[B]_t$, is related to the final concentration of B, $[B]_\infty$ (the endpoint), by

$$[B]_t = [B]_\infty\{1 - \exp(-kt)\} \tag{6·6}$$

(cf. equation 4·7). The value of k is usually obtained from the following semilogarithmic plot of $[B]_\infty - [B]_t$ against t:

$$\ln([B]_\infty - [B]_t) = \ln[B]_\infty - kt \tag{6·7}$$

It is important to determine the endpoint $[B]_\infty$ accurately, since errors here cause serious errors in the derived rate constant. A least squares method is often used to fit the data directly to the theoretical equation. It is better to use the observed endpoint with this method rather than treating it as a variable parameter, since the deviations caused by small changes in the endpoint are usually insignificant compared to the "noise" in the data.

b. The Guggenheim method

In cases in which the endpoint cannot be determined, the Guggenheim method may be used.[2] The differences between pairs of readings at t and $t + \Delta t$ (where Δt is a constant time that must be at least 2 or 3 times the half-time) are plotted against t in the semilogarithmic plot, since it may be shown that

$$\ln([B]_{t+\Delta t} - [B]_t) = \text{constant} - kt \tag{6·8}$$

c. Consecutive exponentials

A series of exponentials of the form

$$[B] = X\{\exp(-k_1 t) - \exp(-k_2 t)\} \tag{6·9}$$

(cf. equation 4·31) are relatively straightforward to solve if one of the rate constants is more than 5 or 10 times faster than the other. The slower process is plotted as a simple semilogarithmic plot by using the data from the tail end of the curve after the first process has died out (that is, after 5 to 10 half-lives of the first process have occurred). The data may then be fed back into equation 6·9 to give the faster rate constant. This is often done graphically, as shown in Figure 6·2. If the rate constants are not separated by this factor, it is simpler to try to change their ratio by a change in the reaction conditions than to use a least squares method.

2. Second-order reactions

$$A + B \xrightarrow{k_2} C \tag{6·10}$$

Second-order kinetics are best dealt with by converting them to pseudo-first-order kinetics by using one of the reagents in large excess over the

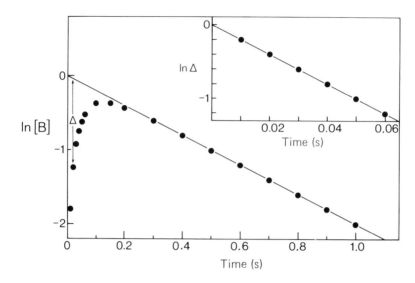

FIGURE 6·2. Graphical method of analyzing the kinetics of a reaction obeying equation 6·9. The logarithm of [B] is plotted against time. The rate constant for the slower process is obtained from the slope of the linear region after the faster process has died out. The rate constant for the faster process is obtained by plotting the logarithm of Δ (the difference between the value of [B] at a particular time and the value of [B] extrapolated back from the linear portion of the plot) against time for the earlier points. The rate constants for this example are 20 and 2 s^{-1}, respectively.

other. If $[A] \gg [B]$ in equation 6·10, [A] changes little during the reaction and

$$\frac{d[C]}{dt} = -\frac{d[B]}{dt} = k_2[A]_0[B] \tag{6·11}$$

The disappearance of B and the appearance of C follow exponential first-order kinetics with a pseudo-first-order rate constant of $k_2[A]_0$. Plotting a series of such reactions with varying $[A]_0$, the initial concentration of A, gives k_2.

The analytical solution for the reaction when $[A]_0$ is similar to $[B]_0$ is

$$\frac{\ln ([A]_0[B]/[B]_0[A])}{[B]_0 - [A]_0} = k_2 t \tag{6·12}$$

—an equation that is tedious to plot. However, the equation is greatly simplified when $[A]_0 = [B]_0$. The analytical solution becomes

$$\frac{1}{[A]_0 - [C]} - \frac{1}{[A]_0} = k_2 t \tag{6·13}$$

This simple equation holds to a good approximation even when the initial concentrations are not exactly equal. In this case, the average of $[A]_0$ and $[B]_0$ should be used in equation 6·13 instead of $[A]_0$.

3. Michaelis-Menten kinetics

Although it has become fashionable to fit data directly to the integrated form of the Michaelis-Menten equation by using a computer, the most satisfactory method of determining k_{cat}, k_{cat}/K_M, and K_M is to use the classic approach of measuring initial rates (i.e., the first 5% or less of the reaction) and making a plot such as that of Eadie and Hofstee (equation 3·29).[3-5] The following multiples of the K_M are a good range of substrate concentrations: 8; 4; 2; 1; 0.5; 0.25; 0.125.

C. Determination of enzyme–ligand dissociation constants

1. Kinetics

As we discussed in Chapter 3, the K_M for an enzymatic reaction is not always equal to the dissociation constant of the enzyme–substrate complex, but may be lower or higher depending on whether or not intermediates accumulate or Briggs-Haldane kinetics hold. Enzyme–substrate dissociation constants cannot be derived from steady state kinetics unless mechanistic assumptions are made or there is corroborative evidence. Pre–steady state kinetics are more powerful, since the chemical steps may often be separated from those for binding.

The *dissociation constants of competitive inhibitors* are readily determined from inhibition studies by using equation 3·32. This equation holds whether or not the K_M for the substrate is a true dissociation constant. The inhibition must first be shown to be competitive by determining the apparent K_M for the substrate at different concentrations of the inhibitor, and calculating the K_I from the apparent K_M and equation 3·32. Significant changes in K_M are obtained only at relatively high values of the inhibitor concentration.

2. Equilibrium dialysis

This is a method of directly measuring the concentrations of free and enzyme-bound ligand. A solution of the enzyme and ligand is separated from a solution of the ligand by a semipermeable membrane across which only the small ligand may equilibrate (Figure 6·3). After equilibration, a sample from the chamber containing protein gives the sum of the concentrations of free and bound ligand, whereas a sample from the other chamber gives the concentration of free ligand. Measurements may be made by using radioactively labeled ligands in chamber volumes of only

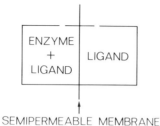

SEMIPERMEABLE MEMBRANE

FIGURE 6·3. The principle of equilibrium dialysis.

20 μL and sampling triplicate aliquots of 5 μL. However, since the apparatus requires at least 1 to 2 hours of equilibration with even the most porous membranes, the method cannot be used with unstable ligands or enzymes. *Nonequilibrium dialysis* has been used to make rapid measurements.[6,7] In this technique, the *rate* of diffusion of the ligand across the membrane from the side containing enzyme is measured as a function of concentration; binding slows down the rate.

3. Equilibrium gel filtration[8]

Certain gels, such as the commercial products Sephadex, Sephacryl, and Biogel, are made with pores that are large enough to be occupied by small ligands but not by proteins. If a chromatography column is packed with one of these gels and a solution of protein and ligand is applied, the protein travels faster through the column than the ligand does, since the ligand has to pass through the volume of solution surrounding the beads of gel and in the pores, whereas the protein has only to travel through the surrounding water. In equilibrium gel filtration, such a column is equilibrated with a solution of the ligand, and a sample of the enzyme in the equilibrating buffer is applied to the column. As the protein travels through the column it drags any bound ligand with it at the same flow rate. The result is that a peak of ligand travels through the column in the position of the protein, and a trough follows at the position normally occupied by a small ligand. The area under the peak should be the same as that in the trough, and should be equal to the amount of bound ligand.

One advantage of this method is that the enzyme is in contact with any particular substrate molecule for a short time only, and so can be used in cases in which the enzyme slowly hydrolyzes the ligand. Another advantage is that some available gels are able to distinguish between the size of one polymer and another, so that, for example, the binding of a tRNA (M_r = 25 000) to an aminoacyl-tRNA synthetase (M_r = 100 000) may be measured.[9]

The method may be scaled down. A convenient size for the binding of small ligands is a 1-mL tuberculin syringe packed with Sephadex G-25. A sample volume of 100 μL is applied, and individual drops (about 35 to 40 μL) are collected in siliconized tubes. The volume of each drop is measured by drawing it into a syringe. However, where the gel has to distinguish between a protein and a large ligand, much larger columns have to be used (a typical example being a 0.7- by 18-cm column for the case of tRNA/aminoacyl-tRNA synthetase).

a. Some possible errors

The criteria for an accurate experiment are that (1) the base line must be constant, (2) there must be a region of base line between the peak and the trough to show that the ligand and the protein have sufficiently different mobilities, and (3) the area of the peak must equal that of the trough. (Since it is very difficult to make up an enzyme–ligand solution that is

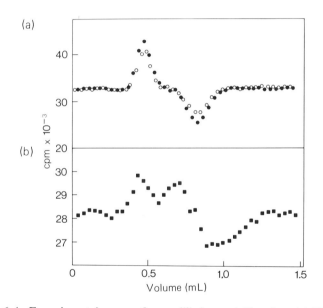

FIGURE 6·4. Experimental curves for equilibrium gel filtration. (a) The 1-mL tuberculin syringe was incubated in 109-μM [^{14}C]valine (○) or 109-μM [^{14}C]valine and 4-mM ATP (●). Then 100 μL of a solution of 26-μM valyl-tRNA synthetase was added to the same solution. Stoichiometries of 0.8 and 1.1, respectively, were found for the binding of the amino acid. Note the return to baseline between the peak and the trough—the mark of a good equilibrium gel filtration experiment. (b) An artifact-induced double peak obtained from the binding of [γ-^{32}P]ATP and valine to the enzyme. Some of the labeled ATP hydrolyzed to [^{32}P]orthophosphate, which traveled down the column faster than the [γ-^{32}P]ATP did.

exactly the same as the solution of the ligand in the equilibrating buffer, it is advisable to assay an aliquot of the enzyme solution before it is added to the column, in order to make any necessary correction to the area of the trough.)

Artifacts can occur due to the specific retardation of a ligand. ATP, ADP, and AMP, for example, bind to Sephadex and are retarded. The faster mobility of phosphate and pyrophosphate produced in reactions in which $[\gamma\text{-}^{32}P]ATP$ is hydrolyzed causes additional peaks and troughs (Figure 6·4).

4. Ultracentrifugation

The binding of a small polymer to a larger one, such as the binding of tRNA to an aminoacyl-tRNA synthetase, cannot be determined by equilibrium dialysis, and also requires relatively large volumes for equilibrium gel filtration. In this case, the binding may be measured on a 100- to 200-µL scale by using the analytical ultracentrifuge. The cell is filled with a mixture of, say, the tRNA and aminoacyl-tRNA synthetase, and the absorbances of the bound tRNA and the free tRNA are directly measured by the ultraviolet optics during sedimentation. The higher-molecular-weight complex of the enzyme and the tRNA sediments faster than the free tRNA, and there is a sharp moving boundary of absorbance as the complex moves down the cell (Figure 6·5).[10]

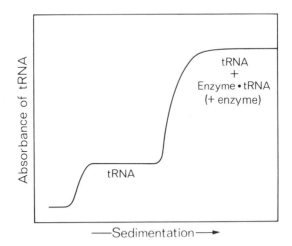

FIGURE 6·5. Sedimentation of a mixture of tRNA and aminoacyl-tRNA synthetase. The complex of tRNA and enzyme sediments faster than the free tRNA. The free enzyme and its complex with tRNA migrate together since they are in rapid equilibrium, the tRNA exchanging between the two.

5. Filter assays[11]

Many proteins are adsorbed (along with any slowly dissociating bound ligands) on nitrocellulose filters, while the free ligands are not retained. In special cases, this provides a very economical procedure for assaying binding. In particular, much useful data on the binding of nucleic acids to proteins has resulted. Care must be taken to avoid possible errors. Binding is often less than 100% efficient, and it may be variable from batch to batch of filters. The filters also saturate at fairly low concentrations of protein.

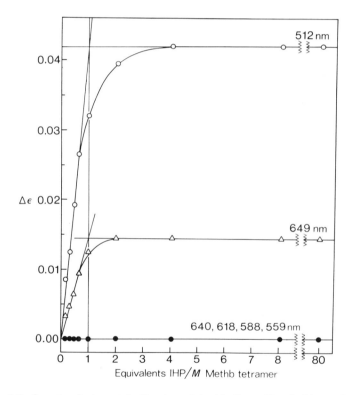

FIGURE 6·6. Spectrophotometric titration of the binding of inositol hexaphosphate (IHP) to methemoglobin (Methb). The complex has an increased absorbance at 512 and 649 nm, and no increases at 640, 618, 588, and 559 nm. The concentration of methemoglobin (20 μM) is about 14 times higher than the dissociation constant of 1.4 μM for the complex. The intersection of the slope of the increase in absorbance with the maximum value gives the stoichiometry (1, in this case). Note that this simple procedure cannot be used if the protein is not initially present at such a high concentration relative to the dissociation constant, since the assumption is that all the added ligand is bound to the protein for the early additions.

6. Spectroscopic methods

Except in NMR experiments, in which the concentrations of free and bound ligand may often be individually measured, spectroscopic methods do not usually give a direct measurement of the number of bound ligand molecules. On the binding of the ligand there is indeed a change in spectroscopic signal that is related to the fraction of the protein that is binding. But without additional evidence, the number of ligand molecules binding cannot be known. In these spectroscopic methods, the binding of a ligand to an enzyme typically causes either a change in the fluorescence of the protein, or a change in the fluorescence or the optical spectrum of the ligand. The usual procedure is to add increasing amounts of the ligand to a relatively dilute solution of the protein, and to plot the change in the spectroscopic signal by one of the procedures given in the following section. A variation of the procedure for ligands that are not chromophores is to measure their competitive inhibition of the binding of chromophoric ligands.

7. Titration procedures

If the dissociation constant of the complex is sufficiently low, it may be possible to determine the number of equivalents of ligand that are required to give the maximum spectral change that occurs when all the binding sites are occupied. For example, increasing amounts of the ligand are added to a solution in which the concentration of the protein is at least 10 times higher than the dissociation constant. The results are plotted as in Figure 6·6 to give the stoichiometry.

Another procedure is that of P. Job.[12]

D. Plotting binding data

1. The single binding site

The binding of a ligand to a single site on a protein is described by the following equations:

$$EL \overset{K_S}{\rightleftharpoons} E + L \tag{6·14}$$

$$K_S = \frac{[E][L]}{[EL]} \tag{6·15}$$

where [E] and [L] are the concentrations of the unbound enzyme and ligand.

In terms of the total enzyme concentration $[E]_0$,

$$[EL] = \frac{[E]_0[L]}{[L] + K_S} \tag{6·16}$$

Equation 6·16 is in the same form as the Michaelis-Menten equation, and may be manipulated in the same way. A good strategy in plotting the data is to use the equivalent of the Eadie plot:

$$[EL] = [E]_0 - K_S \frac{[EL]}{[L]} \tag{6·17}$$

A plot of [EL] against [EL]/[L] gives K_S.

Equation 6·17 cannot be used directly with spectroscopic data since [EL] is not known. However, because [EL] is usually directly proportional to the change in the spectroscopic signal being observed, we have

$$\Delta F = \Delta F_{max} - K_S \frac{\Delta F}{[L]} \tag{6·18}$$

where ΔF is the change in spectroscopic signal when [L] is added to the protein solution. A plot of ΔF against $\Delta F/[L]$ gives K_S and ΔF_{max}, the change in signal when all the protein is converted into complex.

2. Multiple binding sites

a. Identical

If there are n identical noninteracting sites on the protein, equation 6·17 may be modified to the *Scatchard plot*,

$$v = n - K_S \frac{v}{[L]} \tag{6·19}$$

where v is the number of moles of ligand bound per mole of protein. The stoichiometry n and K_S are obtained from the plot of v against $v/[L]$.

b. Nonidentical

If there are two classes of sites, one weak and the other strong, the Scatchard plot will be biphasic and composed of the sum of two different Scatchard plots. The determination of the values of K_S from such plots is satisfactory only when they differ by at least a factor of 10.

Appendix: Two convenient scintillants

1. BBOT

Although this water-miscible scintillant does not have the capacity of Bray's solution,[13] it is simpler to make up and does not suffer from the effects of strip lighting. It can take up to about 1% water. Two liters of scintillant contain 9 g of 2,5-bis[5'-*tert*-butylbenzoxazolyl-(2')]thiophene (BBOT), 1.5 L of toluene, and 0.5 L of methoxyethanol.

2. PPO/POPOP

Nitrocellulose disks may be suspended in this water-immiscible scintillant, which may be prepared from 12.5 g of 2,5-diphenyloxazole (PPO) and 0.75 g of 1,4-bis[2(5-phenyloxazolyl)]benzene (POPOP) dissolved in 2.5 L of toluene.

References

1   M. Ehrenberg, E. Cronvall, and R. Rigler, *FEBS Letts.* **18**, 199 (1971).
2   E. A. Guggenheim, *Phil. Mag.* **2**, 538 (1926).
3   I. A. Nimmo and G. L. Atkins, *Biochem. J.* **141**, 913 (1974).
4   R. R. Jennings and C. Niemann, *J. Am. Chem. Soc.* **77**, 5432 (1955).
5   A. Cornish-Bowden, *Biochem. J.* **149**, 305 (1975).
6   S. P. Colowick and F. C. Womack, *J. Biol. Chem.* **244**, 774 (1969).
7   N. A. Sparrow, A. E. Russell, and L. Glasser, *Analyt. Biochem.* **123**, 255 (1982).
8   J. P. Hummel and W. J. Dreyer, *Biochim. Biophys. Acta* **63**, 530 (1962).
9   R. M. Waterson, S. J. Clarke, F. Kalousek, and W. H. Konigsberg, *J. Biol. Chem.* **248**, 181 (1973).
10 G. Krauss, A. Pingoud, D. Boehme, D. Riesner, F. Peters, and G. Maass, *Eur. J. Biochem.* **55**, 517 (1975).
11 M. Yarus and P. Berg, *Analyt. Biochem.* **35**, 450 (1970).
12 P. Job, *Annls. Chim.* **9** (Ser. 10), 113 (1928).
13 G. A. Bray, *Analyt. Biochem.* **1**, 279 (1960).

7

Detection of intermediates in reactions by kinetics

The mechanism of an enzymatic reaction is ultimately defined when all the intermediates, complexes, and conformational states of the enzyme are characterized and the rate constants for their interconversion are determined. The task of the kineticist in this elucidation is to detect the number and sequence of these intermediates and processes, define their approximate nature (that is, whether covalent intermediates are formed or conformational changes occur), measure the rate constants, and, from studying pH dependence, search for the participation of acidic and basic groups. The chemist seeks to identify the chemical nature of the intermediates, by what chemical paths they form and decay, and the types of catalysis that are involved. These results can then be combined with those from x-ray diffraction and calculations by theoretical chemists to give a complete description of the mechanism.

We shall now discuss some of the techniques that have been used to detect intermediates and delineate reaction pathways, using some well-known enzymes as examples.

A. Pre–steady state vs. steady state kinetics

It is often said that kinetics can never prove mechanisms but can only rule out alternatives. Although this is certainly true of steady state kinetics, in which the only measurements made are those of the rate of appearance of products or disappearance of reagents, it is not true of pre–steady state kinetics. If the intermediates on a reaction pathway are directly observed and their rates of formation and decay are measured, kinetics can prove a particular mechanism. This is the basic strength of pre–steady state kinetics: the enzyme is used in *substrate* quantities and the events on the enzyme are directly observed. There may, of course,

be intermediates that remain undetected because they do not accumulate
or give rise to spectral signals, or simply because they are beyond the
time scale of the measurements, but an overall reaction pathway may be
proved in a scientifically acceptable manner by pre–steady state kinetics.
The basic weakness of steady state kinetics is that the evidence is always
ambiguous. No direct information is obtained about the number of intermediates, so the minimum number is always assumed. This is not to say
that pre–steady state kinetics should be performed to the exclusion of
steady state kinetics, but rather that a combination of the two approaches
should be used. Once pre–steady state kinetics has given information
about the intermediates on the pathway, steady state kinetics becomes
much more powerful.

There are certain practical advantages of pre–steady state kinetics.
Processes that are essentially very simple may be measured, among them
the stoichiometry of a burst process, the rate constant for the transfer of
an enzyme-bound intermediate to a second substrate, and a ligand-induced conformational change. Also, the first-order processes that are usually measured are independent of enzyme concentration, unlike the rate
constants of steady state measurements. Although very high concentrations of enzymes are required for the rapid reaction measurements, these
are usually close to those that occur *in vivo*. Furthermore, the high enzyme
concentrations are usually similar to those used for making direct measurements on the physical state of the protein, so that data may be obtained
for the state of aggregation, etc., under the reaction conditions.

Pre–steady state kinetics involves direct measurements, and direct
measurements are always preferable, especially considering the tendency
of enzymes to "misbehave" (section B5).

1. Detection of intermediates: What is "proof"?

Much of the following discussion centers on determining the chemical
pathways of enzymatic reactions. This usually requires the detection of
the chemical intermediates involved, since they give *positive* evidence.
We shall consider that an intermediate is "proved" to be on a reaction
pathway if the following criteria are satisfied:

1. The intermediate is isolated and characterized.
2. The intermediate is formed sufficiently rapidly to be on the reaction
   pathway.
3. The intermediate reacts sufficiently rapidly to be on the reaction pathway.

These criteria require that pre–steady state kinetics be used at some stage
in order to measure the relevant formation and decomposition rate constants of the intermediate. But the rapid reaction measurements are not

sufficient by themselves, since the rate constants must be shown to be consistent with the activity of the enzyme under steady state conditions. Hence the power, and the necessity, of combining the two approaches.

It must always be borne in mind that an intermediate that has been isolated could be the result of a rearrangement of another intermediate, and might not itself be on the reaction pathway. This is why criteria 2 and 3 are needed. It is also possible that a genuine intermediate is isolated but that during the experimental work-up the enzyme takes on a different, low-activity conformation. In this case, criteria 2 and 3 are not met, even though the chemical nature of the intermediate is correct.

B. Chymotrypsin: Detection of intermediates by stopped-flow spectrophotometry, steady state kinetics, and product partitioning

The currently accepted mechanism for the hydrolysis of amides and esters catalyzed by the archetypal serine protease chymotrypsin involves the initial formation of a Michaelis complex followed by the acylation of Ser-195 to give an acylenzyme (Chapter 1) (equation 7·1). Much of the kinetic work with the enzyme has been directed toward detecting the acylenzyme. This work can be used to illustrate the available methods that are based on pre–steady state and steady state kinetics. The

$$ \text{E} + \text{RCO—X} \overset{K_S}{\rightleftharpoons} \text{RCO—X·E} \xrightarrow{k_2} \underset{+\,\text{XH}}{\text{RCO—}E} \xrightarrow{k_3} \text{RCO}_2\text{H} + \text{E} \qquad (7\cdot1) $$

acylenzyme accumulates in the hydrolysis of activated or specific ester substrates ($k_2 > k_3$), so that the detection is relatively straightforward. Accumulation does not occur with the physiologically relevant peptides ($k_2 < k_3$), and detection is difficult.

1. Detection of intermediates from a "burst" of product release

In 1954, B. S. Hartley and B. A. Kilby[1] examined the reaction of substrate quantities of chymotrypsin with excess p-nitrophenyl acetate or p-nitrophenyl ethyl carbonate. They noted that the release of p-nitrophenol did not extrapolate back to zero but instead involved an initial "burst," equal in magnitude to the concentration of the enzyme (Chapter 4, Figure 4·10). They postulated that initially the ester rapidly acylated the enzyme in a mole-to-mole ratio, and that the subsequent turnover of the substrate involved the relatively slow hydrolysis of the acylenzyme as the rate-determining step. This was later verified by the stopped-flow experiments described in section B2.

Such burst experiments have since been performed on many other enzymes. However, bursts may be due to effects other than the accumulation of intermediates, and artifacts can occur. Some examples are the following:

1. The enzyme is converted to a less active conformational state on combination with the first mole of substrate.
2. The dissociation of the product is rate-determining.
3. There is severe product inhibition.

It is not a trivial matter to eliminate possibilities 1 and 2. But this has been done for chymotrypsin in the following series of stopped-flow experiments.

2. Proof of formation of an intermediate from pre–steady state kinetics under single-turnover conditions

The strategy is to measure the rate constants k_2 and k_3 of the acylenzyme mechanism (equation 7·1) and to show that each of these is either greater than or equal to the value of k_{cat} for the overall reaction in the steady state (i.e., apply rules 2 and 3 of section A1). This requires: (1) choosing a substrate (e.g., an ester of phenylalanine, tyrosine, or tryptophan) that leads to accumulation of the acylenzyme, (2) choosing reaction conditions under which the acylation and deacylation steps may be studied separately, and (3) finding an assay that is convenient for use in pre–steady state kinetics. The experiments chosen here illustrate stopped-flow spectrophotometry and chromophoric procedures.

a. Measurement of the rate constant for acylation, k_2
The step k_2 is isolated by mixing an excess of ester substrate with the enzyme. Under this condition, the acylenzyme is formed; it accumulates and then remains at a constant concentration over an extended period of time, as long as there is enough substrate present to ensure that the enzyme remains acylated (Chapter 4, equations 4·37 to 4·46). The process of the E·S complex giving the acylenzyme is thus isolated. The rate constant for this step is said to be measured under *single-turnover* conditions, since a single turnover of the enzyme from the E·S to the acylenzyme is measured, and not the recycling of enzyme as in the steady state.

Three different chromophoric procedures may be used. The first two depend on the synthesis of chromophoric substrates. The third utilizes an independent probe that can be applied to a wide range of substrates.

1. *Chromophoric leaving group.*[2–4] The original work on p-nitrophenyl acetate has been extended by synthesizing p-nitrophenyl esters of specific acyl groups, such as acetyl-L-phenylalanine, -tyrosine, and -tryptophan. The rate of acylation of the enzyme is determined from the rate of appearance of the nitrophenol or nitrophenolate ion, which absorbs at a different wavelength from the parent ester.

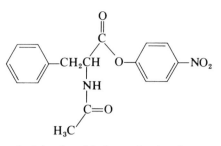

Acetyl-L-phenylalanine *p*-nitrophenyl ester

When the ester is mixed with the enzyme, there is a rapid exponential phase followed by a linear increase in the absorbance due to the nitrophenol. The rate constant for acylation and the dissociation constant of the enzyme–substrate complex may be calculated from the concentration dependence of the rate constant for the exponential phases (Chapter 4, equation 4·46). (The rate constant of the linear portion gives the deacylation rate, but this is a steady state measurement.) Unfortunately, nitrophenyl esters are often so reactive that the acylation rate is too fast for stopped-flow measurement.

2. *Chromophoric acyl group.*[4,5] The spectrum of the furylacryloyl group depends on the polarity of the surrounding medium, and also on the nature of the moiety to which it is attached. The spectrum of furylacryloyl-L-tyrosine ethyl ester changes slightly when it is bound to chymotrypsin. There are also further changes on formation of the acylenzyme and on the subsequent hydrolysis. The rate constants for acylation and deacylation and the dissociation constant of the Michaelis complex may be measured by the appropriate experiments.

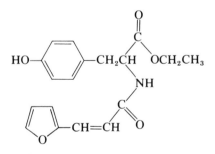

Furylacryloyl-L-tyrosine ethyl ester

When the ester is mixed with the enzyme, there is an initial change in absorbance that is due to the formation of the Michaelis complex. The rate constant for this is beyond the time scale of stopped flow, but the magnitude of the change can be used to calculate the dissociation constant. The absorbance then changes exponentially as the acylenzyme accumulates. There are further changes in the spectrum of the furylacryloyl group as the ester is gradually hydrolyzed to the free acid.

3. *Chromophoric inhibitor displacement.*[6,7] The spectrum of the dye proflavin changes significantly with solvent polarity. It is a competitive

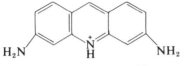

Proflavin (3,6-diaminoacridine)

inhibitor of chymotrypsin, trypsin, and thrombin, and it undergoes a large increase in absorbance at 465 nm ($\Delta\epsilon \approx 2 \times 10^4 \, M^{-1} \, cm^{-1}$) on binding (Figure 7·1).

When an ester such as acetyl-L-phenylalanine ethyl ester is mixed with a solution of chymotrypsin and proflavin, the following events occur. There is a rapid displacement of some of the proflavin from the active site as the substrate combines with the enzyme, leading to a decrease in A_{465}. (This is complete in the dead time of the apparatus.) Then, as the acylenzyme is formed, the binding equilibrium between the ester and the

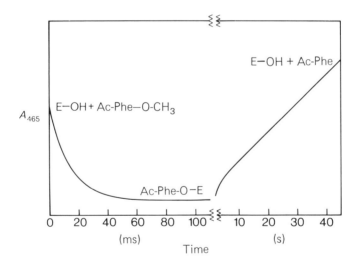

FIGURE 7·1. The proflavin displacement method. A solution of chymotrypsin (10 μM), proflavin (50 μM), and Ac-Phe-OCH$_3$ (2mM) is mixed at pH 6 and 25°C in a stopped-flow spectrophotometer. The substrate-binding step is too fast to be observed. The rapid exponential decrease in absorbance at 465 nm is caused by the displacement of proflavin from the enzyme on formation of the acylenzyme. The slow increase in absorbance is due to the depletion of the substrate and the consequent decrease in the steady state concentration of the acylenzyme. [From A. Himoe, K. G. Brandt, R. J. DeSa, and G. P. Hess, *J. Biol. Chem.* **244**, 3483 (1969).]

dye is displaced, leading to the displacement of all the proflavin. The absorbance remains constant until the ester is depleted and the acylenzyme disappears. The dissociation constant of the enzyme–substrate complex may be calculated from the magnitude of the initial rapid displacement, whereas the rate constant for acylation may be obtained from the exponential second phase.

Use of the proflavin displacement method is far more convenient than use of the furylacryloyl group, since no special substrates have to be synthesized and one readily available compound can be used with all substrates. In general, it is better not to use modified substrates: not only are they chemically inconvenient to synthesize, but they are always open to criticism on the grounds that the results could be artifacts.

b. Measurement of the rate constant for deacylation, k_3

The thematic approach to isolating the deacylation step is to generate the acylenzyme *in situ* in the stopped-flow spectrophotometer by mixing a substrate that acylates very rapidly with an excess or *stoichiometric* amount of the enzyme. The acylenzyme is formed in a rapid step that consumes all the substrate. This is then followed by relatively slow hydrolysis under single-turnover conditions. For example, acetyl-L-phenylalanine *p*-nitrophenyl ester may be mixed with chymotrypsin in a stopped-flow spectrophotometer in which the enzyme is acylated in the dead time. The subsequent deacylation may be monitored by the binding of proflavin to the free enzyme as it is produced in the reaction.[8]

There are also nonthematic methods that allow the formation of acylenzymes under conditions where they are stable, so that they can be stored in a syringe in a stopped-flow spectrophotometer. For example, it is possible to synthesize certain nonspecific acylenzymes and store them at low pH.[9-12] When they are restored to high pH, they are found to deacylate at the rate expected from the steady state kinetics. This approach has been extended to cover specific acylenzymes. When acyl-L-tryptophan derivatives are incubated with chymotrypsin at pH 3 to 4, the acylenzyme accumulates. The solution may then be "pH-jumped" by mixing it with a concentrated high-pH buffer in the stopped-flow spectrophotometer.[13,14] The deacylation rate has been measured by the proflavin displacement method and by using furylacryloyl compounds.

c. Characterization of the intermediate

The preceding experiments prove that there is an intermediate on the reaction pathway: in each case, the measured rate constants for the formation and decay of the intermediate are at least as high as the value of k_{cat} for the hydrolysis of the ester in the steady state. They do not, however, prove what the intermediate is. The evidence for covalent modification of Ser-195 of the enzyme stems from the early experiments on the irreversible inhibition of the enzyme by organophosphates such as diisopropyl fluorophosphate: the inhibited protein was subjected to partial hydrolysis, and the peptide containing the phosphate ester was isolated and shown to be esterified on Ser-195.[15,16] The ultimate characterization of acylenzymes has come from x-ray diffraction studies of nonspecific acylenzymes at low pH, where they are stable (e.g., indolylacryloyl-chymotrypsin),[17] and of specific acylenzymes at subzero temperatures and at low pH.[18] When stable solutions of acylenzymes are restored to conditions under which they are unstable, they are found to react at the required rate. These experiments thus prove that the acylenzyme does occur on the reaction pathway. They do not rule out, however, the possibility that there are further intermediates. For example, they do not rule out an initial acylation on His-57 followed by rapid intramolecular transfer. Evidence concerning this and any other hypothetical intermediates must come from additional kinetic experiments and examination of the crystal structure of the enzyme.

3. Detection of the acylenzyme in the hydrolysis of esters by steady state kinetics and partitioning experiments

In the last section we saw that stopped-flow kinetics can detect intermediates that *accumulate*. Detection of these intermediates by steady state kinetics is of necessity indirect and relies on inference. Proof depends ultimately on relating the results to the direct observations of the pre–steady state kinetics. But steady state kinetics can also detect intermediates that do not accumulate, and, by extrapolation from the cases in which accumulation occurs, can *prove* their existence and nature.

Detection of intermediates by steady state kinetics depends on:

1. The accumulation of an intermediate that is able to react either with an acceptor whose concentration may be varied, or, preferably, with several different acceptors.
2. The generation of a common intermediate E—R by a series of different substrates all containing the structure R. This intermediate must be able to react with different acceptors.

The hydrolysis of esters (and amides) by chymotrypsin satisfies these criteria. The hydrolysis of, say, acetyl-L-tryptophan *p*-nitrophenyl ester forms an acylenzyme that reacts with various amines such as hydroxylamine, alaninamide, hydrazine, etc., and also with alcohols such as methanol, to give the hydroxamic acid, dipeptide, hydrazide, and methyl ester, respectively, of acetyl-L-tryptophan. The same acylenzyme is generated in the hydrolysis of the phenyl, methyl, ethyl, etc. esters of the amino acid (and also during the hydrolysis of amides).

The kinetic consequences of the common intermediate can be used to diagnose its presence.

a. The rate-determining breakdown of a common intermediate implies a common value of V_{max} or k_{cat}

If several different substrates generate the same intermediate and if its breakdown is rate-determining, then they should all hydrolyze with the same value of k_{cat}:

$$RCOX \atop RCOY \atop RCOZ \longrightarrow RCO-E \xrightarrow{\text{slow}} RCO_2H + E \qquad (7\cdot2)$$

This has been found for many series of ester substrates of chymotrypsin (Table 7·1) since the original study of H. Gutfreund and B. R. Hammond in 1959.[19] For weakly activated esters, the value of k_{cat} decreases to below that of k_3 because k_2 becomes partly rate-determining (equation 7·1). With

TABLE 7·1. *Comparison of values of k_{cat} for the hydrolysis of substrates by α-chymotrypsin at pH 7.0 and 26°C*

Derivative	k_{cat} (s^{-1})	K_M (mM)	Rate-determining step	Ref.
N-Acetyl-L-tryptophan derivatives				
Amide	0.026	7.3	Acylation	1
Ethyl ester	27	0.1	Deacylation	1
Methyl ester	28	0.1	Deacylation	1
p-Nitrophenyl ester	30	0.002	Deacylation	1
N-Acetyl-L-phenylalanine derivatives				
Amide	0.039	37	Acylation	1
Ethyl ester	63	0.09	Deacylation	1
Methyl ester	58	0.15	Deacylation	1
p-Nitrophenyl ester	77	0.02	Deacylation	1
N-Benzoylglycine derivatives				
Ethyl ester	0.1	2.3	Mainly acylation	2
Methyl ester	0.14	2.4	Mainly acylation	2
Isopropyl ester	0.05	2.3	Acylation	2
Isobutyl ester	0.17	2.4	Mainly acylation	2
Choline ester	0.43	1.2	Deacylation	2
4-Pyridinemethyl ester	0.51	0.092	Deacylation	2
p-Methoxyphenyl ester	0.61	0.1	Deacylation	3
Phenyl ester	0.54	0.14	Deacylation	3
p-Nitrophenyl ester	0.54	0.03	Deacylation	3

1 B. Zerner, R. P. M. Bond, and M. L. Bender, *J. Am. Chem. Soc.* **86**, 3674 (1964).
2 R. M. Epand and I. B. Wilson, *J. Biol. Chem.* **238**, 1718 (1963).
3 A. Williams, *Biochemistry* **9**, 3383 (1970). (The data are corrected to pH 7.0 from pH 6.91, assuming a pK_a of 6.8.)

amides, k_{cat} is very low and k_2 is completely rate-determining. The steady state analysis of k_{cat} in relation to k_2 and k_3 was presented in Chapter 3 [equation 3·22, where $k_{cat} = k_2 k_3/(k_2 + k_3)$]. Also given there was the relationship between K_M and K_S [equation 3·21, where $K_M = K_S k_3/(k_2 + k_3)$]. This latter relationship is shown very clearly in Table 7·1 in the reactions of the derivatives of acetyl-L-tryptophan and acetyl-L-phenylalanine. For $k_2 \ll k_3$—i.e., the situation for the amide substrates—$K_M = K_S$. But for the ester substrates, $k_2 > k_3$ and so $K_M \doteqdot K_S k_3/k_2$. Since k_3 is the same for the derivatives of any one acylamino acid, K_M should decrease with increasing k_2, as found. The enhanced reactivity of the activated p-nitrophenyl ester is manifested in a low value for K_M. The trend of K_M values decreasing with increasing reactivity of the leaving group is seen also for the hydrolysis of benzoylglycine esters.

The occurrence of a common value of k_{cat} in the reaction of a series of substrates is not sufficient evidence for the accumulation of a common covalent intermediate whose breakdown is rate-determining. The value of k_{cat} is constant for the hydrolysis of a wide range of phosphate esters by alkaline phosphatase.[20,21] This was once interpreted as evidence for the rate-determining hydrolysis of a phosphorylenzyme. But it now seems likely that, at alkaline pH, dephosphorylation is rapid[22] and there is a rate-determining dissociation of inorganic phosphate from the enzyme–product complex (E·P_i).[23,24] The common value of k_{cat} is caused by a common intermediate, but it is a noncovalent one.

b. Partitioning of the intermediate between competing acceptors
If an intermediate that may react with different acceptors is generated, two procedures may be used for its detection.

The first involves determining product ratios. For example, the hydrolysis of a series of esters of hippuric acid by chymotrypsin in solutions containing hydroxylamine leads to the formation of the free hippuric acid and hippurylhydroxamic acid in a constant ratio (Table 7·2):

TABLE 7·2. *Product ratios in the hydrolysis of N-benzoylglycine esters by α-chymotrypsin in 0.1-M hydroxylamine*[a]

Ester	Hydroxylaminolysis/hydrolysis	
	Enzymatic, pH 6.6–6.8	Nonenzymatic, pH 12
Methyl	0.37	0.99
Isopropyl	0.38	0.29
Homocholine	0.37	1.73
4-Pyridinemethyl	0.37	3.03

[a] From R. M. Epand and I B. Wilson, *J. Biol. Chem.* **238**, 1718 (1963); **240**, 1104 (1965).

$$E + RCOOR' \longrightarrow RCO\text{—}E \begin{cases} \xrightarrow{\text{NH}_2\text{OH}} RCONHOH \\ \xrightarrow{\text{H}_2\text{O}} RCO_2H \end{cases} \tag{7.3}$$

The nonenzymatic hydrolysis under the same conditions leads to variable product ratios.[25] This is good evidence for a common intermediate.

The second procedure involves measuring the rates of formation of the products for a particular substrate at varying acceptor concentrations. This gives information on the rate-determining step of the reaction as well as detecting the intermediate. Suppose that the rate-determining step is the formation of the acylenzyme. Then, since the acceptor reacts with the acylenzyme after the rate-determining step, it cannot increase the rate of destruction of the ester. The overall formation rates will be as in Figure 7·2. If the rate-determining step is the hydrolysis of the acylenzyme, the acceptor increases the rate of its reaction and hence increases the overall reaction rate. The product formation rates will be as in Figure 7·3.

The steady state kinetics for partition may be calculated from

$$E + RCOOR \underset{}{\overset{K_S}{\rightleftharpoons}} E\cdot RCOOR' \xrightarrow{k_2} RCO\text{—}E \begin{cases} \xrightarrow{k_3'[\text{H}_2\text{O}]} RCO_2H \\ \xrightarrow{k_4[N]} RCO\text{-}N \end{cases} \tag{7.4}$$
$$\downarrow$$
$$R'OH$$

The following expressions for k_{cat} and K_M may be derived by the usual procedures for the reactions of chymotrypsin.[26] The kinetics are

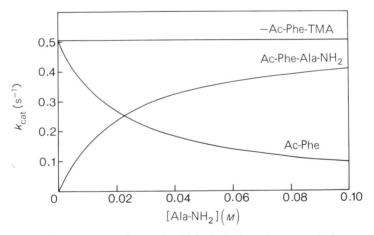

FIGURE 7·2. The chymotrypsin-catalyzed hydrolysis and transacylation reactions of acetylphenylalanine p-trimethylammoniumanilide (Ac-Phe-TMA) in the presence of various concentrations of Ala-NH$_2$. The values of k_{cat} for the depletion of Ac-Phe-TMA and the production of Ac-Phe and Ac-Phe-Ala-NH$_2$ are calculated from equation 7·3 by using $k_2 = 0.504$ s^{-1}, $k_3'[\text{H}_2\text{O}] = 144$ s^{-1}, and $k_4 = 6340$ s^{-1} M^{-1}. [From J. Fastrez and A. R. Fersht, *Biochemistry* **12**, 2025 (1973).]

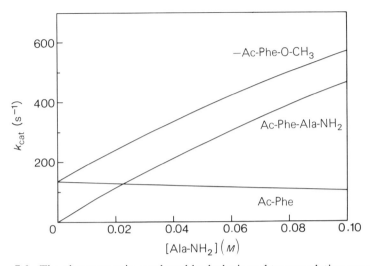

FIGURE 7·3. The chymotrypsin-catalyzed hydrolysis and transacylation reactions of Ac-Phe-OCH$_3$ in the presence of various concentrations of Ala-NH$_2$. The values of k_{cat} for the depletion of Ac-Phe-OCH$_3$ and the production of Ac-Phe and Ac-Phe-Ala-NH$_2$ are calculated from equation 7·3 by using $k_2 = 2200$ s^{-1}, $k_3'[H_2O] = 144$ s^{-1}, and $k_4 = 6340$ s^{-1} M^{-1}. [From J. Fastrez and A. R. Fersht, *Biochemistry* **12**, 2025 (1973).]

simplified in this example because the acceptor N does not bind to the enzyme:

$$K_M = K_S \frac{k_3'[H_2O] + k_4[N]}{k_2 + k_3'[H_2O] + k_4[N]} \tag{7·5}$$

For the formation of RCO$_2$H,

$$k_{cat} = \frac{k_2 k_3'[H_2O]}{k_2 + k_3'[H_2O] + k_4[N]} \tag{7·6}$$

For the formation of RCO-N,

$$k_{cat} = \frac{k_2 k_4[N]}{k_2 + k_3'[H_2O] + k_4[N]} \tag{7·7}$$

Note that k_{cat}/K_M for the rate of disappearance of RCO$_2$R' is equal to k_2/K_S, and is obtained by dividing the sum of the two values of k_{cat} from equations 7·6 and 7·7 by the K_M. It may be recalled from Chapter 3, section F, that this is always true for such a series of sequential reactions following an equilibrium binding step. Thus another criterion for the detection of an intermediate is provided. If the reaction of the nucleophiles involves the direct attack on the Michaelis complex, as in equation 7·8, k_{cat}/K_M will be a function of the concentration of N.[8] For example, if

$$E + RCOOR' \xrightleftharpoons{K_S} E{\cdot}RCOOR' \begin{array}{c} \xrightarrow{k_2[H_2O]} RCO_2H \\ \\ \xrightarrow{k_3[N]} RCO\text{-}N \end{array} \qquad (7{\cdot}8)$$

then, for the disappearance of ester,

$$\frac{k_{cat}}{K_M} = \frac{k_2[H_2O] + k_3[N]}{K_S} \qquad (7{\cdot}9)$$

(Complications will arise, of course, if the nucleophiles bind to the enzyme and both compete for a single site.)

The speeding up of the deacylation rate by adding nucleophiles has been used to give the rate constant k_2 for deacylation with substrates for which k_3 is normally rate-determining. 1,4-Butanediol is a sufficiently good nucleophile that moderate concentrations cause the deacylation rate to become faster than k_2.[27] Values of some rate constants obtained by this method are listed in Table 7·3.

Partition experiments provide a very powerful approach for the detection of intermediates.

4. Detection of the acylenzyme in the hydrolysis of amides and peptides[8]

The acylenzyme does not accumulate in the hydrolysis of amides, so detection is indirect and difficult. Fortunately, the direct detection of the acylenzyme in ester substrates can be used to provide a rigorous proof of the acylenzyme with amides.

The acylenzyme mechanism was proved for derivatives of acetyl-L-phenylalanine (Ac-Phe) as follows:

1. The hydrolysis of amides in the presence of acceptor nucleophiles gives the same product ratios as those found for the hydrolysis of the methyl ester (Ac-Phe-OCH$_3$) under the same conditions (Table 7·4). Furthermore, these product ratios are the same as those expected from direct rate measurements of the attack of the nucleophiles on Ac-Phe-chymotrypsin, generated *in situ* in the stopped-flow spectrophotometer (Table 7·5).
2. Under conditions where over 94% of the amide that is reacting in the presence of the acceptor nucleophiles forms Ac-Phe-nucleophile, there is no significant increase in the rate of disappearance of the amide. This is consistent with attack by the nucleophile after the rate-determining step—that is, after the formation of an intermediate, with at least 94% of the reaction going through this intermediate.
3. The final *proof* of the acylenzyme route comes from calculating the

TABLE 7·3. *Kinetic constants for the hydrolysis of N-acyl-L-amino acid esters by α-chymotrypsin at 25°C, pH 7.8, and ionic strength 0.1, determined by partitioning experiments*[a]

Acyl	Amino acid	Ester	k_{cat} (s^{-1})	K_M (mM)	k_2 (s^{-1})	k_3 (s^{-1})	K_S (mM)
Acetyl	Gly	OCH$_3$	0.109	862	0.49	0.14	3380
Acetyl	Gly	OC$_2$H$_5$	0.051	445	0.094	0.11	823
Benzoyl	Gly	OCH$_3$	0.31	4.24	0.42	1.17	5.78
Acetyl	But	OCH$_3$	1.41	66.7	8.81	1.68	417
Benzoyl	But	OCH$_3$	0.32	1.41	0.41	1.52	1.79
Benzoyl	Ala	OC$_2$H$_5$	0.069	5.97	0.069	0.6	5.97
Acetyl	Norval	OCH$_3$	5.08	14.3	35.6	5.93	100
Benzoyl	Norval	OCH$_3$	2.45	0.85	4.16	5.93	1.45
Acetyl	Val	OCH$_3$	0.173	87.7	0.98	0.21	500
Acetyl	Val	OC$_2$H$_5$	0.152	110	0.55	0.21	398
Acetyl	Val	i-OC$_3$H$_7$	0.096	177	0.178	0.21	327
Chloroacetyl	Val	OCH$_3$	0.127	43	0.32	0.21	108.8
Benzoyl	Val	OCH$_3$	0.064	4.17	0.09	0.22	5.84
Acetyl	Norleu	OCH$_3$	16.1	5.37	103	19.1	34.4
Acetyl	Phe	OCH$_3$	97.1	0.93	796	111	7.63
Acetyl	Phe	OC$_2$H$_5$	68.6	1.85	265	92.7	7.14
Acetylala (L)	Phe	OCH$_3$	57.3	0.296	176	85	0.909
Benzoyl	Phe	OCH$_3$	30.7	0.0349	45.8	91.6	0.0524
Acetyl	Tyr	OC$_2$H$_5$	192	0.663	5000	200	17.2
Benzoyl	Tyr	OCH$_3$	90.9	0.018	364	121	0.072
Benzoyl	Tyr	OC$_2$H$_5$	85.9	0.022	249	131	0.0638
Acetylleu (L)	Tyr	OCH$_3$	65.7	0.0192	158	113	0.0461
Furoyl	Tyr	OCH$_3$	50	0.417	66.7	200	0.56

[a] From I. V. Berezin, N. F. Kazanskaya, and A. A. Klyosov, *FEBS Lett.* **15**, 121 (1971).

rate constant for the hydrolytic reaction from the rate constant for the reverse reaction (the synthesis of the substrate by the acylenzyme route) and the *Haldane equation* (Chapter 3, section H). It is found that amines will react with the acylenzyme to produce amides and peptides. Hence, by the principle of microscopic reversibility, the re-

TABLE 7·4. *Product ratios in the hydrolysis of substrates by δ-chymotrypsin*[a]

Substrate	Transacylation/hydrolysis (M^{-1})		
	Acceptor: Ala-NH$_2$	Gly-NH$_2$	H$_2$NNH$_2$
Ac-Phe-OCH$_3$	43	13	2.2
Ac-Phe—NH—⟨⟩—$\overset{+}{N}$(CH$_3$)$_3$	45	11	1.8
Ac-Phe-Ala-NH$_2$	43	9	—

[a] At 25°C, pH 9.3. [From J. Fastrez and A. R. Fersht, *Biochemistry* **12**, 2025 (1973).]

TABLE 7·5. *Rate constants for the attack of nucleophiles on Ac-Phe-δ-chymotrypsin*[a]

| Nucleophile | $k(s^{-1} M^{-1})$ | |
	Direct kinetic measurement	Calculated from product ratios[b]
Ala-NH$_2$	4800	6200
Gly-NH$_2$	1500	1600
H$_2$NNH$_2$	330	280
(H$_2$O	142 s^{-1})	

[a] At 25° C, pH 9.3. [From J. Fastrez and A. R. Fersht, *Biochemistry* **12**, 2025 (1973).]
[b] From Table 7·4.

verse reaction (the hydrolysis of peptides by the acylenzyme mechanism) must also occur. The question is whether or not this reaction is rapid enough to account for the observed hydrolysis rate. This can be answered by measuring $(k_{cat}/K_M)_S$ for the synthesis of a peptide by the acylenzyme route, and K_{eq} for the hydrolysis of the peptide; $(k_{cat}/K_M)_H$ for the hydrolytic reaction can then be calculated from the Haldane equation,

$$K_{eq} = \frac{(k_{cat}/K_M)_H}{(k_{cat}/K_M)_S}$$

The calculated value is close to the experimental value.

5. The validity of partitioning experiments and some possible experimental errors

The occurrence of neither a constant value of V_{max} nor a constant product ratio is sufficient proof of the presence of an intermediate. It was seen for alkaline phosphatase that a constant value for V_{max} is an artifact, and also that there is no *a priori* reason why the attack of acceptors on a Michaelis complex should not also give constant product ratios. In order for partitioning experiments to provide a satisfactory proof of the presence of an intermediate, they must be linked with rate measurements. When the rate measurements are restricted to steady state kinetics, the most favorable situation is when the intermediate accumulates. If the kinetics of equations 7·5 to 7·7 hold, it may be concluded beyond a reasonable doubt that an intermediate occurs. The ideal situation is a combination of partitioning experiments with pre–steady state studies, as described for chymotrypsin and amides.

Errors can arise when the enzymes "misbehave." Chymotrypsin is often treated as a solution of imidazole and serine. But proteins are quite sensitive to their environment; they often bind organic molecules and ions nonspecifically to alter their kinetic properties slightly. The first experimental rule is that reactions should be carried out, to the extent possible, at the same concentration of enzyme. Many proteins aggregate somewhat, and this can cause changes in rate constants. The second rule is that product ratios should be determined by direct analysis of the products rather than by indirect measurements. For example, the rate of attack of Gly-NH$_2$ on the acylenzyme Bz-Tyr-chymotrypsin was once measured from the decrease in k_{cat} for the hydrolysis of Bz-Tyr-Gly-NH$_2$ on addition of Gly-NH$_2$. (The Gly-NH$_2$ inhibits the reaction, since it reacts with the acylenzyme to regenerate the Bz-Tyr-Gly-NH$_2$.) Unfortunately, it was subsequently found that amines bind to chymotrypsin, causing increases of up to 30% in the k_{cat}.[8] This increase is on the same order as the expected decreases due to the reversal of the reaction. In general, small changes of rate are not reliable in enzymatic reactions under circumstances that would give reliable results in chemical kinetics.

There are circumstances in which the simple rules for the partition of intermediates break down. If the acceptor nucleophile reacts with the acylenzyme before the leaving group has diffused away from the enzyme-bound intermediate, the partition ratio could depend on the nature of the leaving group (e.g., due to steric hindrance of attack, etc.). Also, the measurement of rate constants for the attack of the nucleophiles on the intermediate could be in slight error due to the nonspecific binding effects mentioned above.

C. Further examples of detection of intermediates by partition and kinetic experiments

1. Alkaline phosphatase

There is little doubt that a phosphorylenzyme is formed during the hydrolysis of phosphate esters by alkaline phosphatase.[28,29] The phosphorylenzyme is stable at low pH, and it may be isolated.[30,31] The phos-

TABLE 7·6. *Relative values of V_{max} for the hydrolysis of phosphate esters by alkaline phosphatase*[a]

Phosphate	V_{max}	Phosphate	V_{max}
5'-AMP	1	dCTP	1.05
Pyrophosphate	1	Ribose 5-phosphate	0.7
3'-AMP	0.9	β-Glycerol phosphate	0.9
ApAp	0.6	Ethanolamine phosphate	0.7
ATP	1.05	Glucose 1-phosphate	0.8
dATP	1.05	Glucose 6-phosphate	0.9
dGTP	1.05	Histidinol phosphate	0.8
UDP	1.0	p-Nitrophenyl phosphate	1.0
5'-UMP	0.85		

[a] From L. A. Heppel, D. R. Harkness, and R. J. Hilmoe, *J. Biol. Chem.* **237**, 841 (1962).

TABLE 7·7. *Product ratios in the hydrolysis of phosphate esters and phosphoramidates by alkaline phosphatase*[a]

Phosphate	Transphosphorylation/ hydrolysis
Acceptor = 1-M tris, pH 8	
Phenyl	1.42
Cresyl	1.41
Chlorophenyl	1.38
p-tert-Butylphenyl	1.37
p-Nitrophenyl	1.37
o-Methoxy-*p*-methylphenyl	1.40
α-Naphthyl	1.40
β-Naphthyl	1.40
Acceptor = 2-M tris, pH 8.2	
p-Nitrophenyl	1.2
Phosphoramidates	1.1[b]

[a] From H. Barrett, R. Butler, and I. B. Wilson, *Biochemistry* **8**, 1042 (1969); S. Snyder and I. B. Wilson, *Biochemistry* **11**, 3220 (1972).
[b] Average for several phosphoramidates.

phorylenzyme has been detected by gel electrophoresis.[28] Partition experiments using tris buffer as a phosphate acceptor give a constant product ratio (Table 7·6).[32,33] Earlier kinetic experiments have to be evaluated in light of the recent findings that the enzyme as isolated may contain tightly, although noncovalently, bound phosphate,[34] and that the rate-determining step in the reaction at high pH is the dissociation of the tightly bound phosphate[23,24] (although there is no unanimity on the kinetic points or on the stoichiometry of binding).[35,36] As mentioned earlier, it seems that the constant value of V_{max} for the hydrolysis of a wide series of phosphate esters (Table 7·7)[20,21] is the result of a slow, rate-determining dissociation of the enzyme–product complex, $E \cdot P_i$.

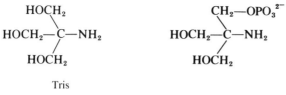

Tris

[*tris*-(hydroxymethyl)-aminomethane]   *O*-Phosphoryl-tris

2. Acid phosphatase

Acid phosphatase gives a straightforward example of the accumulation of a phosphorylenzyme intermediate with rate-determining breakdown:

$$E\text{—}OH + RO\overline{P}O_3H \rightleftharpoons E\text{—}OH \cdot RO\overline{P}O_3H \xrightarrow{\text{fast}} E\text{—}O\overline{P}O_3H \xrightarrow{\text{slow}}$$

$$\searrow$$

$$ROH$$

$$E\text{—}OH + H_2PO_4^- \qquad (7\cdot10)$$

A wide variety of esters are hydrolyzed with the same $V_{\max}$ (Table 7·8);[37] constant product ratios are found (Table 7·9); stopped-flow studies using p-nitrophenyl phosphate find a burst of 1 mol of p-nitrophenolate ion released per enzyme subunit; and the enzyme is covalently labeled by diisopropyl fluorophosphate.[38]

3. β-Galactosidase

Chapter 8 (section C3) presents stereochemical evidence that the hydrolysis of β-D-galactosides catalyzed by β-galactosidase (equation 7·11)

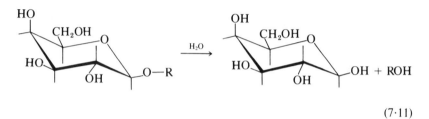

$$(7\cdot11)$$

involves two successive displacements on the C-1 carbon—i.e., involves an intermediate. Further evidence for an intermediate from partitioning experiments is presented in Table 7·10. There is constant partitioning between water and methanol.[39–41] Examination of $V_{\max}$ suggests that formation of the intermediate is rate-determining for the weakly activated substrates since $V_{\max}$ is variable, but that hydrolysis of the intermediate is rate-determining for the highly activated dinitro compounds because the rate levels off. Consistent with this is the observation that the rate of disappearance of the 2,4-dinitro and 3,5-dinitro substrates is increased by

TABLE 7·8. *Hydrolysis of phosphate esters by prostatic acid phosphatase at pH 5 and 37°C[a]*

Phosphate ester	K_M (mM)	$V_{\max}$ (relative)
β-Glycerol phosphate	1.1	1
2′-AMP	0.28	1
Acetyl phosphate	0.17	1
3′-AMP	0.068	1
p-Nitrophenyl phosphate	0.034	1

[a] From G. S. Kilsheimer and B. Axelrod, *J. Biol. Chem.* **227**, 879 (1957).

TABLE 7·9. *Product ratios in the hydrolysis of phosphate esters by prostatic acid phosphatase*[a]

Phosphate ester	Transphosphorylation/hydrolysis	
	Acceptor: Ethanol	Ethanolamine
p-Nitrophenyl phosphate	0.29	0.044
Phenyl phosphate	0.26	0.044
3'-UMP	0.28	—
3'-AMP	0.30	0.046
β-Glycerol phosphate	0.28	0.041

[a] From W. Ostrowski and E. A. Barnard, *Biochemistry* **12**, 3893 (1973).

added methanol, but not the rate of disappearance of the less reactive substrates (see Figures 7·2 and 7·3).

A change in rate-determining step is also indicated from $^{16}O/^{18}O$ kinetic isotope effects.[42] V_{max} for the hydrolysis of 4-nitrophenyl β-D-galactoside exhibits a $^{16}O/^{18}O$ effect of 1.022 (for the phenolic oxygen) compared with an expected value of 1.042 for complete C—O bond fission in the transition state. The 2,4-dinitrophenyl derivative for which degalactosylation is thought to be rate-determining has a negligible isotope effect on V_{max} (= 1.002). (Additional evidence suggests that the enzyme has an S_N2 reaction with the C-1 carbon to give a covalent intermediate for most substrates. The reaction of the activated dinitrophenyl derivative, however, possibly has a contribution from an S_N1 pathway.[42])

TABLE 7·10. *Product ratios and relative values of V_{max} for the hydrolysis of β-galactosides by β-galactosidase at 25°C and pH 7.0–7.5*[a]

β-Galactoside	Methanolysis/ hydrolysis (M^{-1})	V_{max} (relative)	Rate-determining step
2,4-Dinitrophenyl	—	1.3	Degalactosylation
3,5-Dinitrophenyl	—	1.1	Degalactosylation
2,5-Dinitrophenyl	—	1.1	Degalactosylation
2-Nitrophenyl	1.97	1.0	—
3-Nitrophenyl	1.96	0.9	—
3-Chlorophenyl	2.08	0.5	Galactosylation
4-Nitrophenyl	1.99	0.2	Galactosylation
Phenyl	1.94	0.1	Galactosylation
4-Methoxyphenyl	2.14	0.1	Galactosylation
4-Chlorophenyl	2.13	0.02	Galactosylation
4-Bromophenyl	2.02	0.02	Galactosylation
Methyl	2.2	0.06	Galactosylation

[a] From T. M. Stokes and I. B. Wilson, *Biochemistry* **11**, 1061 (1972); M. L. Sinnott and O. M. Viratelle, *Biochem. J.* **133**, 81 (1973); M. L. Sinnott and I. J. Souchard, *Biochem. J.* **133**, 89 (1973).

D. Aminoacyl-tRNA synthetases: Detection of intermediates by quenched flow, steady state kinetics, and isotope exchange

1. The reaction mechanism

The aminoacyl-tRNA synthetases catalyze the formation of aminoacyl-tRNA from the free amino acid (AA) and ATP:

$$AA + ATP + tRNA \xrightarrow{E} AA\text{-}tRNA + AMP + PP_i \qquad (7\cdot12)$$

In the absence of tRNA, the enzymes will, with a few exceptions, activate amino acids to the attack of nucleophiles, and ATP to the attack of pyrophosphate.[43-46] This is done by forming a tightly bound complex with the aminoacyl adenylate, the mixed anhydride of the amino acid, and AMP. (The chemistry of activation is discussed in Chapter 2, section D2c.)

The activation to the attack of pyrophosphate is measured by the pyrophosphate exchange technique. The enzyme, the amino acid, and ATP are incubated with [^{32}P]-labeled pyrophosphate so that β,γ-labeled ATP is formed by the continuous recycling of the E·AA-AMP complex. The complex is formed as in equation 7·13, and the reaction is reversed by the attack of labeled pyrophosphate to generate labeled ATP. This process is repeated until the isotopic label is uniformly distributed among all the reagents.

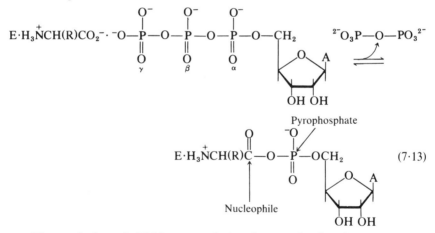

$$(7\cdot13)$$

The acylation of tRNA proceeds by the attack of a ribose hydroxyl of the terminal adenosine of the tRNA on the carbonyl group, as indicated in equations 7·13 and 7·14.

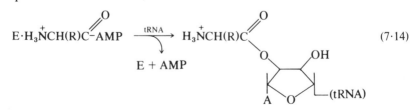

$$(7\cdot14)$$

There is no doubt that the enzyme-bound aminoacyl adenylate is formed in the absence of tRNA. It may be isolated by chromatography and the free aminoacyl adenylate obtained by precipitation of the enzyme with acid.[47,48] Furthermore, the isolated complex will transfer its amino acid to tRNA.

The following mechanism is derived logically from these observations:

$$E + ATP + AA \xrightarrow[PP_i]{} E \cdot AA\text{-}AMP \xrightarrow{tRNA} AA\text{-}tRNA + AMP + E \quad (7 \cdot 15)$$

Despite this, it seemed at one stage that not all the evidence was consistent with the aminoacyl adenylate pathway. An alternative mechanism appeared possible: in the presence of tRNA, perhaps an aminoacyl adenylate was *not* formed, and the reaction occurred instead by the simultaneous reaction of the tRNA, the amino acid, and ATP.[49] As is shown below, this mechanism is not correct, but at the time it was suggested it was a valid possibility. Furthermore, it made an important point: the finding of a partial reaction in the absence of one of the substrates (for example, the formation of the aminoacyl adenylate from the amino acid and ATP in the absence of tRNA) does not mean that the same reaction occurs in the presence of *all* the substrates. Indeed, there are examples in which such partial reactions have been found to be artifacts.

The aminoacyl adenylate pathway is proved very simply from three quenched-flow experiments by using the three criteria for proof: the intermediate is isolated; it is formed fast enough; and it reacts fast enough to be on the reaction pathway.[50] The following is found for the isoleucyl-tRNA synthetase (IRS):

1. When the preformed and isolated IRS·[^{14}C]Ile-AMP is mixed with tRNA in the pulsed quenched-flow apparatus (Figure 7·4), the first-order rate constant for the transfer of the [^{14}C]Ile to the tRNA is measured to be the same as the k_{cat} for the steady state aminoacylation of the tRNA under the same reaction conditions. The rate constant for the reaction of the intermediate is thus fast enough to be on the reaction pathway; furthermore, reaction of the intermediate appears to be the rate-determining step.

2. When IRS, isoleucine, tRNA, and [γ-^{32}P]ATP (labeled in the terminal phosphate) are mixed in the pulsed quenched-flow apparatus (Figure 7·5), there is a burst of release of labeled pyrophosphate before the steady state rate of aminoacylation of tRNA is reached. This means either that the aminoacyl adenylate is formed before the aminoacylation of tRNA, thus proving the mechanism, *or* that there could be a pathway that involves the formation of aminoacyl-tRNA in a rapid process followed by a subsequent slow step, such as the dissociation of the IRS·Ile-tRNA complex:

$$E + AA + ATP + tRNA \xrightarrow{\text{fast}} E \cdot AA\text{-}tRNA \xrightarrow{\text{slow}} E + AA\text{-}tRNA \qquad (7\cdot16)$$
$$AMP + PP_i$$

This reaction, which was thought to occur by many workers, is disproved by a third quenching experiment. Equation 7·16 predicts that there should be a burst of charging of tRNA, since 1 mol of enzyme-bound aminoacyl-tRNA is formed rapidly whereas the subsequent turnover is slow.

3. When IRS, [^{14}C]Ile, tRNA, and ATP are mixed in the quenched-flow apparatus (Figure 7·6), the initial rate of charging of tRNA extrapolates back through the origin without any indication of a burst of charging. The burst of pyrophosphate release is due to the formation of the aminoacyl adenylate before the transfer of the amino acid to tRNA.

2. The editing mechanism

Chapter 13 points out that during protein biosynthesis, the cell distinguishes between certain amino acids with an accuracy far greater than would be expected from their differences in structure. This sensitivity could be caused either by the specifically catalyzed hydrolysis of the aminoacyl adenylate complex of the "wrong" amino acid by the aminoacyl-tRNA synthetase, or by hydrolysis of the mischarged tRNA.[48] One example is the rejection of threonine by the valyl-tRNA synthetase (VRS).[51] This enzyme catalyzes the pyrophosphate exchange reaction in the presence of threonine, and also forms a stable VRS·Thr-AMP complex. In the presence of tRNA and threonine, the VRS acts as an ATP pyrophosphatase, hydrolyzing ATP to AMP and pyrophosphate, and does

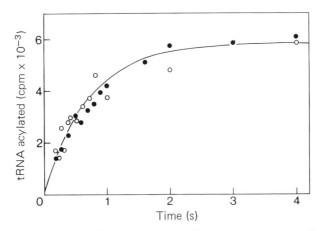

FIGURE 7·4. Transfer of [^{14}C]Ile from IRS·[^{14}C]Ile-AMP to tRNA^{Ile} when the complex is mixed with excess tRNA in the pulsed quenched-flow apparatus.

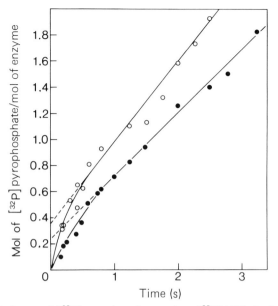

FIGURE 7·5. Release of [³²P]pyrophosphate when [³²P]ATP, isoleucine, tRNA, and enzyme are mixed in the pulsed quenched-flow apparatus. The extrapolated burst of product formation is below 1 mol per mole of enzyme because the concentrations of ATP are not saturating (Chapter 4, section D). Open circles (○) are for $[ATP] = 2 \times K_M$; filled circles (●) are for $[ATP] = 1 \times K_M$.

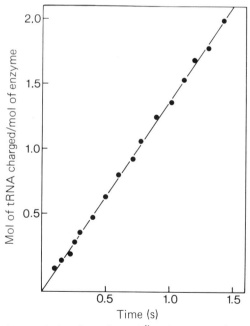

FIGURE 7·6. Initial rate of charging of tRNA^Ile when saturating concentrations of [¹⁴C]Ile, ATP, and tRNA are mixed with the enzyme.

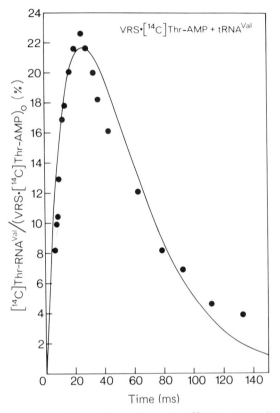

FIGURE 7·7. The transient formation of $[^{14}C]$Thr-tRNA^{Val} when VRS·$[^{14}C]$Thr-AMP is mixed with excess tRNA in the quenched-flow apparatus.

not catalyze the net formation of Thr-tRNA^{Val}. During this reaction there is the intermediate formation of VRS·Thr-AMP shown by the occurrence of the pyrophosphate exchange reaction.

Rapid quenching experiments show that the editing mechanism for the rejection of threonine involves the mischarging of tRNA followed by its rapid hydrolysis. The transiently mischarged tRNA may be trapped, isolated, and found to be hydrolyzed at the necessary rate. When the VRS·$[^{14}C]$Thr-AMP complex is mixed with tRNA^{Val} in the quenched-flow apparatus, $[^{14}C]$Thr-tRNA^{Val} is transiently formed (Figure 7·7). This may be isolated by rapidly quenching the reaction with phenol and precipitating the mischarged tRNA from the aqueous layer. The mischarged tRNA is hydrolyzed by the VRS with a rate constant of 40 s^{-1} (Figure 7·8). The rate of transfer of the threonine from the VRS·Thr-AMP complex may be measured independently from the rate of liberation of AMP

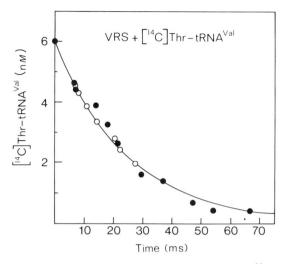

FIGURE 7·8. The VRS-catalyzed hydrolysis of mischarged [^{14}C]Thr-tRNA^{Val}.

(using the VRS·Thr-[^{32}P]AMP compound: Figure 7·9). The solid curve in Figure 7·8 is calculated from the independently measured formation and hydrolysis rate constants given in equation 7·17. Kinetic data thus obtained are consistent with the levels of the mischarged tRNA in the steady state.[52]

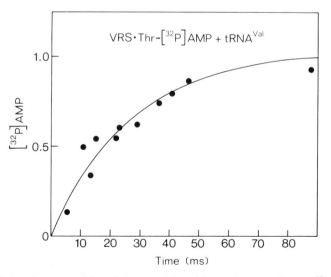

FIGURE 7·9. The rate of breakdown of Thr-AMP when VRS·Thr-[^{32}P]AMP is mixed with excess tRNA^{Val}.

$$VRS \cdot Thr\text{-}AMP \cdot tRNA^{Val} \xrightarrow{36 \text{ s}^{-1}} VRS \cdot Thr\text{-}tRNA^{Val} + AMP \qquad (7 \cdot 17)$$

$$\downarrow {}_{40 \text{ s}^{-1}}$$

$$VRS + Thr + tRNA^{Val}$$

One experimental point worth noting in regard to Figure 7·8 is that the rate constant for the deacylation of the mischarged tRNA can be measured by the pre–steady state kinetics even in the presence of a large fraction of uncharged tRNA. This is not easily done by steady state kinetics, because of the competitive inhibition by the uncharged material. But in the rapid quenching experiment, an excess of enzyme over total tRNA—i.e., a single turnover—is used.

E. Detection of conformational changes

The examples of intermediates discussed so far have been relatively clear cut, since real chemical changes that generate either covalent complexes with the enzyme or chemical products have been involved. A more difficult problem to work with is the detection and analysis of conformational changes of the enzyme on the binding of substrates and also on their interconversion to products. The approach to this problem was outlined in Chapter 4, section D7. A systematic analysis of the relaxation times on the binding of substrates or analogues is usually essential. In addition, since the property of the protein that is being monitored in the rapid reaction study is usually poorly defined (e.g., a change of tryptophan fluorescence), as much independent corroborative evidence as possible must be gathered and more than one probe used if available (Chapter 4, section D7). Because the nature of the problem is less concrete than for the detection of chemical intermediates, the examples are more different to analyze.

Frequently, the existence of a conformational change is inferred because rate constants measured for events in a reaction are inadequate to describe the overall scheme, or because they appear "suspicious." An example of the latter is illustrated in Chapter 4, Figure 4·9: the second-order rate constant measured for the binding of an inhibitor to lysozyme is too low for a diffusion-controlled reaction. Application of a full relaxation-time analysis to the binding revealed a second step that is presumably related to the small conformational change described in Chapter 1, section D3. Another example is illustrated in Figure 4·6, the binding of tyrosine to the tyrosyl-tRNA synthetase. The second-order rate constant is 2 to 3 orders of magnitude below that of diffusion control, so it is most likely that at least one conformational change is involved although not directly detected. There is independent evidence for this, in that although the enzyme is a dimer in which the two active sites are spatially separate, the binding of 1 mol of tyrosine prevents binding at the second site. As will be discussed in Chapter 10, the binding of tyrosine to one of the sites must cause a conformational change to be propagated through the molecule. Another step in the reaction of that enzyme, the formation of the E·Tyr-AMP complex from E·Tyr·ATP, probably involves a further conformational change: stopped-flow fluorescence has detected a large change in tryptophan fluorescence occurring with the same rate constant as that for the chemical step that was monitored independently by rapid quenching. The temporal coupling between the conformational change and the chemical step has not been shown by these studies. The kinetics just indicate a parallel change, but there could be distinct steps. For example, there could be a slow rate-determining conformational change followed by a rapid chemical step, or *vice versa*.

References

1 B. S. Hartley and B. A. Kilby, *Biochem. J.* **56**, 288 (1954).
2 H. Gutfreund, *Discuss. Faraday Soc.* **20**, 167 (1955).
3 H. Gutfreund and J. M. Sturtevant, *Biochem. J.* **63**, 656 (1956).
4 A. Himoe, K. G. Brandt, R. J. DeSa, and G. P. Hess, *J. Biol. Chem.* **244**, 3483 (1969).
5 T. E. Barman and H. Gutfreund, *Biochem. J.* **101**, 411 (1966).
6 S. A. Bernhard and H. Gutfreund, *Proc. Natn. Acad. Sci. USA* **53**, 1238 (1965).
7 J. McConn, E. Ku, A. Himoe, K. G. Brandt, and G. P. Hess, *J. Biol. Chem.* **246**, 2918 (1971).
8 J. Fastrez and A. R. Fersht, *Biochemistry* **12**, 2025 (1973).
9 M. L. Bender, G. R. Schonbaum, and B. Zerner, *J. Am. Chem. Soc.* **84**, 2540 (1962).
10 M. Caplow and W. P. Jencks, *Biochemistry* **1**, 883 (1962).
11 S. A. Bernhard, S. J. Lau, and H. Noller, *Biochemistry* **4**, 1108 (1965).
12 J. De Jersey, D. T. Keough, J. K. Stoops, and B. Zerner, *Eur. J. Biochem.* **42**, 237 (1974).
13 C. G. Miller and M. L. Bender, *J. Am. Chem. Soc.* **90**, 6850 (1968).
14 A. R. Fersht, D. M. Blow, and J. Fastrez, *Biochemistry* **12**, 2035 (1973).
15 E. F. Jansen, M. D. Nutting, and A. K. Balls, *J. Biol. Chem.* **179**, 201 (1949).
16 N. K. Schaffer, S. C. May, and W. H. Summeson, *J. Biol. Chem.* **202**, 67 (1953).
17 R. Henderson, *J. Molec. Biol.* **54**, 341 (1970).
18 T. Alber, G. A. Petsko, and D. Tsernoglou, *Nature, Lond.* **263**, 297 (1976).
19 H. Gutfreund and B. R. Hammond, *Biochem. J.* **73**, 526 (1959).
20 A. Garen and C. Levinthal, *Biochim. Biophys. Acta* **38**, 470 (1960).
21 L. A. Heppel, D. R. Harkness, and R. J. Hilmoe, *J. Biol. Chem.* **237**, 841 (1962).
22 W. N. Aldridge, T. E. Barman, and H. Gutfreund, *Biochem. J.* **92**, 23C (1964).
23 W. E. Hull and B. D. Sykes, *Biochemistry* **15**, 1535 (1976).
24 W. E. Hull, S. E. Halford, H. Gutfreund, and B. D. Sykes, *Biochemistry* **15**, 1547 (1976).
25 R. M. Epand and I. B. Wilson, *J. Biol. Chem.* **238**, 1718 (1963); **240**, 1104 (1965).

26 M. L. Bender, G. E. Clement, C. R. Gunter, and F. J. Kézdy, *J. Am. Chem. Soc.* **86**, 3697 (1964).
27 I. V. Berezin, N. F. Kazanskaya, and A. A. Klyosov, *FEBS Lett.* **15**, 121 (1971).
28 M. Cocivera, J. McManaman, and I. B. Wilson, *Biochemistry* **19**, 2901 (1980).
29 M. Caswell and M. Caplow, *Biochemistry* **19**, 2907 (1980).
30 L. Engström, *Biochim. Biophys. Acta* **54**, 179 (1961); **56**, 606 (1962).
31 J. H. Schwartz and F. Lipmann, *Proc. Natn. Acad. Sci. USA* **47**, 1996 (1961).
32 H. Barrett, R. Butler, and I. B. Wilson, *Biochemistry* **8**, 1042 (1969).
33 S. L. Snyder and I. B. Wilson, *Biochemistry* **11**, 3220 (1972).
34 W. Bloch and M. J. Schlesinger, *J. Biol. Chem.* **248**, 5794 (1973).
35 J. F. Chlebowski, I. M. Armitage, P. P. Tusa, and J. E. Coleman, *J. Biol. Chem.* **251**, 1207 (1976).
36 D. Chappelet-Tordo, M. Iwatsubo, and M. Lazdundski, *Biochemistry* **13**, 3754 (1974).
37 G. S. Kilsheimer and B. Axelrod, *J. Biol. Chem.* **227**, 879 (1957).
38 W. Ostrowski and E. A. Barnard, *Biochemistry* **12**, 3893 (1973).
39 T. M. Stokes and I. B. Wilson, *Biochemistry* **11**, 1061 (1972).
40 M. L. Sinnott and O. M. Viratelle, *Biochem. J.* **133**, 81 (1973).
41 M. L. Sinnott and I. J. L. Souchard, *Biochem. J.* **133**, 89 (1973).
42 S. Rosenberg and J. F. Kirsch, *Biochemistry* **20**, 3189 (1981).
43 M. B. Hoagland, *Biochim. Biophys. Acta* **16**, 288 (1955).
44 P. Berg, *J. Biol. Chem.* **222**, 1025 (1956).
45 M. B. Hoagland, E. B. Keller, and P. C. Zamecnik, *J. Biol. Chem.* **218**, 345 (1956).
46 P. R. Schimmel and D. Söll, *Ann. Rev. Biochem.* **48**, 601 (1979).
47 A. Norris and P. Berg, *Proc. Natn. Acad. Sci. USA* **52**, 330 (1964).
48 A. N. Baldwin and P. Berg, *J. Biol. Chem.* **241**, 839 (1966).
49 R. B. Loftfield, *Progr. Nucl. Acid Res. (& Mol. Biol.)* **12**, 87 (1972).
50 A. R. Fersht and M. M. Kaethner, *Biochemistry* **15**, 818 (1976).
51 A. R. Fersht and M. M. Kaethner, *Biochemistry* **15**, 3342 (1976).
52 A. R. Fersht and C. Dingwall, *Biochemistry* **18**, 1238 (1979).

8

Stereochemistry of enzymatic reactions

Stereospecificity is the hallmark of enzyme catalysis, so a knowledge of the basic principles of stereochemistry is essential for appreciating enzyme mechanisms. Stereochemical evidence can provide important information about the topology of enzyme–substrate complexes. In particular, the positions of catalytic groups on the enzyme relative to the substrate may often be indicated, as may be the conformation or configuration of a substrate or intermediate during the reaction. Further, comparison of the stereochemistry of the substrates and products may reveal the likelihood of intermediates during the reaction.

A. Optical activity and chirality

A compound is optically active, rotating the plane of polarization of plane-polarized light, if it is not superimposable on its mirror image, i.e., on its enantiomer. A simple diagnostic test for superimposability is to determine whether there is a plane or center of symmetry. The presence of such symmetry indicates a lack of activity; the absence indicates activity. As with many organic molecules of biological interest, the optical activity in the amino acids is caused by an asymmetric carbon atom with four different groups around it. This type of carbon is now called a *chiral* center (from Greek for hand). A carbon atom of the form $CR_2R'R''$ is called *prochiral*. Although it is not optically active because it is bound to two identical groups (and thus has a plane of symmetry, making it superimposable with its mirror image), it is *potentially* chiral since it can be made chiral by any operation that produces a difference between the two R groups. The idea that enzymes may be specific for only one enantiomer of a pair of optically active substrates is as old as the study of stereochemistry itself. L. Pasteur, that towering genius of chemistry and biochemistry, reported in 1858 a form of yeast that fermented dextrorotatory tartaric acid but not levorotatory. Early work on the proteases showed that derivatives of L-amino acids and not those of D-amino acids are hydrolyzed.

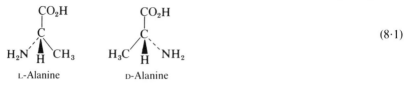

L-Alanine   D-Alanine

(8·1)

1. Notation[1,2]

The letters D and L, which are often used to denote configuration, have the drawback that they are not absolute but are relative to a reference compound. A more useful notation, denoting *absolute* configuration of a chiral center, is the *RS* convention. The groups around the chiral carbon are assigned an order of "priority" based on a series of rules that depend on atomic number and mass. The atom directly attached to the chiral carbon is considered first; the higher its atomic number, the higher its priority. For isotopes, the higher mass number has priority. For groups that have the same type of atom attached to the chiral carbon, the atomic numbers of the next atoms out are considered. This is best illustrated by the following list of the most commonly found groups: $—SH > —OR >$ $—OH > —NHCOR > —NH_2 > —CO_2R > —CO_2H > —CHO >$ $—CH_2OH > —C_6H_5 > —CH_3 > —T > —D > —H.$ (Note: $—CHO$ has priority over $—CH_2OH$ because a $C{=}O$ carbon is counted as being bonded to *two* oxygen atoms.)

A chiral carbon is designated as being R or S as follows. The carbon is viewed from the direction opposite to the ligand of lowest priority. If the priority order of the remaining three ligands decreases in a clockwise direction, the absolute configuration of the molecule is said to be R (Latin, *rectus* = right). If it decreases in the counterclockwise direction it is said to be S (Latin, *sinister* = left). For example:

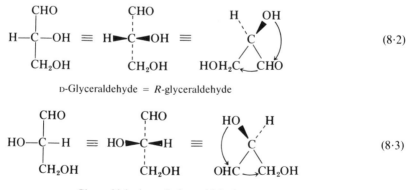

D-Glyceraldehyde = R-glyceraldehyde

(8·2)

L-Glyceraldehyde = S-glyceraldehyde

(8·3)

(Note: The formulas for glyceraldehyde on the left-hand sides of equations 8·2 and 8·3 are written according to the Fischer projection notation. Bonds in the "east-west" direction come up out of the page and those running "north-south" go into the page, as represented in the middle formulas in the equations.)

$$\tag{8·4}$$

L-Alanine = *S*-alanine

If compounds contain different isotopes, e.g., hydrogen and deuterium, the priority rules are applied first to the atomic numbers. If this does not give an unambiguous assignment, the higher-mass-number isotope is given priority.

There is also a *prochirality* rule. For example, in ethanol, if we label the two protons by the subscripts a and b, and *arbitrarily* give H_a priority over H_b as in structure 8·5, H_a is said to be pro-*R* because of the clockwise

$$\tag{8·5}$$

order of priority. Conversely, H_b is pro-*S*. Note that if we repeat the treatment but give H_b priority over H_a, H_a is still found to be pro-*R* and H_b pro-*S*. Prochirality is thus absolute and does not depend on whether H_a or H_b is given priority.

The two faces of a compound containing a trigonal carbon atom are described as *re* (rectus) and *si* (sinister) by a complicated set of rules. Two simple cases are illustrated by the faces presented to the reader by structures 8·6.

$$\tag{8·6}$$

      re   *si*

2. Differences between the stereochemistries of enzymatic and nonenzymatic reactions

The crucial difference between nonenzymatic and enzymatic reactions is that the former generally take place in a homogeneous solution, whereas the latter occur on the surface of a protein that is asymmetric. Because of this, an asymmetric enzyme is able to confer asymmetry on the reactions of symmetric substrates. For example, although the two hydrogen atoms in CH_2RR' are equivalent in simple chemical reactions, the equivalence may be lost when the compound binds to the asymmetric active site of an enzyme.[3] The attachment to the enzyme by R and R' in structure 8·7 causes the two hydrogen atoms to be exposed to different environ-

$$(8·7)$$

ments: H_a may be next to a catalytic base, whereas H_b may be in an inert position.

Another example of this is found in the reactions of carbonyl compounds and carbon-carbon double bonds. An optically active compound is formed if a reagent attacks just one side of a planar trigonal carbon. For example, if in equation 8·8 the nucleophile attacks acetaldehyde from

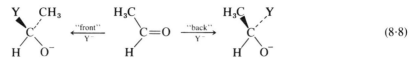

$$(8·8)$$

the "front" side, the product on the left is formed, whereas attack from the "back" of the page gives the enantiomer on the right. (Note: Attack on a trigonal carbon always occurs perpendicular to the plane of the double bond.) In a simple chemical reaction in solution, there is an equal probability of attack at either face of the trigonal carbon in equation 8·8, so that a racemic mixture of 50% of each enantiomer is formed. But in an enzymatic reaction, attack may occur on one face only, because the substrate is firmly held at an asymmetric active site with only one face exposed to the attacking group.

It was pointed out some 50 years ago that the recognition of a chiral (or, as subsequently realized, a prochiral) carbon by an enzyme implies that at least three of the groups surrounding the carbon atom must interact with the enzyme. This is the *multi-point attachment theory*.[4] If only two of the groups interact, the other two may be interchanged without affecting the binding of the substrate (structures 8·9).

$$(8·9)$$

The structural basis of one of the classic examples of such stereospecificity, that of chymotrypsin for L-amino acid derivatives, is immediately obvious on examination of the crystal structure of the enzyme. D-Amino acid derivatives differ from those of L-amino acids by having the H atom and the side chain attached to the chiral carbon interchanged (structure 8·10). The D derivatives cannot bind because of steric hindrance

(8·10)

between the side chain and the walls of enzyme around the position normally occupied by the H atom of L derivatives.

3. Conformation and configuration

The terms conformation and configuration, although often used interchangeably in biochemistry, have precise and different meanings: conformation refers to any one of a molecule's instantaneous orientations in space caused by free rotation about its single bonds; configuration refers to the geometry about a rigid or dissymmetric part of a molecule (e.g., about a double bond, a chiral center, or a single bond where there is steric hindrance to free rotation). In other words, a change of conformation requires just that single bonds rotate, whereas, in general, a change of configuration requires that covalent bonds be broken. The distinction between conformation and configuration becomes blurred in examples in which there is restricted rotation about single bonds: at low temperatures rotation is slow, so different conformational isomers are technically in different configurations; whereas at high temperatures the rate of rotation may be increased so that there is a conformational equilibration. It is important to preserve the distinction between the terms for the clear-cut examples.

B. Examples of stereospecific enzymatic reactions

1. NAD⁺- and NADP⁺-dependent oxidation and reduction

NAD^+ (structure 8·11, R = H) and $NADP^+$ (structure 8·11, R = PO_3^-) function as coenzymes in redox reactions by reversibly accepting hydrogen at the 4 position of the nicotinamide ring (equation 8·12).

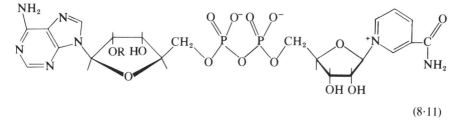

(8·11)

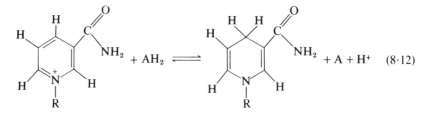

The 4 position in the dihydronicotinamide ring is prochiral. The faces of the nicotinamide ring and the C-4 protons of the dihydronicotinamide rings may be labeled according to the RS convention by giving the portion of the ring containing the —$CONH_2$ group priority over the other portion:

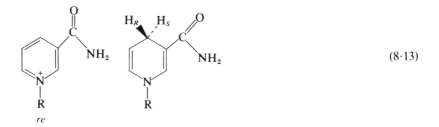

It was discovered in a historically important series of experiments that there is direct and stereospecific transfer between the substrate and NAD^+.[5,6] Yeast alcohol dehydrogenase transfers 1 mol of deuterium from CH_3CD_2OH to NAD^+. When the NADD formed with the enzyme and unlabeled acetaldehyde is incubated, *all* the deuterium is lost from the NADD and incorporated in the alcohol that is formed. The deuterium or hydrogen is transferred stereospecifically to one face of the NAD^+ and then transferred back from the same face. A nonspecific transfer to both faces would lead to a transfer of 50% of the deuterium back to the acetaldehyde (or considerably less than 50% because of the kinetic isotope effect slowing down deuterium transfer). Some dehydrogenases transfer to the same face as does the alcohol dehydrogenase (Class A dehydrogenases), others to the opposite face (Class B); these are listed in Chapter 15. It has subsequently been found that the Class A enzymes transfer to the *re* face of NAD^+ (or $NADP^+$) and use the pro-R hydrogen of NADH:

$$NAD^+ + CH_3CD_2OH \longrightarrow NADD + CH_3CDO + H^+ \qquad (8 \cdot 14)$$

$$NADD + CH_3CHO + H^+ \longrightarrow NAD^+ + CH_3CDHOH \qquad (8 \cdot 15)$$

The transfer to the aldehyde is also stereospecific. The alcohol formed is the R enantiomer. The S enantiomer (structure 8·16), formed from CH_3CDO and NADH, has a specific rotation of $-0.28 \pm 0.03°$.[7]

$$(8 \cdot 16)$$

2. Stereochemistry of the fumarase-catalyzed hydration of fumarate

Fumarase catalyzes the addition of the elements of water across the double bond of fumarate to form malate (equation 8·17). An NMR analysis

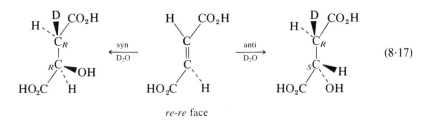

$$(8 \cdot 17)$$

re-re face

of the stereochemistry of the malic acid formed shows that the addition of D_2O is anti (D^+ adding to the *re* face at the top and OD^- adding to the *si* face at the bottom) rather than syn (both adding to the *re-re* face). Similarly, other enzymes in the citric acid cycle catalyze anti hydrations of double bonds.[8] It should be noted that, from the principle of microscopic reversibility, the dehydration reaction of malate must occur in a conformation in which the two carboxyl groups are anti, as are the elements of water that are to be eliminated. It will be seen below that this observation is critical in the analysis of the stereochemistry of the chiral methyl group.

3. Demonstration that the enediol intermediate in aldose–ketose isomerase reactions is syn

As will be discussed in Chapter 15, the catalysis of a reaction by aldose–ketose isomerases involves an enediol intermediate in which the transferred proton (T in equation 8·18) remains on the *same* face of the intermediate. The stereochemistry of the products shows that the intermediate is syn rather than anti.[9]

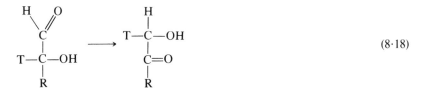

$$(8 \cdot 18)$$

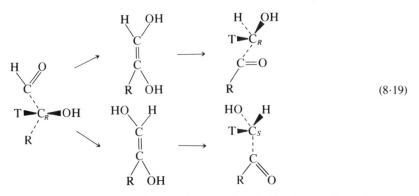

(8·19)

Aldoses that are R at C-2, as in equation 8·19, always form ketoses that are R at C-1. This implies the *syn*-enediol in the upper branch of 8·19 rather than the anti intermediate of the lower branch.

4. Use of locked substrates to determine the anomeric specificity of phosphofructokinase

Phosphofructokinase catalyzes the phosphorylation of fructose 6-phosphate to fructose 1,6-diphosphate (Chapter 10, section G):

> D-fructose 6-phosphate + ATP ⟶
>
> > D-fructose 1,6-diphosphate + ADP   (8·20)

In solution, the substrate exists as about 80% β anomer and 20% α anomer. The two forms rapidly equilibrate in solution via the open-chain keto form (equation 8·21). The enzyme has been shown to be specific for the β form

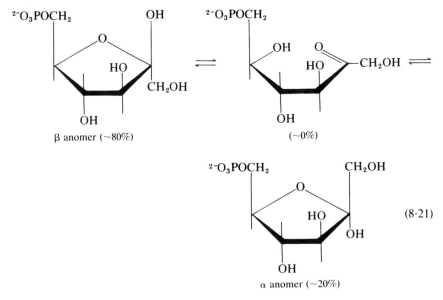

(8·21)

by rapid reaction measurements on a time scale faster than that for the interconversion of the anomers, and also by determination of the activity toward model substrates that are locked in either of the configurations. By using sufficient enzyme to phosphorylate all the active anomer of the substrate before the two forms can re-equilibrate, it is found that 80% of the substrate reacts rapidly, and that the remaining 20% reacts at the rate constant for the anomerization. The kinetics were followed both by quenched flow using $[\gamma\text{-}^{32}P]ATP$[10] and by the coupled spectrophotometric assay of equation 6·4.[11] The other evidence comes from the steady state data on the following substrates:[12]

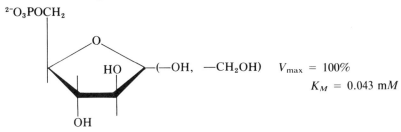

$^{2-}O_3POCH_2$

$(-OH, -CH_2OH)$ $V_{max} = 100\%$
$K_M = 0.043$ mM

α,β-D-Fructose 6-phosphate

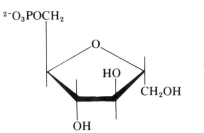

$^{2-}O_3POCH_2$

$V_{max} = 87\%$ $K_M = 0.41$ mM

2,5-Anhydro-D-mannitol 1-phosphate
(locked in β configuration)

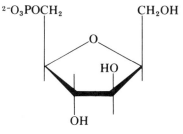

$^{2-}O_3POCH_2$ CH_2OH

$V_{max} = 0\%$ $K_i = 0.34$ mM
(Competitive inhibitor)

2,5-Anhydro-D-glucitol 1-phosphate
(locked in α configuration)

2,5-Anhydro-D-mannitol 1-phosphate is locked in a configuration that is equivalent to the β anomer of D-fructose: it lacks the 2-hydroxyl group and cannot undergo mutarotation to the equivalent of the α anomer because the ring cannot open. The glucitol derivative, on the other hand, is locked into the equivalent of the configuration of the α anomer. It is seen from the values of V_{max} and K_M (or K_i) that although both bind, only the β anomer is bound productively and is phosphorylated.

C. Detection of intermediates from retention or inversion of configuration at chiral centers

1. Stereochemistry of nucleophilic reactions

The reaction that usually begins most elementary courses on mechanistic organic chemistry is nucleophilic substitution at saturated carbon. There are two extreme forms of the mechanism. The bimolecular S_N2 reaction of a nucleophile (Y^-) and an alkyl halide leads to inversion of the chiral carbon (equation 8·22). The unimolecular S_N1 reaction, on the other hand,

$$\text{R'}-\underset{\underset{Y^-}{\overset{|}{R''}}}{\overset{R}{C}}-X \longrightarrow Y-\underset{\overset{}{R''}}{\overset{R}{C}}\text{---R'} + X^- \tag{8·22}$$

generates a planar carbonium ion that reacts randomly at each face to generate a racemic product (equation 8·23). The inversion of configuration

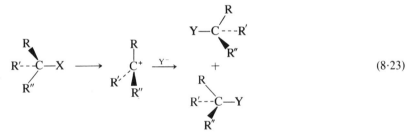

$$\tag{8·23}$$

during an S_N2 reaction has led to certain criteria for the detection of intermediates in enzymatic reactions. For example, suppose that an enzyme catalyzes the nucleophilic reaction 8·22. Then a direct attack of Y^- on the alkyl halide will lead to inversion. But if the substrate first reacts with a nucleophilic group on the enzyme to give an intermediate that then reacts with Y^- (equation 8·24), there will be two successive inversions,

$$\underset{\overset{}{E}}{\overset{R}{\underset{R''}{\text{R'}-C-X}}} \longrightarrow \underset{\underset{+X^-}{\overset{}{R''}}}{\overset{R}{E-C\text{---R'}}} \longrightarrow \underset{\underset{+E}{\overset{}{R''}}}{\overset{R}{\text{R'---C}-Y}} \tag{8·24}$$

and hence an overall retention of configuration. Thus, retention in such reactions is generally taken as evidence for the presence of an intermediate on a reaction pathway, whereas inversion of configuration is interpreted as evidence for the direct reaction of one substrate with the other.[13] How valid are these and other stereochemical arguments?

2. The validity of stereochemical arguments

Stereochemical criteria by themselves have never solved a reaction mechanism. This is because there is the same basic philosophical problem in analyzing stereochemistry as there is in analyzing steady state kinetics (Chapter 7). Only the products and the starting materials are being directly observed; the information regarding the intermediates is indirect and inferential only. For example, although inversion of configuration can arise from a single nucleophilic displacement reaction, it can also arise from three successive displacements, or from five or from any *odd* number of successive displacements. Similarly, retention of configuration implies not just two successive displacement reactions but any *even* number. Proof, as discussed in Chapter 7, requires direct observation of intermediates. Stereochemical criteria do, however, place constraints on the range of possible mechanisms. Thus, as was true for steady state kinetics, *stereochemical evidence per se can never prove mechanisms but can only rule out alternative pathways.*

3. Intermediates in reactions of lysozyme and β-galactosidase

Lysozyme and β-galactosidase, which are both glycosidases, catalyze very similar reactions. Both enzymes are found to catalyze the alcoholysis of their polysaccharide substrates with retention of configuration at the C-1 carbon (equation 8·25).[14-17] This is consistent with the evidence pre-

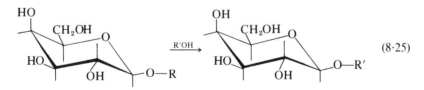

$$(8\cdot25)$$

sented in Chapter 7, section C3, that there is at least one (but probably only one) intermediate on the reaction pathway. However, kinetic isotope data are consistent with the interpretation that the intermediate in the reaction of β-galactosidase is covalent and that there are two successive S_N2 displacements, whereas the intermediate with lysozyme is a bound carbonium ion formed in an S_N1 reaction (Chapter 15). The carbonium ion, unlike an analogous one in solution, reacts stereospecifically on the enzyme. Thus, the stereochemical evidence by itself has given no indication of the *nature* of the intermediate.

The above stereochemical experiments were relatively easy to perform because the natural substrates are chiral. We shall examine two areas in which clever chemistry was required to build chiral substrates.

D. The chiral methyl group

Many enzymatic reactions involve the conversion of methylene groups (CH_2XY or $CH_2=$) to methyl (CH_3-), and *vice versa*. The stereochemistry of these reactions may be studied by synthesizing a methyl group that is chiral. This is possible by using all three isotopes of hydrogen to form a classical asymmetric carbon atom (equation 8·26). The chemistry

$$(8·26)$$

R-Methyl   *S*-Methyl

($D = {}^2H$; $T = {}^3H$)

involved is most elegant, being based on a combination of organic and physical organic chemistry and enzymology.[18,19]

1. The fundamental difference between generating a chiral methyl group from a methylene group and converting a chiral methyl group into methylene

There is a crucial difference between generating a chiral methyl group and converting the same group to methylene; this distinction lies at the heart of much of the experimentation.[20] A methylene group CH_2XY is prochiral, and when it interacts with an enzyme by a three-point contact the two hydrogen atoms are nonequivalent. Their reactivity is thus governed by stereospecificity. A methyl group in CH_3X, on the other hand, rotates freely about the C—X bond, both in solution and when the group is bound to an enzyme (Chapter 1, section E2d). The reactivities of the hydrogen isotopes in CHDTX are thus governed by the *kinetic isotope effect* on the reaction. (Recall from Chapter 2, section G1, that, in general, H is transferred faster than D, which reacts faster than T: that is, $k_H > k_D > k_T$.) Accordingly, whereas a prochiral methylene compound may be transformed by a stereospecific reaction into a homogeneous chiral product, a chiral methyl group reacts to give a mixture of products, the distribution of which depends on the magnitude of the kinetic isotope effect. Further, because of the possibility that a chiral methyl group is gradually exchanging isotopes by recycling in a reversible reaction, it is important that either the reaction being studied is irreversible, or that the products are rapidly converted into stable form by a subsequent reaction so that the isotopes do not become scrambled by equilibration.

2. The chirality assay

a. The assay depends on radioactivity and not on optical measurements

It is clearly not practical to measure the optical activities of chiral methyl compounds because of the vast amounts of radioactive tritium that would have to be handled. Fortunately, as discussed in the next section, it is not necessary to produce substrates that are 100% [HDT]methyl. It is sufficient that practically all tritium-containing methyl groups also contain one deuterium, and that all, or nearly all, such methyl groups have the same chirality. There is thus a small amount of chiral, but also *radioactive*, methyl in a sea of nonradioactive material. By using a radioactivity assay on a mixture of unlabeled and labeled reagents, the only compounds that are measured are those that are labeled and hence chiral.

b. The assay uses the stereochemistry of known enzymatic reactions

R and *S* isomers of [HDT]acetic acid were synthesized by chemical and enzymatic methods that yield products of known stereochemistry.[18,19] The two isomers were then distinguished by using the following ingenious enzymatic assays. The acetic acid was first converted to acetyl–coenzyme A (by a reaction of the carboxyl group—and not the methyl—of acetic acid). The acetyl–coenzyme A was then condensed with glyoxylate to form malate in an essentially *irreversible* reaction catalyzed by malate synthase (equation 8·27). The crucial feature of this reaction is that it is

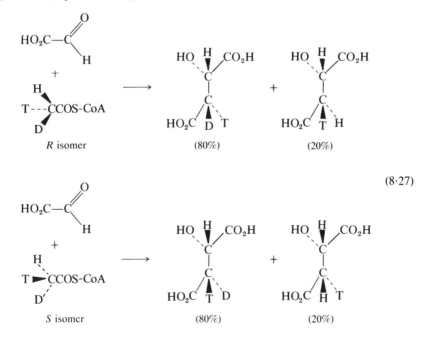

(8·27)

subject to a normal kinetic isotope effect, so that more H than D is lost on condensation. In fact, there is a k_H/k_D of 4 so that 80% of the radioactive product has lost H and 20% has lost D.[21] The stereochemistry may now be resolved by using the enzyme fumarase. The crucial feature of the fumarase reaction (equation 8·28) is that, as described previously

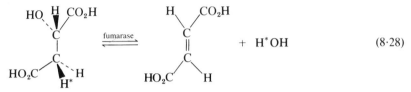

$$+ \ H^*OH \qquad\qquad (8·28)$$

(equation 8·17), the stereochemistry of dehydration of malate is controlled stereospecifically and is not influenced by a kinetic isotope effect: dehydration occurs from a malate conformation in which the two carboxyl groups are anti, as are the H and OH groups that are eliminated (equation 8·28).[8] It is found that the fumarate eventually formed from the R isomer of acetic acid gives a product containing 80% of the tritium and that the fumarate from the S isomer contains 20%, compared with a control reaction of TCH_2CO_2H, which gives a product containing 50%. This clearly provides an assay for the chirality of unknown acetyl groups: R should retain 80% of T, and S 20%.

It should be noted that although the stereochemistry of the products of the malate synthase reaction is now known (see the next section), with the kinetic isotope effect taken into account, this information is not necessary for the chirality assay: all that is essential is that R and S isomers of [HDT]acetic acids give different yields of radioactive fumarate.[20]

3. Stereochemistry of the malate synthase reaction[18–21]

The stereospecificity of the malate synthase reaction was inferred from prior knowledge that the $—C(H)(OH)(CO_2H)$ moiety of malic acid is S and the specificity of fumarase is anti, and assuming a normal k_H/k_D effect of greater than 1. Given, then, that the malic acid has lost more H than D, inversion of the configuration of acetate will yield the products shown in equations 8·27, where the D-containing form is greater than 50% and the H-containing form less than 50%. Retention of configuration will give products with the opposite configurations (equations 8·29). Equations 8·29 predict that the fumarate formed from R-acetate in the condensation of glyoxylate with acetyl–coenzyme A will contain greater than 50% of the T if inversion has occurred, and less than 50% with retention. The observed figure of 80% is therefore consistent with inversion of configuration. Similarly, it is predicted, as found, that inversion of the configuration of the S isomer will give fumarate containing less than 50% of the T. It is seen that even though the rotation of a chiral methyl group is not

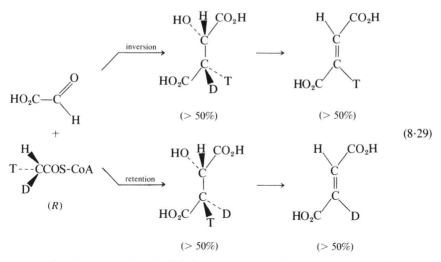

$$(8 \cdot 29)$$

constrained, stereochemical information may be derived, provided that there is a kinetic isotope effect. The absence of an effect would give 50% of each product in equations 8·27 and 8·29, and hence no information on stereospecificity.

It has since been proved that malate synthase proceeds with inversion independently of any assumption about isotope effect. This has been done via a lyase system that splits malate to acetate[21]—a methylene-to-methyl conversion. By using methylene-labeled malate and generating chiral acetate, it could be shown unambiguously that this cleavage is an inversion. When the lyase reaction was applied to malate formed from chiral acetate by malate synthase, it was found that the acetate thus generated was of the same chirality as the starting material. This proved that there is also inversion in the reaction of malate synthase.

The stereospecificity of other methyl-methylene transformations has been assigned by converting the reaction products to acetyl–coenzyme A and using the malate synthase/fumarase assay. Similarly, the stereochemistry of transfer of chiral methyl groups between two acceptors has also been studied by converting the methyl group into chiral acetate. The stereochemistry of dozens of such reactions has been investigated and reviewed in detail.[22,23]

E. Chiral phosphate[24–27]

Phosphate esters play a central and ubiquitous role in biochemistry: the genetic information of all living organisms is stored in the phosphodiester polymers DNA and RNA; the genetic information is translated into protein via RNA; the di- and triphosphates of nucleosides are important energy carriers, as are other phosphorylated compounds; groups are often activated by phosphorylation; energy transduction (work and movement) is generally mediated via the hydrolysis of a nucleoside triphosphate; phosphorylation of proteins and enzymes is typically an important part of control mechanisms; phosphorylation of metabolites is often invoked to increase their solubility or prevent their passage across membranes. Phosphate transfer reactions are accordingly most common. The chemistry of phosphoryl transfer is far more complicated than that of acyl transfer. A very important piece of information in the elucidation of the mechanism of phosphoryl transfer is the stereochemistry.

1. A preview of phosphoryl transfer chemistry

Unlike carbon, which forms only stable tetravalent compounds, phosphorus forms stable trivalent, tetravalent, and pentavalent compounds. A phosphoryl transfer reaction such as

$$R'OPO_3^{2-} + ROH \longrightarrow ROPO_3^{2-} + R'OH \qquad (8\cdot30)$$

can proceed by either of two general types of mechanisms, analogous to the S_N1 and S_N2 mechanisms of carbon chemistry. In a *dissociative* mechanism, there is first the elimination of the leaving group to produce an unstable *metaphosphate* intermediate, and then the rapid addition of the nucleophile:

$$R'OPO_3^{2-} \longrightarrow R'O^- + PO_3^- \xrightarrow{ROH} ROPO_3^{2-} + R'OH \qquad (8\cdot31)$$

In an *associative* mechanism, there is first the addition of the nucleophile to give a pentacovalent intermediate, followed by the elimination of the leaving group. This mechanism is subdivided further. There is an *in-line* mechanism, in which the attacking nucleophile enters opposite the leaving group (equation 8·32); and there is an *adjacent* mechanism, in which the

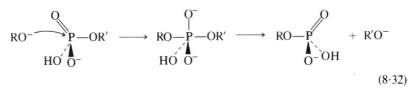

$$(8\cdot32)$$

nucleophile enters on the same side as the leaving group (equation 8·33).

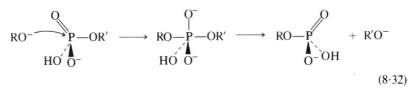

$$(8\cdot33)$$

The adjacent mechanism involves an additional step that is a consequence of the symmetry of the intermediate. The pentacovalent intermediate is a trigonal bipyramid in which groups assume either of two topologically different positions, *equatorial* and *apical* (equation 8·34). In the in-line

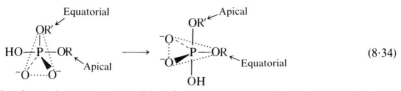

$$(8·34)$$

mechanism, the entering and leaving groups naturally take up apical positions. In the adjacent mechanism, however, the entering nucleophile is apical whereas the leaving group is equatorial. For the leaving group to be expelled, it must move to an apical position. (This is required by the principle of microscopic reversibility; groups enter apically and so must leave apically.) The movement occurs by a process termed *pseudorotation* (equation 8·34).

The stereochemical consequences of the three mechanisms in enzymatic reactions are illustrated in Figure 8·1. The phosphoryl group has the same tetrahedral geometry as saturated carbon compounds. A direct in-line associative reaction between two substrates leads to inversion (Figure 8·1a), as would an in-line dissociative mechanism (Figure 8·1b). Although a dissociative reaction in solution would generally lead to racemization, the stereochemistry in an enzymatic reaction is governed by the spatial arrangements of the substrates (cf. retention of configuration of the carbonium ion in lysozyme catalysis, section C3). The adjacent associative mechanism leads to retention of configuration (Figure 8·1c). In an enzymatic reaction there is the fourth possibility that a covalent intermediate is formed with a nucleophilic group on the enzyme (Figure 8·1d). The stereochemistry should obey the rules that normally apply when such intermediates occur: an odd number of intermediates with inversion at each step will lead to retention of configuration, whereas an even number will lead to inversion. *But*, the adjacent mechanism always leads to retention, since configuration is retained at each step.

2. Chirality of phosphoryl derivatives

A phosphate diester of the form $ROPO_2OR'$ is prochiral, since one of the nonalkylated oxygen (^{16}O) atoms must be replaced to produce chirality. A monoester $ROPO_3{}^{2-}$ is *pro*-prochiral, since two oxygen atoms must be substituted to produce a chiral compound. Chiral phosphates have been synthesized *de novo* by using stereospecific chemical and enzymatic reactions with isotopic and/or atomic substitutions. For example, a chiral phosphorothioate may be synthesized from a prochiral phosphate by replacing an oxygen atom with a sulfur atom. Similarly, what would otherwise be a pro-prochiral phosphate has been synthesized as a chiral product by replacing one oxygen atom with sulfur and another with

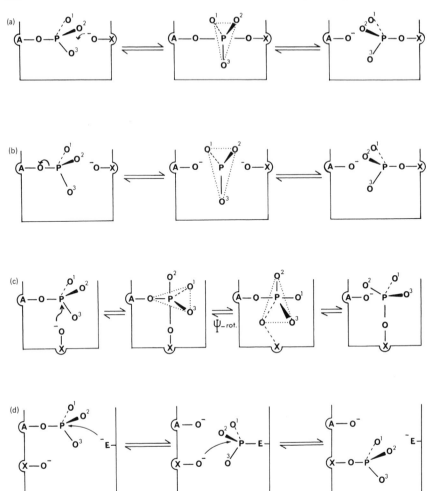

FIGURE 8·1. The four mechanisms of phosphoryl transfer to which phosphokin-ases are subject. The three peripheral oxygen atoms of the phosphoryl group are each labeled to illustrate the stereochemical consequences of the mechanisms. [From G. Lowe, P. M. Cullis, R. L. Jarvest, B. V. Potter, and B. S. Sproat, *Phil. Trans. R. Soc.* **B293**, 75 (1981).]

^{17}O or ^{18}O, or by replacing two oxygen atoms with both isotopes to give [^{16}O,^{17}O,^{18}O]phosphate.

Assays of isotopic and chiral phosphoryl compounds have been greatly facilitated by ^{31}P NMR: substitution with ^{18}O causes a shift to higher field,[28,29] the magnitude of the isotope shift being related to the bond order;[30] ^{17}O, with its spin of $\frac{5}{2}$ and nuclear electronic quadrupole moment, rapidly relaxes the ^{31}P resonance and hence washes out its signal.[30–32] The methods will not be detailed in this chapter. Many of the assignments have involved some complex chemical and/or enzymatic derivatization.

The different degrees of substitution required to generate chirality have led to different methods for studying the stereochemistry of the various reactions. These reactions may be divided into four categories:

1. Prochiral substrate giving prochiral product (i.e., diester → diester).
2. Prochiral substrate giving pro-prochiral product (i.e., diester → monoester).
3. Pro-prochiral substrate giving pro-prochiral product (i.e., monoester → monoester).
4. Pro-prochiral substrate giving pro-pro-prochiral product (i.e., monoester → phosphate).

An example of each type of reaction follows.

3. Examples of chiral phosphoryl transfer

a. Prochiral substrate giving prochiral product
These reactions are the easiest to tackle, since they require only one phosphoryl oxygen to be substituted in both the substrate and the product. The classic example of this experiment is the first step in the hydrolysis of RNA catalyzed by bovine pancreatic ribonuclease. As discussed in detail in Chapter 15, ribonuclease catalyzes the hydrolysis of RNA by a two-step reaction in which a cyclic intermediate is formed. The stereochemistry of the first step (cyclization) (equation 8·35) was solved by some

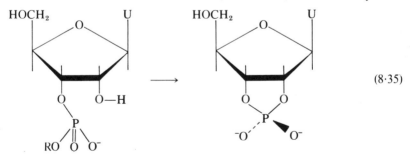

(8·35)

powerful, elegant experiments with the chiral substrate uridine 2',3'-cyclic phosphorothioate (structure 8·36).[33,34] This compound was crystallized and its structure and absolute stereochemistry were determined by x-ray diffraction. Incubation with ribonuclease in aqueous methanol solution formed a methyl ester by the reverse of mechanism 8·35 (equation 8·37). The methyl ester was crystallized and its absolute stereochemistry was determined by x-ray diffraction to be as in equation 8·37. This product

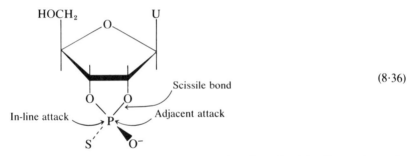

(8·36)

corresponds to an in-line attack. When incubated with ribonuclease in aqueous solution, the methyl ester re-forms the original cyclic phosphorothioate (structure 8·36). This result is expected from the principle of microscopic reversibility, since the forward and reverse reactions must

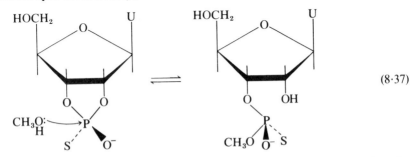

(8·37)

go through the same transition state. But it does show directly that the cyclization step involves an in-line attack: an adjacent attack of the ribose hydroxyl in the cyclization of the methyl ester as in the right-hand structure 8·38 would give the enantiomer of structure 8·36.

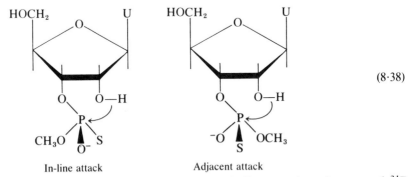

(8·38)

More recent experiments involving similar reactions have used ^{31}P NMR on the products (or on chemical derivatives) to assign configuration.

b. Prochiral substrate giving pro-prochiral product
The second step of the ribonuclease reaction, ring opening (equation 8·39),

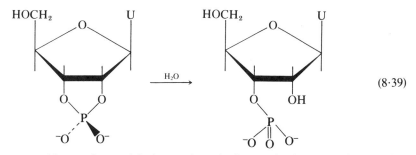

(8·39)

is expected by analogy with the methanolysis step in the preceding section to be an in-line attack of water. This was demonstrated directly, however, by using a technique generally applied to studying the reaction of prochiral phosphate giving pro-prochiral phosphate: the chiral phosphorothioate was hydrolyzed in $H_2{}^{18}O$ to give a chiral product whose configuration could be assigned (equation 8·40).[35]

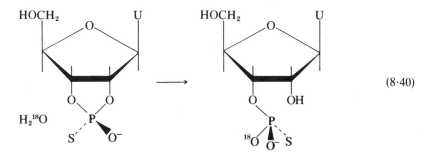

(8·40)

c. Pro-prochiral substrate giving pro-prochiral product
These reactions involve the transfer of the phosphoryl moiety itself and require that two of the three phosphoryl oxygen atoms be tagged in the substrate. This can be achieved by synthesizing an ^{18}O-substituted phosphorothioate[36,37] or $[^{16}O,^{17}O,^{18}O]$phosphomonoester[38,39] substrate. An example of the latter is the demonstration that the hexokinase reaction

$$\beta\text{-D-glucose} + ATP \rightleftharpoons \text{glucose 6-phosphate} + ADP \qquad (8\cdot41)$$

proceeds with inversion of configuration.[25,40] The products were derivatized and their configurations were determined by NMR.[25]

d. Pro-prochiral substrate giving pro-pro-prochiral product
Experiments conducted so far for studying the stereochemistry of hydrolysis of phosphomonoesters to inorganic phosphate have used $[^{16}O,-{}^{17}O,^{18}O]$thiophosphate.[41,42] For example, the hydrolysis of ATP catalyzed by the myosin ATPase (equation 8·42) has been shown to proceed with inversion of configuration.

(8·42)

4. Positional isotope exchange

None of the experiments just discussed gives information on whether or not the reactions are associative or dissociative, since both mechanisms predict the same configurational changes. This aspect of the mechanism may be probed, however, by a different approach which is based on the procedure of *molecular isotope exchange.*[43] This may be illustrated by reactions of ATP. The dissociative mechanism is stepwise with prior formation of the metaphosphate intermediate. Thus, a kinase that catalyzes the transfer of the terminal phosphoryl group of ATP to an acceptor may, in the absence of the acceptor molecule, generate metaphosphate ion and ADP in a rapid and reversible reaction. If the β,γ-bridge oxygen is tagged as in equation 8·43, then because of the torsional symmetry of the phos-

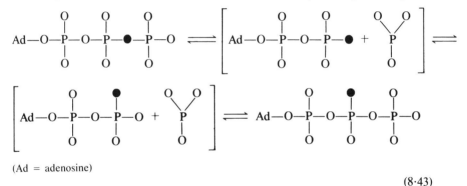

(Ad = adenosine)

(8·43)

phate group, the bridge oxygen may become scrambled.[44] There are several "may"s in this argument, so what is the status of data from such positional isotope exchange experiments? The rules of proof discussed at the beginning of Chapter 7 for the detection of intermediates apply equally well here. If scrambling occurs at a rate that is consistent with the value of k_{cat} for the transfer of the phosphoryl group to the acceptor in the steady state, this is very good evidence for the importance of the dissociative route. However, for the following reasons, the absence of scrambling does not necessarily rule out the dissociative mechanism. It is possible that:

1. There is no scrambling in the absence of the acceptor because it is required to cause a conformational change in the enzyme, or to exert an electrostatic effect on the reaction (i.e., the acceptor is a promoter).

2. The lifetime of the intermediate is very short compared with that for rotation of the phosphate, so that the intermediate reverts to substrate faster than it can rotate.
3. Rotation of the intermediate may be restricted because an oxyanion is bound by a group on the enzyme (e.g., via a bridging metal ion).

Positional isotope exchange experiments conducted on creatine kinase,[45] hexokinase,[46] and pyruvate kinase[47] have provided no evidence in favor of a dissociative mechanism. In each case, the data are consistent with a straightforward nucleophilic attack on phosphorus to generate a trigonal bipyramidal intermediate.[47]

5. A summary of the stereochemistry of enzymatic phosphoryl transfers

So far, all evidence is consistent with the interpretation that enzymatic reactions at phosphorus proceed with inversion by an in-line associative mechanism. There has been no need to invoke adjacent mechanisms, metaphosphate intermediates, or pseudorotation.

Phosphokinases proceed with inversion at phosphorus, and it is thus usually assumed that these reactions involve direct transfer between the two substrates. This has been challenged, however, for acetate kinase: it has been argued that the inversion results from a triple-displacement mechanism with two phosphorylenzyme intermediates and hence three transfers.[48]

Mutases are, in effect, "internal kinases," in that they transfer a phosphoryl group from one hydroxyl to another in the molecule. Unlike kinases, however, their reactions proceed with retention. Since there is generally good corroborative evidence for a phosphorylenzyme in mutase reaction pathways, double-displacement reactions are most likely.

Most phosphodiesterases catalyze transfer of the phosphoryl group to water with retention of configuration. Again, in keeping with other evidence for phosphorylenzyme intermediates, these probably represent double-displacement reactions. One exception occurs with the $3' \rightarrow 5'$ exonuclease of T4 DNA polymerase (Chapter 14); its reactions proceed with inversion.[49]

F. Stereoelectronic control of enzymatic reactions

One of the developments in organic chemistry in recent years is the realization that many reactions are under stereoelectronic control. That is, there is a relationship between the energetics of the electronic changes that occur in bond making and breaking and the conformation or configuration of the reactants. Molecules may thus have an optimal conformation for a particular reaction. This could be of importance in enzymatic reactions for controlling product formation and minimizing side reactions. Stereoelectronic control helps explain the specificity of the reactions of pyridoxal phosphate derivatives. These reactions nicely illustrate the principles involved in the addition or elimination of groups in molecules containing conjugated double bonds.

1. Pyridoxal phosphate reactivity

As discussed in Chapter 2, section C2, pyridoxal phosphate condenses with amino acids to form a Schiff base (structure 8·44). Each of the three groups around the chiral carbon at the top of structure 8·44 may be cleaved

$$(8\cdot44)$$

to give an anion that is stabilized by delocalization of the electrons over the π orbitals. An essential feature of such stabilization is that the atoms in the π system are planar. The extended molecular orbital is constructed from atomic orbitals that are perpendicular to the plane. Thus, for the electrons involved in any bond making or breaking processes to be stabilized by delocalization, the bonds that are being made or broken must also be perpendicular to the plane. This criterion may be used by pyridoxal phosphate–utilizing enzymes in choosing which bond to cleave, as may be seen when the intermediate 8·44 is redrawn so that it is perpendicular to the plane of the paper (structures 8·45; the pyridine ring is represented

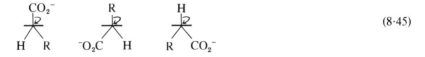

$$(8\cdot45)$$

as a solid bar). In each case, the bond that is broken is the one at the top, so that the electrons may be fed into the π system.

The same principle could well be involved in stabilizing other intermediates against unwanted side reactions. For example, any tendency to eliminate phosphate from an enolate anion intermediate of dihydroxyacetone phosphate (structure 8·46) in the reactions of triosephosphate

(8·46)

isomerase will be minimized if the P—O bond is coplanar with the carbon skeleton.

2. Stereoelectronic effects in reactions of proteases

A different type of stereoelectronic control has been found in the breakdown in solution of tetrahedral addition intermediates that arise in ester and amide hydrolysis and other reactions of carboxyl and carbonyl groups. In the case of an intermediate such as 8·47, in which there are

R O⁻
  \ /
   C
  / \
 X   OR′

(8·47)

two atoms with nonbonded electrons (generally O or N), the lowest-energy transition state for breakdown is a conformation in which nonbonded electrons of each are anti to the group being expelled (structures 8·48).[50]

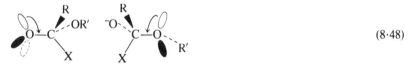

(8·48)

Two views of lone-pair electron orbitals
on —O⁻ and —OR′ anti to X

It is expected from the principle of microscopic reversibility that during the *formation* of the tetrahedral intermediate, the electrons developing on the two heteroatoms must be anti to the bond being made. This has implications in the mechanism of the serine proteases (Chapters 1 and 15).[51-53] A tetrahedral intermediate is formed during proteolysis by the attack of the hydroxyl of Ser-195 on the amide bond. According to the stereoelectronic theory, the lone pairs produced in the tetrahedral intermediate are anti to the incoming hydroxyl group, so the intermediate is set up to expel Ser-195. Expulsion of the amino moiety of the peptide requires proton transfer from His-57 to the N atom, but its lone pair is seen on examination of the structure of the enzyme–substrate complex to be pointing away from the important general-acid-base catalyst. However, if an inversion of the nitrogen switches around its lone pair and its bonded hydrogen (equation 8·49), not only is the N able to accept the

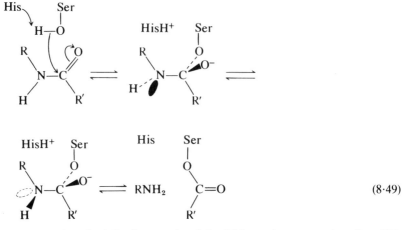

$$(8 \cdot 49)$$

necessary proton, but the lone pair of the N is no longer anti to Ser-195. Perhaps inversion is an essential step.

References

1 G. Popjak, *The Enzymes* **2**, 115 (1970).
2 K. R. Hanson, *J. Am. Chem. Soc.* **88**, 2731 (1966).
3 A. Ogston, *Nature, Lond.* **162**, 963 (1948).
4 M. Bergmann and J. S. Fruton, *J. Biol. Chem.* **117**, 189 (1937).
5 F. H. Westheimer, H. Fisher, E. E. Conn, and B. Vennesland, *J. Am. Chem. Soc.* **73**, 2043 (1951).
6 H. F. Fisher, E. E. Conn, B. Vennesland, and F. H. Westheimer, *J. Biol. Chem.* **202**, 687 (1953).
7 H. R. Levy, F. A. Loewus, and B. Vennesland, *J. Am. Chem. Soc.* **79**, 2949 (1957).
8 O. Gawron, A. J. Glaid III, and T. P. Fondy, *J. Am. Chem. Soc.* **83**, 3634 (1981).
9 I. A. Rose and E. L. O'Connell, *Biochim. Biophys. Acta* **42**, 159 (1960).
10 R. Fishbein, P. A. Benkovic, K. J. Schray, I. J. Siewers, J. J. Steffens, and S. J. Benkovic, *J. Biol. Chem.* **249**, 6047 (1974).
11 B. Wurster and B. Hess, *FEBS Lett.* **38**, 257 (1974).
12 T. A. W. Koerner, Jr., E. S. Younathan, A.-L. E. Ashour, and R. J. Voll, *J. Biol. Chem.* **249**, 5749 (1974).
13 D. E. Koshland, Jr., in *Mechanisms of enzyme action* (W. D. McElroy and B. Glass, Eds.), Johns Hopkins University Press, p. 608 (1954).
14 K. Wallenfels and G. Kurz, *Biochem. Z.* **335**, 559 (1962).
15 K. Wallenfels and O. P. Malhotra, *Adv. Carbohyd. Chem.* **16**, 239 (1961).
16 J. A. Rupley and V. Gates, *Proc. Natn. Acad. Sci. USA* **57**, 496 (1967).
17 J. J. Pollock, D. M. Chipman, and N. Sharon, *Archs. Biochem. Biophys.* **120**, 235 (1967).
18 J. W. Cornforth, J. W. Redmond, H. Eggerer, W. Buckel, and C. Gutschow, *Nature, Lond.* **221**, 1212 (1969).
19 J. Lüthey, J. Rétey, and D. Arigoni, *Nature, Lond.* **221**, 1213 (1969).
20 J. W. Cornforth, *Chem. Brit.* **6**, 431 (1970).

21 H. Lenz and H. Eggerer, *Eur. J. Biochem.* **65**, 237 (1976).
22 H. H. Floss and M. D. Tsai, *Adv. Enzymol.* **50**, 243 (1979).
23 J. W. Cornforth, in *Structural and functional aspects of enzyme catalysis* (H. Eggerer and R. Huber, Eds.), Springer-Verlag, p. 3 (1981).
24 J. R. Knowles, *Ann. Rev. Biochem.* **49**, 877 (1980).
25 G. Lowe, P. M. Cullis, R. L. Jarvest, B. V. Potter, and B. S. Sproat, *Phil. Trans. R. Soc.* **B293**, 75 (1981).
26 P. A. Frey, in *New comprehensive biochemistry,* Vol. 3 (Ch. Tamm, Ed.), Elsevier, p. 201 (1982).
27 G. Lowe, *Accts. Chem. Res.* **17**, 244 (1983).
28 M. Cohn and A. Hu, *Proc. Natn. Acad. Sci. USA* **75**, 200 (1978).
29 G. Lowe and B. S. Sproat, *J. Chem. Soc. Chem. Commun.* **1978**, 565 (1978).
30 G. Lowe, B. V. L. Potter, B. S. Sproat, and W. E. Hull, *J. Chem. Soc. Chem. Commun.* **1979**, 733 (1979).
31 M. D. Tsai, *Biochemistry* **18**, 1468 (1979).
32 J. J. Villafranca and F. M. Raushel, *Ann. Rev. Biophys. Eng.* **9**, 363 (1980).
33 D. A. Usher, D. I. Richardson, and F. Eckstein, *Nature, Lond.* **288**, 663 (1970).
34 D. A. Usher, E. S. Erenrich, and F. Eckstein, *Proc. Natn. Acad. Sci. USA* **69**, 115 (1972).
35 P. M. J. Burgers, F. Eckstein, D. H. Hunneman, J. Baraniak, R. W. Kinas, K. Lesiak, and W. J. Stec, *J. Biol. Chem.* **254**, 9959 (1979).
36 G. A. Orr, J. Simon, S. R. Jones, G. J. Chin, and J. R. Knowles, *Proc. Natn. Acad. Sci. USA* **75**, 2230 (1978).
37 J. P. Richard, H.-T. Ho, and P. A. Frey, *J. Am. Chem. Soc.* **100**, 7756 (1978).
38 S. J. Abbott, S. R. Jonas, S. A. Weinman, and J. R. Knowles, *J. Am. Chem. Soc.* **100**, 2558 (1978).
39 P. M. Cullis and G. Lowe, *J. Chem. Soc. Chem. Commun.* **1978**, 512 (1978).
40 W. A. Blättler and J. R. Knowles, *Biochemistry* **18**, 3927 (1979).
41 M. R. Webb and D. R. Trentham, *J. Biol. Chem.* **255**, 1775 (1980).
42 M. D. Tsai, *Biochemistry* **19**, 5310 (1980).
43 I. A. Rose, *Adv. Enzymol.* **50**, 361 (1979).
44 C. F. Midelfort and I. A. Rose, *J. Biol. Chem.* **251**, 5881 (1976).
45 G. Lowe and B. S. Sproat, *J. Biol. Chem.* **255**, 3944 (1980).
46 I. A. Rose, *Biochem. Biophys. Res. Commun.* **94**, 573 (1980).
47 A. Hassett, W. Blättler, and J. R. Knowles, *Biochemistry* **21**, 6335 (1982).
48 L. B. Spector, *Proc. Natn. Acad. Sci. USA* **77**, 2625 (1980).
49 A. Gupta, C. DeBrosse, and S. J. Benkovic, *J. Biol. Chem.* **257**, 7689 (1982).
50 P. Deslongchamps, *Tetrahedron* **31**, 2463 (1975).
51 S. A. Bizzozero and B. O. Zweifel, *FEBS Lett.* **59**, 105 (1975).
52 S. A. Bizzozero and H. Dutler, *Bioorg. Chem.* **10**, 46 (1982).
53 M. Fujinaga, R. J. Read, A. Sieleck, W. Ardelt, M. Lastrowski, Jr., and M. N. G. James, *Proc. Natn. Acad. Sci. USA* **79**, 4846 (1982).

Further reading

New comprehensive biochemistry, Vol. 3: *Stereochemistry* (Ch. Tamm, Ed.), Elsevier (1982).

9

Active-site-directed and enzyme-activated irreversible inhibitors: "Affinity labels" and "suicide inhibitors"

Amino acid side chains of proteins react with a variety of chemical reagents to form covalent bonds. In general, the reagents tend to be nonspecific and to react with any accessible amino acid residue that has the appropriate chemical nature, with variable consequences. Covalent modification of an enzyme may lead to an irreversible loss of activity if a catalytically essential residue is blocked, if substrate binding is sterically impeded, or if the protein is distorted or its mobility impaired. Alternatively, modification may not affect activity if an unimportant residue is modified, or the inhibition may be only transitory if the chemical reaction is reversible. In this chapter, we are concerned mainly with specially designed irreversible inhibitors that bind specifically to the active site of an enzyme and utilize components of its catalytic apparatus for their effect. Their mode of inhibition is thus controlled and efficient. Before discussing the design of such highly specific irreversible inhibitors, we first list briefly the types and uses of the chemical modifications most commonly employed, and survey the chemical reactivities expected of amino acid chains.

A. Chemical modification of proteins

Protein chemistry is an extensive and highly developed area of organic chemistry that deals with the chemical reactions of proteins. Much of this chemistry concerns reactions that occur in aqueous solution at ambient temperatures and neutral pH, that is, under conditions where proteins are stable. The objective is to modify residues in proteins chemically, either to provide mechanistic information or to produce useful alterations of activity. Some of the more frequent modification reactions are listed in Table 9·1. Spectroscopic probes may be covalently linked to specific regions of proteins; e.g., fluorescent derivatives such as the dansyl group

248

TABLE 9·1. *Chemical modification reactions referred to in this text*

Residue (reactive ionic form)	Reagents	Products	Comments
$—CO_2H$	$R'—N=C=NR'$ (soluble diimide) $+ RNH_2$	$—CONHR$	Surface carboxyls readily blocked; cf. chymotrypsin, pp. 166, 172.
$—CO_2H$	$N_2CHCONHR$ (diazoamide or -ester)	$—CO_2CH_2CONHR$	Reactive in affinity labels; cf. pepsin, pp. 255, 423.
$—CO_2^-$	(epoxide)[a]	$—CO_2CH_2CH(OH)R$	Reactive in affinity labels; cf. glucose 6-phosphate isomerase, p. 255.
$—CO_2^-$	XCH_2COR[b] $(X = I, Br, Cl;$ haloacetates, etc.)	$—CO_2CH_2COR$	Reactive in affinity labels; cf. triosephosphate isomerase, p. 256, 442.
$—NH_2$	Acetic anhydride	$—NHCOCH_3$	Surface groups readily blocked; see pp. 166, 172.
$—NH_2$	Succinic anhydride	$—NHCOCH_2CH_2CO_2^-$	As above, but more soluble protein; see p. 165.
$—NH_2$	$RR'CO, NaBH_4$ (sodium borohydride)	$—NHCHRR'$	Trapping Schiff base adducts; cf. acetoacetate decarboxylase, p. 71. H_2CO and $NaBH_4$ give reductive methylation of surface amino groups.
$—NH_2$	$CH_3OC(=\overset{+}{N}H_2)CH_3$ (methyl acetimidate)	$—\overset{+}{N}H=C(NH_2)CH_3$	—
$—NH_2,$ $—NH_2$			Cross-linking of proteins by linking $—NH_2$ groups, e.g., dimethylsuberimidate; see p. 251.[c]

TABLE 9·1. *Chemical modification reactions referred to in this text (cont.)*

Residue (reactive ionic form)	Reagents	Products	Comments
—NH₂	Dansyl chloride	—NH-dansyl	Fluorescent label; cf. immunoglobin; see p. 35.
—S⁻	(N-ethylmaleimide)		Selective for —SH. Et may be replaced by spin label; cf. pyruvate dehydrogenase, p. 41.
—S⁻		—SCH₂—CH(OH)R	Reactive in affinity label; cf. glyceraldehyde 3-phosphate dehydrogenase, p. 256.
—S⁻	XCH₂COR (haloacetate, etc.)	—SCH₂COR	S⁻ sufficiently nucleophilic to react rapidly in solution; see p. 252.
	(NO₂)₄C (tetranitromethane)		Chromophore, low pK_a (~7); cf. carboxypeptidase see p. 421.
	I₃⁻	(I)	Radioactive tag; see p. 251.
	(EtOC)₂O (diethylpyrocarbonate)		Relatively stable at neutral pH; cf. lactate dehydrogenase, p. 400.

TABLE 9·1. *Chemical modification reactions referred to in this text (cont.)*

Residue (reactive ionic form)	Reagents	Products	Comments
[imidazole ring, HN⎯N]	XCH_2COR (haloacetate, etc.)	[imidazole ring, N⎯NCH$_2$COR]	$\sim 100 \times$ less reactive than —S$^-$. Useful in affinity labels; cf. chymotrypsin, p. 253.

[a] The —NH$_2$ group also reacts with epoxides.
[b] The α carbon of haloacetates is particularly susceptible to nucleophilic attack. The order of reactivity for different halides is I > Br > Cl.
[c] Cross-linkers have been synthesized with an —SS— bond that may be cleaved by a reducing agent to reverse the cross-linking.

may be linked to amino groups, spin labels such as nitroxide derivatives may be attached to cysteine residues, and tyrosine residues may be modified by nitration. These probes are sometimes termed *reporter groups* since they examine local structure or overall conformation. Other common examples are: the cross-linking of proteins by bifunctional reagents such as dimethylsuberimidate, in order to measure the degree of association or subunit composition of oligomers; the insertion of radioactive tags into molecules, by such procedures as the iodination of tyrosine side chains by radioactive iodine and the reductive methylation of amino groups with formaldehyde and radioactively labeled (^{3}H) borohydride; searching for unusually reactive groups by their high reaction rates; measuring the degree of exposure of groups to solvent from their activity with reagents; measuring the pK_a values and ionic states of reactive groups from the pH dependence of their reactivities (as described in Chapter 5); assessing whether groups are important in catalysis by noting the effect of modification on reaction rates; irreversibly inhibiting activity by chemical modification.

The subject of protein chemistry is too large to be surveyed systematically in one chapter. Various examples of chemical modifications are covered nonsystematically throughout this volume, however, in discussions of individual enzymes and methods. To aid in locating these examples, some are listed in Table 9·1.

1. The chemical reactivity of amino acid side chains

The principal chemically reactive groups in proteins are *nucleophiles*. These are generally the same groups that are found at the active sites of enzymes and that are responsible for catalysis, as discussed in Chapter 2. The nucleophiles that are potent toward "hard" electrophilic centers, such as the carbonyl, phosphoryl, and sulfuryl groups (Chapter 2, section D2a), include: the —OH of serine, threonine, and tyrosine (the alcohols being activated by general-base catalysis, with the phenols reacting via their ionized forms at neutral and higher pH's); the ϵ-NH_2 group of lysine and the α-amino groups of the N-termini; the imidazole ring of histidine; the —S^- of cysteine; and the —CO_2^- of aspartate, glutamate, and the C-termini. These nucleophiles are also reactive in varying degrees toward "soft" electrophiles such as saturated carbon, according to the principles discussed in Chapter 2, section D2b. In addition, the S atom in the side chain of methionine is nucleophilic toward soft electrophiles, as is the aromatic ring of tyrosine. Consequently, the majority of reagents that are used to modify proteins are *electrophiles*.

Two important classes of reagents that are used to modify the nucleophilic side chains are the acylating and similarly reacting agents, and the alkylating agents. For example, acetic anhydride and other activated acyl compounds acylate amino groups to form stable amide derivatives. The phenolic hydroxyl group of tyrosine is also acylated, but the resultant ester is unstable in mild alkali. The acyl derivatives of cysteine (thioester) and histidine (acylimidazole) are also rapidly hydrolyzed. Thus, acylating agents are not generally useful for inhibiting these important catalytic groups irreversibly. The products of alkylation, however, are usually quite stable, and so irreversible inhibitors are often based on alkylating agents.

Alkylating agents are soft electrophiles. As was explained in Chapter 2, section D2b, one of the most reactive common nucleophiles toward soft electrophiles is the —S^- ion. Accordingly, it is found that cysteine residues in proteins are readily alkylated by haloacetates and similar reagents. On the other hand, carboxylate ions are usually so unreactive that they are not modified by simple haloacetates or epoxides. But, as will be seen in the following section, cleverly designed reagents may react specifically and rapidly with otherwise poorly reactive groups.

B. Active-site-directed irreversible inhibitors

An affinity label, or active-site-directed irreversible inhibitor, is a chemically reactive compound that is designed to resemble a substrate of an enzyme, so that it binds specifically to the active site and forms covalent bonds with the protein residues.[1-3] Affinity labels are very useful for identifying catalytically important residues and determining their pK_a values from the pH dependence of the rate of modification.

The reaction of an affinity label with an enzyme involves the initial formation of a reversibly bound enzyme–inhibitor complex followed by covalent modification and hence irreversible inhibition:

$$E + I \underset{}{\overset{K_1}{\rightleftharpoons}} E{\cdot}I \overset{k_1}{\longrightarrow} E{—}I \tag{9·1}$$

This scheme is analogous to that of the Michaelis-Menten mechanism, and the reaction should thus show saturation kinetics with increasing inhibitor concentration. The kinetics were solved in Chapter 4, equation 4·71. For the simple case of pre-equilibrium binding followed by a slow chemical step, the solution reduces to

$$-\frac{d[E]}{dt} = \frac{k_1[E][I]}{K_I + [I]} \tag{9·2}$$

An important consequence of the chemical reaction taking place in the confines of an enzyme–"substrate" complex is that not only is the binding specific, but the rate of the chemical step may be unusually rapid because it is favored entropically over a simple bimolecular reaction in solution, in the same way as is a normal enzymatic reaction. Thus, re-agents that are normally only weakly reactive may become very reactive affinity labels.

The principles of the method are very nicely illustrated by one of the first affinity labeling experiments, the reaction of *tos*-L-phenylalanine chloromethyl ketone (TPCK) with chymotrypsin.[1] TPCK resembles substrates like *tos*-L-phenylalanine methyl ester, but the chloromethyl ketone

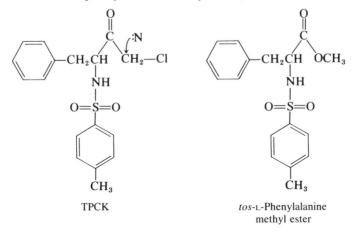

TPCK   *tos*-L-Phenylalanine
                                 methyl ester

group of TPCK is an alkylating reagent. Halomethyl ketones and acids are known to react with thiols and imidazoles. TPCK reacts far more rapidly with chymotrypsin than it does with normal histidine-containing peptides because of its high reactivity as an affinity label. This can be seen in Table 9·2 for an analogous chloromethyl ketone. In addition to this important diagnostic feature, the irreversible inhibition of chymotrypsin by TPCK has four other characteristic features:[1,4]

TABLE 9·2. *High reactivity in affinity labeling*[a]

Reaction	Second-order rate constant ($s^{-1} M^{-1}$)
Cbz-L-phenylalanine chloromethyl ketone + chymotrypsin (alkylation of His-57)	69
Cbz-L-phenylalanine chloromethyl ketone + acetylhistidine	4.5×10^{-5}

[a] From E. N. Shaw and J. Ruscica, *Archs. Biochem. Biophys.* **145**, 484 (1971).

1. The rate of inactivation is inhibited by reversible inhibitors or substrates of the enzyme.
2. The relative rates of inhibition as a function of pH at low inhibitor concentrations ($[I] \ll K_I$) follow a bell-shaped curve, with the same pK_a values as found for the pH dependence of k_{cat}/K_M for the hydrolysis of substrates.
3. The inactivated enzyme has 1 mol of inhibitor covalently bound per mole of active sites. (This was found by using ^{14}C-labeled inhibitor.)
4. The inhibition follows saturation kinetics (with a "K_M" of about 0.3 mM).[5]

(It was later shown that the site of modification is at His-57.)

Characteristics 1, 2, 3, and 4 are diagnostic tests for an affinity label that is modifying the active-site group whose ionization controls activity. The saturation kinetics (feature 4) show that the label binds to the enzyme—although these may not be observed if the K_I is higher than the solubility of the label. Competitive inhibition by substrates, etc. (feature 1) suggests that the binding is at the active site. The 1:1 stoichiometry (feature 3) shows that the modification is selective. The pH dependence (feature 2) gives important evidence. It was shown in Chapter 5 that the pH dependence of V_{max}/K_M or k_{cat}/K_M gives the pK_a's of the free-enzyme groups that are involved in catalysis and binding of the substrate. Similarly, the pH dependence of k_I/K_I (or the relative rates of reaction at $[I]$ $\ll K_I$, since the rate is proportional to k_I/K_I under this condition) gives the pK_a's of the free-enzyme groups that are involved in binding and reacting with the inhibitor. Thus, identical sets of pK_a's should be obtained from the pH dependence of k_{cat}/K_M and k_I/K_I if the groups that are involved in catalysis are also those that react with the affinity label. The pH dependence of k_I gives the pK_a's of the enzyme–inhibitor complex just as k_{cat} gives the pK_a of the enzyme–substrate complex (subject to the provisos of Chapter 5).

Enzymes do have some tricks up their sleeves (or rather, in their pockets) to spring on enzymologists. For example, the affinity label 1,2-

anhydro-D-mannitol 6-phosphate labels a glutamate residue in glucose 6-

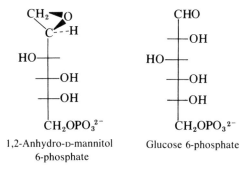

1,2-Anhydro-D-mannitol   Glucose 6-phosphate
6-phosphate

phosphate isomerase.[6] It shows saturation kinetics, competitive inhibition by substrates, and 1:1 stoichiometry, and the pH dependence of k_1 gives pK_a values for the enzyme–inhibitor complex similar to those for enzyme–substrate complexes. However, a crystallographic study has provided some speculative evidence that the group modified was not the one aimed at, but another base that is catalytically important.[7] If this is correct, the unexpected result is more interesting than the expected one, since an important but previously unknown catalytic group has been located.

Another good example of the use of affinity labels involves pepsin, and is illustrated in Chapter 15, equations 15·31 and 15·32. The enzyme has two catalytical important aspartic acid residues, one ionized and the other un-ionized. The ionized carboxylate is trapped with an epoxide, which, of course, requires the reaction of a nucleophilic group. The un-ionized carboxyl is trapped with a diazoacetyl derivative of an amino acid ester:

$$\overset{+}{N_2}CHCONHCH(R)CO_2CH_3 \xrightarrow[N_2]{Cu(II)} Cu(II)\overset{\cdot\cdot}{C}HCONHCH(R)CO_2CH_3$$

(9·3)

The reaction is catalyzed by cupric ions and presumably results from a copper-complexed carbene.[8] The electron-deficient carbene with only six electrons in its outer valence shell is known to add across the O—H bonds of un-ionized carboxyl groups to form the methyl ester.

Another general approach is the use of *photoaffinity labels*.[9-11] A compound that is stable in the absence of light but that is activated by photolysis is reversibly bound to an enzyme and photolyzed. The usual reagents are diazo compounds that when photolyzed give highly reactive carbenes, or azides that give highly reactive nitrenes:

$$
\underset{\text{CHN}_2}{\text{RC}}\diagdown^{\displaystyle\text{O}} \xrightarrow{\ h\nu\ } \underset{\text{CH}}{\text{RC}}\diagdown^{\displaystyle\text{O}} \tag{9$\cdot$4}
$$

$$
\text{R—N}_3 \xrightarrow{\ h\nu\ } \text{R—N:} \tag{9$\cdot$5}
$$

The photoaffinity labels have been useful in mapping out residues at the active sites of enzymes and the binding sites of proteins such as antibodies. The normal affinity labels are more useful for kinetic work, since they are selective for the basic and nucleophilic groups that are so prevalent in catalysis, and since they also give information on the pK_a's of the groups that are modified.

Many hundreds of affinity labels have been synthesized, the majority based on halomethyl ketones or epoxides (Table 9.3). They are listed each year in the *Specialist Periodical Reports: Amino-Acids, Peptides, and Proteins*, published by the Chemical Society (U.K.).

C. Enzyme-activated irreversible inhibitors[12–15]

There is considerable interest in the design of highly specific irreversible enzyme inhibitors because of their potential use as therapeutic agents.

TABLE 9·3. *Some affinity labels[a]*

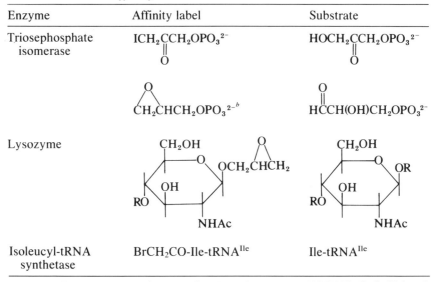

Enzyme	Affinity label	Substrate
Triosephosphate isomerase	$\underset{\displaystyle O}{ICH_2\overset{\parallel}{C}CH_2OPO_3{}^{2-}}$	$\underset{\displaystyle O}{HOCH_2\overset{\parallel}{C}CH_2OPO_3{}^{2-}}$
	$\overset{\displaystyle O}{\overset{\triangle}{CH_2}CHCH_2OPO_3{}^{2-b}}$	$\underset{\displaystyle O}{H\overset{\parallel}{C}CH(OH)CH_2OPO_3{}^{2-}}$
Lysozyme		
Isoleucyl-tRNA synthetase	$BrCH_2CO\text{-Ile-tRNA}^{Ile}$	Ile-tRNA^{Ile}

[a] From F. C. Hartmann, *Biochem. Biophys. Res. Commun.* **33**, 888 (1968); S. G. Waley, J. C. Miller, I. A. Rose, and E. L. O'Connell, *Nature, Lond.* **227**, 181 (1970); E. W. Thomas, J. F. McElvy, and N. Sharon, *Nature, Lond.* **222**, 485 (1969); D. V. Santi and W. Marchant, *Biochem. Biophys. Res. Commun.* **51**, 370 (1973).
[b] Also reacts with glyceraldehyde 3-phosphate dehydrogenase. S. McCaul and L. D. Byers, *Biochem. Biophys. Res. Commun.* **72**, 1028 (1976).

Part of the research program of most pharmaceutical companies is the rational design of drugs based on mechanistic ideas from enzymology, biochemistry, and chemistry. There are serious drawbacks to the use of the affinity labels in this context, since they are, in general, reactive chemical compounds that owe their selectivity to productive binding at the active site of their target enzyme. Although this selectivity is perfectly adequate for specifically modifying the active sites of purified proteins, there are too many possible side reactions for the use of such inhibitors *in vivo*. In particular, the high reactivity of —SH groups in small biological molecules and in other proteins could either destroy an affinity label or cause side effects. There is, however, another class of irreversible inhibitors which, in addition to utilizing the binding specificity of their target enzyme, specifically utilize its catalytic apparatus for chemical activation, with the result that a normally innocuous reversible inhibitor is converted into a powerful irreversible inhibitor. These inhibitors, which are chemically unreactive in the absence of the target enzyme, are commonly called suicide inhibitors[14] (since the enzyme appears to "commit suicide") or k_{cat} inhibitors[13] (since they use the active-site residues for activation). Other terminologies applied to them are mechanism-based, trojan-horse,[16] and enzyme-activated substrate (EASI)[17] inhibitors. They react with the enzyme according to the following scheme:

$$E + I \underset{}{\overset{K_1}{\rightleftharpoons}} E\cdot I \xrightarrow{k_{cat}} E\cdot I^* \begin{array}{c} \xrightarrow{k_I} E\text{—}I^* \\ \xrightarrow{k_{diss}} E + I^* \end{array} \qquad (9\cdot6)$$

An effective suicide inhibitor must have these characteristics:

1. The inhibitor must be chemically unreactive in the absence of enzyme.
2. It must be activated specifically by its target enzyme.
3. It must in its activated form react more rapidly with its target enzyme than it dissociates (i.e., $k_I \gg k_{diss}$). (This criterion may be relaxed if I* is destroyed by water much faster than it reacts with other enzymes.)

In the preceding section, four diagnostic tests of affinity labeling were listed (inactivation inhibited by substrates, pH dependence of inactivation similar to that of catalysis, labeled inhibitor covalently bound in 1:1 stoichiometry, and saturation kinetics obeyed). The same criteria may be used to diagnose suicide inhibition. In addition, tests must be made to detect any diffusion of the activated intermediate I* into solution. For example, the addition of —SH reagents that rapidly react with electrophiles and hence scavenge them should not slow down the rate of reaction. The suicide inhibitor should not, in any case, react with the thiol at an appreciable rate in the absence of enzyme.

    Most suicide inhibitors are based on the generation of an intermediate that has conjugated double bonds and that is susceptible to a Michael addition reaction. A nucleophilic group on the enzyme may then be alkylated by the intermediate (equation 9·7). The conjugated intermediate

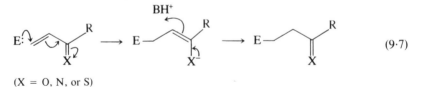

$$(X = O, N, or S)$$

is usually generated by proton abstraction by a basic group on the enzyme. Often, the basic group that is responsible for the proton abstraction is also the nucleophilic group in the Michael addition. Thus, most of the suicide inhibitors made so far have been aimed at enzymes that catalyze the formation of carbanions or carbanion-like intermediates. Suicide inhibitors are typically based on acetylenic compounds (as in equation 9·8), β,γ-unsaturated compounds (as in equation 9·9), or β-halo compounds (as in equation 9·10). (The α protons in such compounds are acidic because

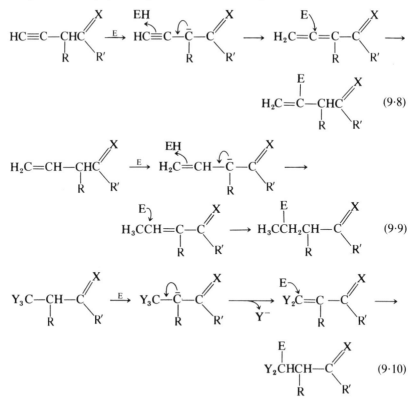

$$(X = O, N, or S; Y = F or Cl)$$

the negative charge in the carbanion is delocalized by the conjugation with X.)

A classic example of suicide inhibition is that of β-hydroxy-decanoyl-dehydrase by 3-decenoyl-N-acetylcysteamine.[12] The enzyme catalyzes the reaction

$$CH_3(CH_2)_6CHOHCH_2CO—NAC \rightleftharpoons$$

$$CH_3(CH_2)_6CH=CHCO—NAC \qquad (9\cdot11)$$
$$\text{(cis and trans)}$$

where $NAC = SCH_2CH_2NHCOCH_3$. The suicide inhibitor is activated by the enzyme, which is then alkylated on the active-site histidine:

$$CH_3(CH_2)_5C\equiv CCH_2CO—NAC \xrightarrow{E}$$

$$CH_3(CH_2)_5CH=C=CHCO—NAC \longrightarrow$$

$$CH_3(CH_2)_5CH=\underset{\underset{E}{|}}{C}CH_2CO—NAC \qquad (9\cdot12)$$

It should be noted that there is a kinetic isotope effect on the normal reaction (9·11) when the α-deuterated compound is used as the substrate. A similar effect is found when the deuterated suicide inhibitor is used. Thus, both reactions involve a proton transfer in the rate-determining step of the reaction. It has also been shown that a sample of the allenic intermediate that is prepared chemically does in fact irreversibly inhibit the enzyme.[18]

1. Pyridoxal phosphate–linked enzymes

Enzymes containing pyridoxal phosphate are prime targets for suicide inhibition because the chemistry is so naturally suitable. As discussed in Chapter 2, section C2, the pyridoxal ring acts as an electron sink that facilitates the formation of carbanions and also forms part of an extended system of conjugated double bonds. For example, vinyl glycine, $CH_2=CHCH(NH_3{}^+)CO_2{}^-$, condenses with the pyridoxal phosphate of aspartate aminotransferase to form a Schiff base, as described in Chapter 2, equation 2·42.[19] The α proton may be abstracted (as in equation 2·43)

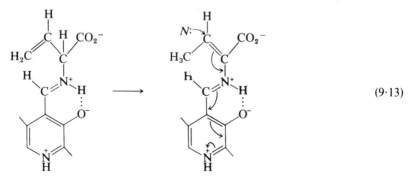

$$(9\cdot13)$$

so that the isomerization shown in equation 9·13 readily occurs. Michael addition may then take place at the terminus of the conjugated system. Lysine-258 at the active site is alkylated by the suicide inhibitor. The reaction is similar to that of equation 9·7, except that in the Michael addition the pyridoxal ring acts as the electron sink, rather than the N atom originating from the vinyl glycine.

The product of certain elimination reactions, e.g., that of serine dehydratase (Chapter 2, equation 2·47), is structure 9·14. This is similar to

$$(9·14)$$

the reactive intermediate of equation 9·13; here, though, the enzyme has evolved to deal with the intermediate and is not inactivated. However, a similar but far more reactive intermediate can be generated from the reaction of the enzyme with β-trifluoroalanine (equation 9·15), which will alkylate the enzyme and inactivate it.[20]

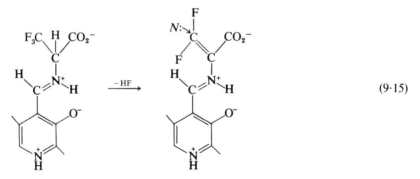

$$(9·15)$$

The intermediate 9·14 is probably generated during the suicide inhibition of β-aspartate decarboxylase,[21] aspartate aminotransferase,[22] and alanine racemase[23,24] by β-chloroalanine. These enzymes are inactivated by the intermediate, since they have not evolved to cope with it during the normal course of reaction.

2. Monoamine oxidases and flavoproteins

Another means of producing Michael acceptors is by the *oxidation* of unsaturated amines of alcohols, such as allyl amine (equation 9·16). These

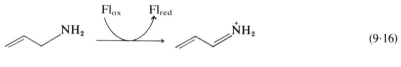

$$(9 \cdot 16)$$

(Fl = flavin)

reactions, which have provided a means of inhibiting the flavin-linked monoamine oxidases, enable us to end on a clinical note. The monoamine oxidases are responsible for the deamination of monoamines such as adrenaline, noradrenaline, dopamine, and serotonin, which act as neurotransmitters. Imbalances in the levels of monoamines cause various psychiatric and neurological disorders: Parkinson's disease is associated with lowered levels of dopamine, and low levels of other monoamines are associated with depression. Inhibitors of monoamine oxidases may consequently be used to treat Parkinson's disease and depression. The flavin moiety is covalently bound to the enzyme by the thiol group of a cysteine residue (equation 9·17). The acetylenic suicide inhibitor N,N-dimethyl-

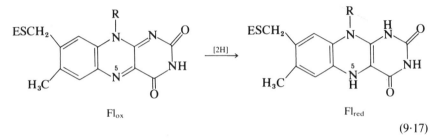

$$(9 \cdot 17)$$

propargylamine inactivates monoamine oxidases by alkylating the flavin on N-5.[25] A likely mechanism for the reaction is the Michael addition of the N-5 of the reduced flavin to the acetylenic carbon:[26]

$$Fl_{ox}$$

$$Fl_{red}$$

$$HC{\equiv}C{-}CH_2{-}N(CH_3)_2 \longrightarrow HC{\equiv}C{-}CH{=}\overset{+}{N}(CH_3)_2 \longrightarrow$$

$$Fl$$
$$|$$
$$HC{=}CH{-}CH{=}\overset{+}{N}(CH_3)_2 \qquad (9 \cdot 18)$$

　　　One drug that is used to treat both Parkinson's disease and depression is (−)deprenyl. It, too, is an acetylenic suicide inhibitor that inhibits the enzyme by binding covalently to the flavin.[27]

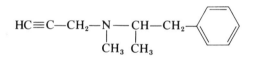

(−)Deprenyl

References

1 G. Schoellmann and E. Shaw, *Biochem. Biophys. Res. Commun.* **7**, 36 (1962); *Biochemistry* **2**, 252 (1963).
2 B. R. Baker, W. W. Lee, E. Tong, and L. O. Ross, *J. Am. Chem. Soc.* **83**, 3713 (1961).
3 L. Wofsy, H. Metzger, and S. J. Singer, *Biochemistry* **1**, 1031 (1961).
4 E. Shaw, *The Enzymes* **1**, 91 (1970).
5 D. Glick, *Biochemistry* **7**, 3391 (1968).
6 E. L. O'Connell and I. A. Rose, *J. Biol. Chem.* **248**, 2225 (1973).
7 P. J. Shaw and H. Muirhead, *FEBS Lett.* **65**, 50 (1976).
8 R. L. Lundblad and W. H. Stein, *J. Biol. Chem.* **244**, 154 (1969).
9 A. Singh, E. R. Thornton, and F. H. Westheimer, *J. Biol. Chem.* **237**, 3006 (1962).
10 V. Chowdhry and F. M. Westheimer, *Ann. Rev. Biochem.* **48**, 293 (1979).
11 J. R. Knowles, *Accts. Chem. Res.* **5**, 155 (1972).
12 K. Bloch, *Accts. Chem. Res.* **2**, 193 (1969).
13 R. R. Rando, *Science* **185**, 320 (1974).
14 R. H. Abeles and A. L. Maycock, *Accts. Chem. Res.* **9**, 313 (1976).
15 C. Walsh, *Tetrahedron* **38**, 871 (1982).
16 F. M. Miesowicz and K. Bloch, *J. Biol. Chem.* **254**, 5868 (1979).
17 E. H. White, L. W. Jelinski, I. R. Politzer, B. R. Branchini, and D. F. Roswell, *J. Am. Chem. Soc.* **103**, 4231 (1981).
18 M. Morisaki and K. Bloch, *Bioorg. Chem.* **1**, 188 (1971).
19 H. Gehring, R. R. Rando, and P. Christen, *Biochemistry* **16**, 4832 (1977).
20 R. B. Silverman and R. H. Abeles, *Biochemistry* **15**, 4718 (1976).
21 E. W. Miles and A. Meister, *Biochemistry* **6**, 1735 (1967).
22 Y. Marin and M. Okamoto, *Biochem. Biophys. Res. Commun.* **50**, 1061 (1973).
23 J. M. Manning, N. E. Merrifield, W. M. Jones, and E. C. Gotschlich, *Proc. Natn. Acad. Sci. USA* **71**, 417 (1974).
24 E. Wang and C. T. Walsh, *Biochemistry* **17**, 1313 (1978).
25 A. L. Maycock, R. H. Abeles, J. L. Salach, and T. P. Singer, *Biochemistry* **15**, 114 (1976).
26 R. H. Abeles, in *Enzyme-activated irreversible inhibitors* (N. Seiler, M. J. Jung, and J. Koch-Weser, Eds.), Elsevier North-Holland, p. 4 (1978).
27 M. B. H. Youdim and J. I. Salach, in *Enzyme-activated irreversible inhibitors* (N. Seiler, M. J. Jung, and J. Koch-Weser, Eds.), Elsevier North-Holland, p. 235 (1978).

Further reading

G. E. Means and R. E. Feeney, *Chemical modification of proteins*, Holden-Day (1971).
Methods in enzymology, Vol. 46 (W. B. Jakoby and M. Wilchek, Eds.), Academic Press (1977).
Theory and practice in affinity techniques (P. V. Dandaram and F. Eckstein, Eds.), Academic Press (1978).
Enzyme-activated irreversible inhibitors (N. Seiler, M. J. Jung, and J. Koch-Weser, Eds.), Elsevier North-Holland (1978).

10

Cooperative ligand binding, allosteric interactions, and regulation

A. Positive cooperativity

Many proteins are composed of subunits and have multiple ligand-binding sites. In some cases the ligand-binding curves do not follow the Michaelis-Menten equation (Chapter 6, section D), but instead are *sigmoid*. This is illustrated in Figure 10·1, where the degree of saturation of hemoglobin with oxygen is plotted against the pressure of oxygen (curve *b*). The sigmoid curve may be compared with the hyperbolic curve that is expected from the Michaelis-Menten equation and that is found for the binding to myoglobin (curve *a*). Sigmoid curves are characteristic of the *cooperative* binding of ligands to proteins that have multiple binding sites. Hemoglobin, for example, is composed of four polypeptide chains, each of which is similar to the single polypeptide chain of myoglobin. The hemoglobin binding curve may be fitted to four successive binding constants, as per the Adair equation (Table 10·1).[1] The surprising feature of this is that the affinity for the fourth oxygen that binds is between a hundred and a thousand times higher than for the first oxygen. The increase in affinity with increasing saturation cannot be explained by four noninteracting sites of differing affinities. If this were the case, the high-affinity sites would fill first, so that the partially ligated molecules would be of lower affinity than the free deoxyhemoglobin. The increase in affinity with increasing saturation is due instead to the sites *interacting*, so that binding at one site causes an increase in affinity at another.

A similar cooperativity of substrate binding is found to occur with some enzymes leading to sigmoid plots of v against [S]. These enzymes are usually the metabolic regulatory or control enzymes whose activities are subject to feedback inhibition or activation. The terminology of cooperative interactions stems from the studies on control.[2] The enzymes are termed *allosteric* (Greek, *allos* = other, *stereos* = solid or space),

263

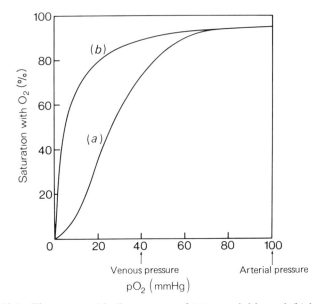

FIGURE 10·1. The oxygen-binding curves of (a) myoglobin and (b) hemoglobin.

since the allosteric effector (the inhibitor or activator) is generally structurally different from the substrates and binds at its own separate site
away from the active site. The term homotropic is sometimes used to
denote the interactions among the identical substrate molecules, whereas
the term heterotropic is used for the interactions of the allosteric effectors
with the substrates.

TABLE 10·1. *Adair constants for the binding of O_2 to hemoglobin*[a]

2,3-Diphosphoglycerate (mM)	K_1	K_2	K_3	K_4
	(mmHg)			
0	0.024	~0.074	~0.086	7.4
2.0	0.01	~0.023	~0.008	11.2

[a] At 25°C, pH 7.4, and 0.1-M NaCl. The Adair equation describes the following:

$$Hb \overset{K_{1,O_2}}{\rightleftharpoons} HbO_2 \overset{K_{2,O_2}}{\rightleftharpoons} Hb(O_2)_2 \overset{K_{3,O_2}}{\rightleftharpoons} Hb(O_2)_3 \overset{K_{4,O_2}}{\rightleftharpoons} Hb(O_2)_4$$

where

$$K_1 = \frac{[HbO_2]}{[Hb][O_2]}, \quad K_2 = \frac{[Hb(O_2)_2]}{[HbO_2][O_2]}, \quad \text{etc.}$$

[From I. Tyuma, K. Imai, and K. Shimizu, *Biochemistry* **12**, 1491 (1973).]

B. Mechanisms of allosteric interactions and cooperativity

The original attempts to explain the mechanism of cooperativity were based on hemoglobin. The best way to understand them is to consider the structures of deoxyhemoglobin and oxyhemoglobin. Hemoglobin is composed of two pairs of very similar chains, α and β, arranged in a symmetrical tetrahedral manner. The oxygen-binding sites, the hemes, are too far apart to interact directly. When deoxyhemoglobin is oxygenated, the overall tetrahedral symmetry is maintained but there are changes in quaternary structure (Figure 10·2).[3] The two α subunits stay together but rotate relative to each other through 15°. There are no gross changes in tertiary structure that can be seen at low resolution. (Later, high-resolution studies revealed the small but important tertiary structural changes caused by ligand binding.) It is known from the binding measurements that deoxyhemoglobin has a low affinity for oxygen, whereas oxyhemoglobin has a high affinity.

1. The Monod-Wyman-Changeux (MWC) concerted mechanism[4]

J. Monod, J. Wyman, and J.-P. Changeux showed that cooperativity can be accounted for in a very simple and elegant manner by assuming that a small fraction of deoxyhemoglobin exists in the quaternary oxy structure that binds oxygen more strongly. On the binding of 1 mol of oxygen, the concentration of the oxy structure is increased, since oxygen binds preferentially to it. When enough oxygen molecules are bound, the oxy form is sufficiently stabilized to be the major structure in solution, so that subsequent binding is strong. In order to simplify the mathematical equations and the physical concepts, the quaternary structure of the molecule is assumed to be always symmetrical; a particular partly ligated molecule is assumed to be either in the oxy state or in the deoxy state, so that mixed states do not occur (Figure 10·3). For this reason, the MWC model is often described as "concerted," "all or none," or "two-state."

The model may be generalized to cover other allosteric proteins if the following assumptions are made:

1. The protein is an oligomer.
2. The protein exists in either of two conformational states, T (= tense), the predominant form when the protein is unligated, and R (= relaxed); the two states are in equilibrium. They differ in the energies and numbers of bonds between the subunits, with the T state being constrained compared with the R state.
3. The T state has a lower affinity for ligands.
4. All binding sites in each state are equivalent and have identical binding constants for ligands (the symmetry assumption).

Oxy-

Deoxy-

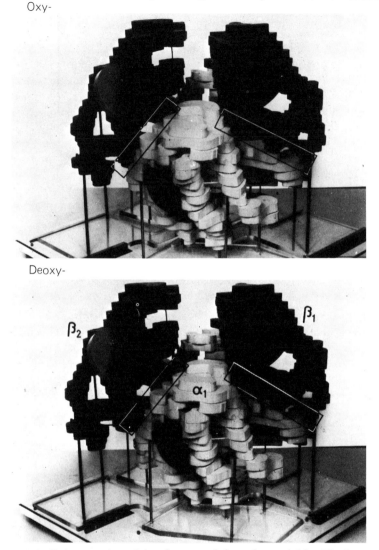

FIGURE 10·2. Balsa wood models of oxy- and deoxyhemoglobin. The hemes are represented by disks. Note the increased separation of the β subunits on deoxygenation.

The sigmoid binding curve of any allosteric protein can be calculated by using just three parameters: L, the allosteric constant, which is equal to the ratio [T]/[R] for the unligated states; and K_T and K_R, the dissociation constants for each site in the T and R states, respectively.

a. An explanation of control through allosteric interactions
The first achievement of the MWC theory was to provide a theoretical curve (based only on L, K_R, and K_T) that fitted the oxygen-binding curve of hemoglobin with high precision. But, even more impressively, the theory provided a very simple explanation for control. Monod, Wyman, and Changeux noted that it is a common feature of allosteric enzymes to exhibit cooperativity in v-vs.-[S] plots in the absence of their allosteric activators or inhibitors. They reasoned that if this cooperativity is due to binding and an R–T equilibrium, then an explanation of control could be that the effector alters the R–T ratio by preferentially stabilizing one of the forms. An activator functions by binding to the R state and increasing its concentration. An inhibitor binds preferentially to the T state and so causes the transition to the R state to be more difficult.

In extreme cases, an activator will displace the R–T equilibrium to such an extent that the R state predominates. Cooperativity is then abolished so that Michaelis-Menten kinetics will hold. This has been verified experimentally for several enzymes (Figure 10·4).

It is also predicted that an allosteric inhibitor should bind noncooperatively to an enzyme that binds its substrates cooperatively, since the inhibitor binds to the predominant T state. The converse should be true for activators binding to multiple binding sites in the R state.

The feature of the MWC model that is most open to criticism is the symmetry assumption. Clearly, changes in the constraints at the subunit

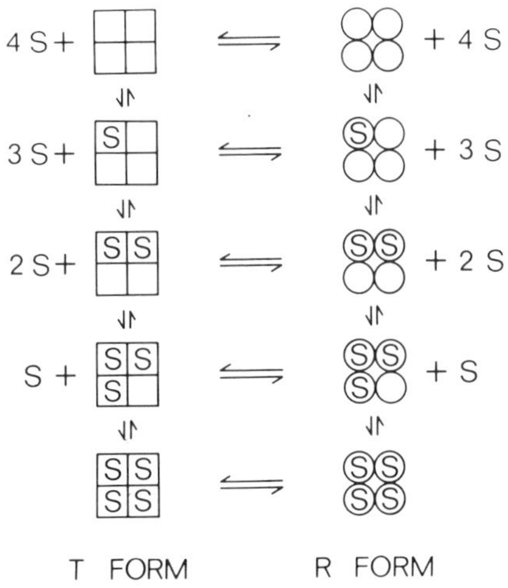

FIGURE 10·3. The MWC model for the binding of ligands to a tetrameric protein. S = substrate; T = tense conformation; and R = relaxed conformation.

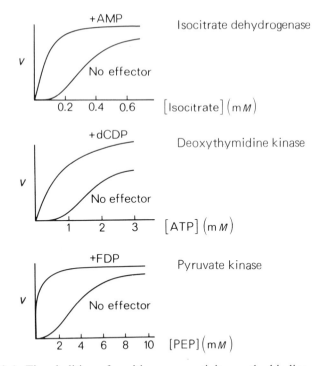

FIGURE 10·4. The abolition of positive cooperativity on the binding of allosteric effectors to some enzymes. Note the dramatic increases in activity at low substrate concentrations on the addition of adenosine monophosphate to isocitrate dehydrogenase, of deoxycytosine diphosphate to deoxythymidine kinase, and of fructose 1,6-diphosphate to pyruvate kinase; this shows how the activity may be "switched on" by an allosteric effector (PEP = phosphoenolpyruvate). [From J. A. Hathaway and D. E. Atkinson, *J. Biol. Chem.* **238**, 2875 (1963); R. Okazaki and A. Kornberg, *J. Biol. Chem.* **239**, 275 (1964); R. Haeckel, B. Hess, W. Lauterhorn, and K.-H. Würster, *Hoppe-Seyler's Z. Physiol. Chem.* **349**, 699 (1968).]

interfaces on ligand binding must be mediated by changes in the tertiary structure of the subunit doing the binding. Because the model minimizes the number of intermediate states, it is only an approximation of reality. But this is also its virtue. The model provides a simple framework in which to rationalize experiments and explain phenomena. The predictions, such as the switch from sigmoid to Michaelis-Menten kinetics in control enzymes at sufficiently high activator concentrations, do not, in any case, depend on the intermediate states. Also, despite its simplicity, the theory accounts for the binding curves of oxygen to a wide series of mutant hemoglobins (see section D2).

The MWC is basically a structural theory. The hypothesis that there are constraints between the subunits in the T state has provided the basis of much of the structural work by M. F. Perutz and others in elucidating the nature and the energies of these constraints in hemoglobin.

b. K systems and V systems

Control in allosteric enzymes may take two extreme forms.[4] In K (= binding) systems, the ones discussed so far, the substrate and the effector molecules have different affinities for the R and T states. The binding of an effector alters the affinity of the enzyme for the substrate, and *vice versa*. The R and the T states can have the same intrinsic value of k_{cat}, and activity is modulated by changes in affinity for the substrate. In V (= rate) systems, the substrate has the same affinity for both states, but one state has a much higher value of k_{cat}. The effector molecule binds preferentially to one of the two states, and so modulates activity by changing the equilibrium position between the two. As the substrate binds equally well to both states, there is no cooperative binding of substrate, and the binding of the allosteric effector does not alter the affinity of the enzyme toward the substrate.

2. The Koshland-Némethy-Filmer (KNF) sequential model[5]

The KNF model avoids the assumption of symmetry but uses another simplifying feature. It assumes that the progress from T to the ligand-bound R state is a sequential process. The conformation of each subunit changes in turn as it binds the ligand, and there is no dramatic switch from one state to another (Figure 10·5). Whereas the MWC model uses a quaternary structural change, the KNF uses a series of tertiary structural changes.

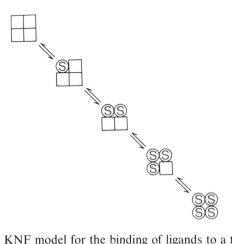

FIGURE 10·5. The KNF model for the binding of ligands to a tetrameric protein.

The two assumptions in the KNF model are:

1. In the absence of ligands, the protein exists in one conformation.
2. Upon binding, the ligand induces a conformational change in the sub-unit to which it is bound. This change may be transmitted to neighboring vacant subunits via the subunit interfaces.

This model embodies Koshland's earlier idea of *induced fit*, according to which the binding of a substrate to an enzyme may cause conformational changes that align the catalytic groups in their correct orientations.

Using these assumptions, it is possible to describe the binding of oxygen by four successive binding constants. This is formally equivalent to the Adair equation; the KNF model may be considered as a molecular interpretation of that equation. In general, the number of constants required is equal to the number of binding sites, unlike the situation in the MWC model, which always uses three.

In sacrificing simplicity, the KNF model is more general and is probably a better description of some proteins than is the MWC model. In turn, the explanation of phenomena is often somewhat more complicated.

3. The general model[6]

M. Eigen has pointed out that the MWC and KNF models are limiting cases of a general scheme involving all possible combinations (Figure 10·6). The scheme is more complicated than the "chessboard" illustrated in the figure, since, for reasons of symmetry, there are 44 possible states

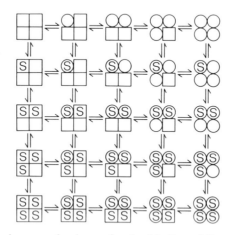

FIGURE 10·6. Eigen's general scheme for the binding of ligands to a tetrameric protein. The columns on the extreme left and right represent the MWC simplification. The diagonal from top left to bottom right represents the KNF simplification.

for the hemoglobin case. The KNF model moves across the chessboard like a bishop confined to the long diagonal, and the MWC like a rook confined to the perimeters. The general model, combining both the MWC and KNF extremes, along with dissociation of the subunits, etc., has been analyzed.[7] But the results are too complex for general use, and it is far more convenient to interpret experiments in terms of the MWC and KNF simplifications.

C. Negative cooperativity and half-of-the-sites reactivity[8,9]

Some enzymes bind successive ligand molecules with decreasing affinity. As pointed out before, this can be explained by the presence of binding sites of differing affinities so that the stronger are occupied first. But in many cases, decreasing affinity is found with oligomeric enzymes composed of identical subunits. In two examples, the tyrosyl-tRNA synthetase[10] and the glyceraldehyde 3-phosphate dehydrogenase[11] from *Bacillus stearothermophilus*, x-ray diffraction studies on the crystalline enzymes show that the subunits are arranged symmetrically so that all sites are initially equivalent. Yet, the dimeric tyrosyl-tRNA synthetase binds only 1 mol of tyrosine tightly; the binding of the second remains undetected even at millimolar concentrations of tyrosine.[12,13] The 4 mol of NAD^+ that bind to the rabbit muscle glyceraldehyde 3-phosphate dehydrogenase do so with increasing dissociation constants of $<10^{-11}$, $<10^{-9}$, 3×10^{-7}, and 3×10^{-5} M.[14,15] Similar changes in affinity are found for the bacterial enzyme. This antagonistic binding of molecules is known as negative cooperativity or anticooperativity.

Negative cooperativity cannot be accounted for by the MWC theory; the binding of the first ligand molecule can only stabilize the high-affinity state and cannot increase the proportion of the T state. The KNF model accounts for negative cooperativity by the binding of the ligand to one site, causing a conformational change that is transmitted to a vacant subunit (assumption 2 of the KNF model). Negative cooperativity is thus a diagnostic test of the KNF model.

A related phenomenon is half-of-the-sites or half-site reactivity, by which an enzyme containing $2n$ sites reacts (rapidly) at only n of them (Table 10.2). This can be detected only by pre–steady state kinetics. The tyrosyl-tRNA synthetase provides a good example, in that it forms 1 mol of enzyme-bound tyrosyl adenylate with a rate constant of 18 s^{-1}, but the second site reacts 10^4 times more slowly.[12,16]

Half-of-the-sites reactivity is inconsistent with the simple MWC theory, since the sites lose their equivalence and symmetry is lost.

A cautionary note must be injected at this point. The diagnosis of half-of-the-sites reactivity depends on knowledge of the exact concentra-

TABLE 10·2. *Some enzymes showing half-of-the-sites reactivity*

Enzyme	Reaction	Number of subunits	Ref.
Acetoacetate decarboxylase	Inactivation of active-site lysine	12	1
Aldolase	Partial reaction with fructose 6-phosphate	2	2
Aminoacyl-tRNA synthetases (some)	Biphasic formation of aminoacyl adenylate	2	3
Cytidine triphosphate synthetase	Stoichiometry of irreversible inhibition	4	4
Glyceraldehyde 3-phosphate dehydrogenase	Reaction with nonphysiological substrates; *but* full site reactivity with physiological	4	5
Aspartate transcarbamoylase	CTP binding Carbamoyl phosphate binding	6 (regulatory) 6 (catalytic)	5
Glutamine synthetase	Irreversible inhibition	8	6

1   D. E. Schmidt, Jr., and F. H. Westheimer, *Biochemistry* **10**, 1249 (1971), and references therein.
2   O. Tsolas and B. L. Horecker, *Archs. Biochem. Biophys.* **173**, 577 (1976).
3   R. S. Mulvey and A. R. Fersht, *Biochemistry* **15**, 243 (1976).
4   A. Levitzki, W. B. Stallcup, and D. E. Koshland, *Biochemistry* **10**, 3371 (1971).
5   F. Seydoux, O. P. Malhotra, and S. A. Bernhard, *C. R. C. Crit. Rev. Biochem.* 227 (1974).
6   S. S. Tate and A. Meister, *Proc. Natn. Acad. Sci. USA* **68**, 781 (1971).

tion of binding sites on the enzyme. This depends on determination of the precise concentration of the protein and its purity. Similarly, a heterogeneous preparation of enzyme containing molecules of differing affinities for the ligand will give a binding curve similar to that of negative cooperativity. Failure to check these points has led to spurious reports about subunit interactions. A further artifact that may be misinterpreted as half-of-the-sites reactivity occurs in the reactions of the lactate and alcohol dehydrogenases, and is discussed in Chapter 15: an enzyme-bound intermediate does not appear to accumulate fully because of an unfavorable equilibrium constant.

D. Quantitative analysis of cooperativity

1. The Hill equation: A measure of cooperativity[17]

Consider a case of completely cooperative binding: An enzyme contains n binding sites and all n are occupied simultaneously with a dissociation constant K. We have

$$E + nS \rightleftharpoons ES_n \qquad (10\cdot1)$$

and

$$K = \frac{[E][S]^n}{[ES_n]} \qquad (10\cdot2)$$

The degree of saturation Y is given by

$$Y = \frac{[ES_n]}{[E]_0} \qquad (10\cdot3)$$

and

$$1 - Y = \frac{[E]}{[E]_0} \qquad (10\cdot4)$$

Equations 10·2 and 10·4 may be manipulated to give

$$\log \frac{Y}{1 - Y} = n \log[S] - \log K \qquad (10\cdot5)$$

A similar equation called the Hill plot (equation 10·6) is found to describe satisfactorily the binding of ligands to allosteric proteins in the region of 50% saturation (10 to 90%) (Figure 10·7). Outside this region, the experimental curve deviates from the straight line. The value of h obtained from the slope of equation 10·6 in the region of 50% saturation is known as the *Hill constant*. It is a measure of cooperativity. The higher h is, the higher the cooperativity. At the upper limit, h is equal to the number of binding sites. If $h = 1$, there is no cooperativity; if $h > 1$, there is positive cooperativity; if $h < 1$, there is negative cooperativity.

$$\log \frac{Y}{1 - Y} = h \log[S] - \log K \qquad (10\cdot6)$$

The Hill equation may be extended to kinetic measurements by replacing Y by v, as in

$$\log \frac{v}{V_{max} - v} = h \log[S] - \log K \qquad (10\cdot7)$$

2. The MWC binding curve[4]

Let the dissociation constant of the ligand from the R state be K_R and that from the T state be K_T. Then c is defined by

$$c = \frac{K_R}{K_T} \qquad (10\cdot8)$$

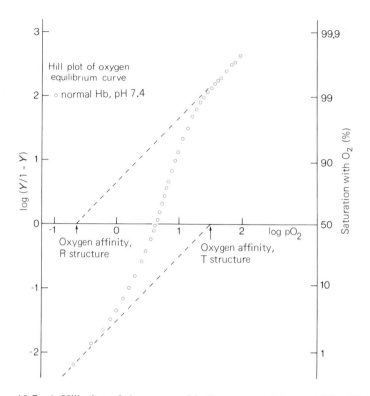

FIGURE 10·7. A Hill plot of the oxygen-binding curve of hemoglobin. [From J. V. Kilmartin, K. Imai, and R. T. Jones, in *Erythrocyte structure and function*, Alan R. Liss, p. 21 (1975).]

The R state with x ligand molecules bound is termed R_x, and the equivalent T state, T_x. The allosteric constant L is then defined by

$$L = \frac{[T_0]}{[R_0]} \qquad (10·9)$$

The fraction of the protein that is bound with ligand, Y, is calculated from the mass balance equations. Recall that it is necessary to "statistically correct" the binding constants. For example, the dissociation constant for the first O_2 binding to hemoglobin is $K_T/4$, since there are four sites to which the ligand may bind, but there is only one site for dissociation when it is bound. Similarly, the dissociation constant for the second molecule that binds is $2K_T/3$, since there are three sites to which it can bind, but once it is bound there are two bound sites that can dissociate.

Y for a protein containing n sites is given by

$$Y = \frac{([R_1] + 2[R_2] + \cdots + n[R_n]) + ([T_1] + 2[T_2] + \cdots + n[T_n])}{n([R_0] + [R_1] + \cdots + [R_n] + [T_0] + [T_1] + \cdots + [T_n])} \quad (10\cdot10)$$

Solving equation $10\cdot10$ in terms of L and c, and using for convenience

$$\alpha = \frac{[S]}{K_R} \quad (10\cdot11)$$

gives

$$Y = \frac{Lc\alpha(1 + c\alpha)^{n-1} + \alpha(1 + \alpha)^{n-1}}{L(1 + c\alpha)^n + (1 + \alpha)^n} \quad (10\cdot12)$$

The fraction of protein in the R state, R, may be similarly derived:

$$R = \frac{(1 + \alpha)^n}{L(1 + c\alpha)^n + (1 + \alpha)^n} \quad (10\cdot13)$$

According to the MWC theory, the saturation curve for any oligo-meric protein composed of n protomers is defined by only three unknown parameters and the concentration of ligand (i.e., L, K_R, K_T, and $[S]$, with the latter three disguised as c and α in equation $10\cdot12$).

Some values of L and c obtained from the computer fitting of equation $10\cdot12$ to the binding curves of some proteins are listed in Table $10\cdot3$. (Note: This does not imply that the structural changes follow the MWC model, since the KNF model also predicts the same binding curve.

a. The dependence of the Hill constant on L and c[18]

The value of h may be calculated from a computer analysis of equation $10\cdot12$. It is found that a plot of h against L at a constant value of c gives

TABLE $10\cdot3$. *Allosteric constants for some proteins*

Protein	Ligand	Number of binding sites (n)	Hill constant (h)	L	c	Ref.
Hemoglobin	O_2	4	2.8	3×10^5	0.01	1
Pyruvate kinase (yeast)	Phosphoenol-pyruvate	4	2.8	9×10^3 [a]	0.01[a]	2
Glyceraldehyde 3-phosphate dehydrogenase (yeast)	NAD$^+$	4	2.3	60	0.04	3

[a] Estimated by the author.

1 S. J. Edelstein, *Nature, Lond.* **230**, 224 (1971).
2 R. Haeckel, B. Hess, W. Lauterhorn, and K.-H. Würster, *Hoppe-Seyler's Z. Physiol. Chem.* **349**, 699 (1968); H. Bischofberger, B. Hess, and P. Röschlau, *Hoppe-Seyler's Z. Physiol. Chem.* **352**, 1139 (1971).
3 K. Kirschner, E. Gallego, I. Schuster, and D. Goodall, *J. Molec. Biol.* **58**, 29 (1971).

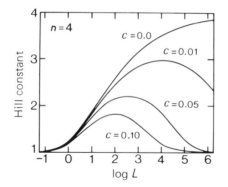

FIGURE 10·8. Variation of the Hill constant with L for a tetrameric protein. [From M. M. Rubin and J.-P. Changeux, *J. Molec. Biol.* **21**, 265 (1966).]

a bell-shaped curve (Figure 10·8). The value of h is equal to 1 for L much greater or lower than c, and is a maximum when

$$L = c^{-n/2} \tag{10·14}$$

(where n is the number of binding sites). The reason for this behavior is that when L is low there is initially sufficient protein in the R state to give good binding, and when L is very high, the concentration of the R state is too small to contribute significantly to binding.

The bell-shaped curve is of particular interest in the analysis of how structural changes in a protein affect L. The Hill constants of a wide series of modified and mutant hemoglobins fit such a curve (Figure 10·9).[19]

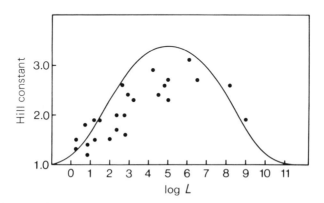

FIGURE 10·9. Variation of the Hill constant with L for the binding of oxygen to various mutant hemoglobins. [From J. M. Baldwin, *Progr. Biophys. Molec. Biol.* **29**, 3 (1975).]

3. The KNF binding curve

The MWC model gives a simple equation for Y because of the assumption of only two dissociation constants. The KNF model requires a different dissociation constant for every intermediate state, and there is no simple general equation for Y. As many variables are required as there are binding sites.

4. Diagnostic tests for cooperativity, and MWC vs. KNF mechanisms

Cooperativity is determined from the value of h in the Hill plot or from the characteristic deviations in the straightforward saturation curves or Scatchard plots (Figure 10·10). The finding of negative cooperativity excludes the simple MWC theory, but positive cooperativity is consistent with both models. Analysis of the shape of the binding curve is generally also ambiguous, since the KNF and MWC models often predict similar shapes. In theory, measurement of the *rates* of ligand binding can distinguish between the two models.[6] The MWC model, for example, predicts fewer relaxation times since fewer states are involved. This approach has been applied with success to the glyceraldehyde 3-phosphate dehydrogenase from yeast, for which it has been shown that the kinetics of NAD^+ binding are consistent with the positive cooperativity due to an MWC model.[20] On the other hand, ligand-binding studies on hemoglobin suggest the importance of intermediate states. For example, the KNF model describes the overall binding curve of oxygen and carp hemoglobin better than does the MWC, since careful analysis shows that there are stages

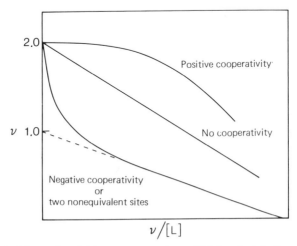

FIGURE 10·10. Plots of stoichiometry ν against $\nu/[L]$ for the binding of ligand (L) to a dimeric protein.

of negative cooperativity.[21] But sophisticated spectroscopic techniques have shown that the β subunits have a slightly higher oxygen affinity than do the α.[22] Such pre-existing asymmetry can, of course, give rise to apparent negative cooperativity without invoking the KNF scheme, as was discussed in section C.

The choice of model often depends on the experiments involved. Workers in the area of, say, the effects of structural changes on the oxygen affinity and Hill constant for hemoglobin prefer the MWC model because it is essentially a *structural* theory. It provides a simple framework for the prediction and interpretation of experiments. Application of the theory to the Hill constant and other *equilibrium* measurements gives very acceptable results. Kineticists prefer the KNF model, since the kinetic measurements are more sensitive to the presence of intermediates. There are more variables in the KNF theory and there is more flexibility in fitting data.

E. Molecular mechanism of cooperative binding to hemoglobin

1. The physiological importance of the cooperative binding of oxygen

Positive cooperativity is not a device for increasing the affinity of hemoglobin for oxygen; the association constant of free hemoglobin chains for oxygen is far higher than that for the binding of the first mole of oxygen to deoxyhemoglobin, and is about the same as for the binding to oxyhemoglobin. It is instead a means of lowering the oxygen affinity over a very narrow range of pressures, so as to allow the hemoglobin to be saturated with oxygen in the lungs and then to unload about 60% of the oxygen to the tissues. This can be done over a relatively narrow range of oxygen pressures, because of the steepness of the curve of saturation with oxygen against oxygen pressure in the region of 50% saturation. A simple Michaelis-Menten curve would require a far greater range of pressures. (The Hill constant is a measure of the steepness of the saturation curve at 50% saturation; h is close to 3 for the sigmoid curve of hemoglobin, but would be 1 for a hyperbolic curve.)

2. Atomic events in the oxygenation of hemoglobin[3,23]

The two pairs of α and β subunits are arranged tetrahedrally around a twofold axis of symmetry (see Figure 10·2). The two α subunits make few contacts with each other, as is also the case with the two β subunits. The main interactions are across the $\alpha_1\beta_2$ and the $\alpha_1\beta_1$ interfaces. There is a cavity at the center of the molecule through which the axis passes. Organic

phosphates such as 2,3-diphosphoglycerate, which is an allosteric effector, bind in this cavity in deoxyhemoglobin in a 1 : 1 stoichiometry (Figure
10·11).[24-26] The negatively charged organic phosphate sits between the
two β subunits, forming four salt linkages with each (with Val-1, His-2,
Lys-82, and His-143). In addition, deoxyhemoglobin is stabilized by four
pairs of salt bridges formed by the C-terminal residues of the α and β
chains: two pairs between the interfaces of the two α chains and two pairs
between the α and β chains.

On oxygenation, the subunits rotate relative to each other by about
15°. The hemes change their angles of tilt by a few degrees and the helical
regions move 2 to 3 Å (0.2 to 0.3 nm) relative to each other. The salt
bridges between the subunits are broken. The α-amino groups of Val-1
of each β subunit in the central cavity move apart by 4 Å as the cavity
narrows, expelling any organic phosphate that is bound. There is also a
decrease in the hydrophobic area that is in contact at the interfaces of
the subunits.[27] The "constraints" that were predicted between the subunits in the T state by Monod, Wyman, and Changeux are thus the salt
bridges and additional hydrophobic interactions.

This explains why deoxyhemoglobin has a low affinity for oxygen

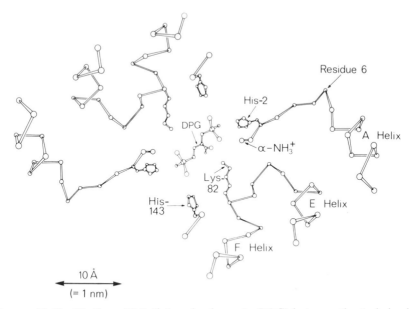

FIGURE 10·11. Binding of 2,3-diphosphoglycerate (DPG) between the β chains in
the central cavity of human hemoglobin. [From A. Arnone and M. F. Perutz,
Nature, Lond. **249**, 34 (1974).]

compared with myoglobin or artificially prepared single chains.[3] On the binding of oxygen, energy must be expended to disrupt favorable interactions in the quaternary T structure. Salt bridges are broken and hydrophobic surfaces are exposed. The mode of action of the allosteric effectors is also clear. Organic phosphates bind strongly to a well-defined site in the T state and stabilize it relative to the R state.

The breaking of the salt bridges on oxygenation also explains the Bohr effect.[28] On oxygenation at physiological pH, there is a release of about 0.7 proton per heme. This is caused by the raised pK_a values of weak bases such as imidazoles and α-amino groups when they are bound to carboxylate ions (Chapter 4). On disruption of the bridges, the pK_a's drop to the normal values, releasing protons.

It seems remarkable that the binding of O_2 causes such extensive structural changes, but nature has provided an ingenious trigger mechanism, based mainly on a small movement of the iron relative to the plane of the porphyrin.[29–32] The movement is a result of the change from 5 to 6 coordination and the accompanying change in the spin of the iron atom (Figure 10·12). Oxyhemoglobin is diamagnetic, with a short Fe—N_P (P = porphyrin) bond distance of 1.99 Å; van der Waals repulsion between the porphyrin nitrogens and the proximal imidazole (His-F8 in Figure 10·12) is balanced by the repulsion between those nitrogens and the bound oxygen. As a result, the iron atom is kept close to or in the porphyrin plane. Deoxyhemoglobin is paramagnetic with a spin of S = 2. Repulsion between the occupied $d_{x^2-y^2}$ orbitals of the iron and the π orbitals of the porphyrin lengthens Fe—N_P to 2.07 Å, longer by 0.3 to 0.4 Å than the distance of the porphyrin nitrogens from the center of the ring. Moreover, van der Waals repulsion between the porphyrin nitrogens and the proximal imidazole is no longer balanced by a sixth ligand. Both factors conspire to pull the iron atoms out of the porphyrin plane by 0.4 Å and to dome the porphyrins toward the proximal histidines. The movements of the iron atoms cause movements of the helixes to which they are linked, and these in turn are transmitted to the nearby subunit boundaries, thus triggering the change in quaternary structure. A second trigger is provided by a valine that obstructs the oxygen-binding site in the β subunits of the T structure. When this valine gives way, it sets in motion changes in tertiary structure that are concerted with those caused by the movement of the iron relative to the porphyrin plane.

The relationship between the structural and the energetic changes has been summarized.[23] It is in remarkable agreement with the original MWC hypothesis.[4] The oligomer has two different modes of close packing of subunit interfaces that can lead to two different quaternary structures. Deoxyhemoglobin is more stable in the T state because there are more hydrophobic, ionic, and van der Waals interactions. On the binding of

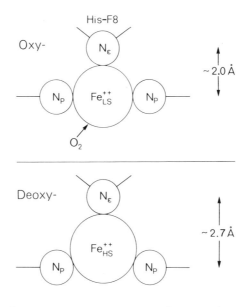

Coordination of iron in oxy- (low-spin)
and deoxy- (high-spin) hemoglobin

FIGURE 10·12. Movement of the imidazole ring of His-F8 on the binding of oxygen to deoxyhemoglobin. Steric repulsion between Nê and the porphyrin nitrogens (N_P), and the slightly too large radius of Fe^{2+} (high-spin), force it out of the porphyrin plane. On the binding of O_2, the steric repulsion between it and the porphyrin in the opposite direction to that of Nê, and the decrease in the radius of Fe^{2+}, force the iron atom to move close to the plane of the porphyrin ring, dragging the His-F8 with it.

ligands, strains are set up in the T state that eventually tip the energy balance in favor of the R state.

3. Chemical models of hemes

Chemical model systems, in particular the "picket fence"[33] (Figure 10·13), have been particularly useful in the studies on hemoglobin. The small chemical models may be crystallized and their structures determined with far higher precision than the structures of the actual proteins can be. Binding measurements may be made on the models without the complications that arise from the protein structure. The precise displacement of the iron atom from the heme and the geometry of iron–ligand bonds were first measured in the models.[29,33,34] The Fe—O_2 bond is bent, whereas the Fe—CO bond is linear (structures 10·15). The Fe—O_2 bond

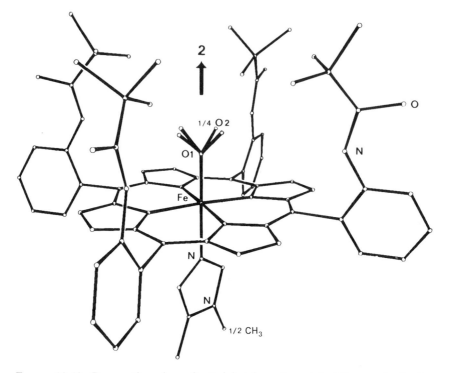

FIGURE 10·13. Perspective view of a "picket fence" model of the porphyrin ring system of hemoglobin. Note that the oxygen molecule is bound at an angle (there is a four-fold statistical distribution of the terminal O because of rotation about the Fe—O bond). [From J. P. Collman, R. R. Gagne, C. A. Reed, W. T. Robinson, and G. A. Rodley, *Proc. Natn. Acad. Sci. USA* **71**, 1326 (1974).]

is bent at 156° in hemoglobin.[35] Binding studies on models show a far

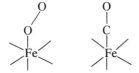   (10·15)

higher affinity for CO than is found for myoglobin or hemoglobin.[36] The proteins have evolved to discriminate against the binding of CO by a combination of two factors. (1) There is steric hindrance against the binding of linearly bound ligands such as CO (more so in myoglobin than in hemoglobin, where the binding pocket is more open).[35] (2) Neutron diffraction studies on oxymyoglobin have located a hydrogen bond between the terminal oxygen of the Fe—O_2 (which is formally the superoxide

$Fe^{3+}O_2^-$) and a histidine.[37] Hemoglobin also has the correct geometry for forming this bond.[35]

F. Regulation of metabolic pathways

The rate of chemical flux through a metabolic pathway is often regulated by controlling the activity of a key enzyme in the pathway. The natural point of control is the enzyme whose catalytic activity is the *rate-limiting step* in the pathway. Such an enzyme may be identified by examination in the steady state *in vivo* of the ratio of the concentrations of its products to those of its substrates—the *mass action ratio*. Since the earlier enzymes in the pathway are producing the substrates of the rate-limiting enzyme faster than it can cope with them, and since its products are being rapidly consumed by the following enzymes in the pathway, the rate-limiting reaction is not in equilibrium. The mass action ratio is therefore less than that expected from the equilibrium concentrations, often by 2 to 4 orders of magnitude. In contrast, the mass action ratios of metabolic enzymes whose activities are not rate-limiting tend to be close to the equilibrium values: enzymes that have very high activities catalyze the forward and reverse reactions of their steps at rates far higher than that of the rate-limiting enzyme, but the *net* rate through each step—i.e., the rate of the forward reaction minus the rate of the reverse reaction—is the same for each step.

The activity of an enzyme can be controlled by regulation of its *concentration*: for example, by induction and repression of its synthesis in prokaryotes, by compartmentation of enzymes in organs or organelles, and by activation of a zymogen by covalent modification (as in the chymotrypsinogen-to-chymotrypsin conversion described in Chapter 15, section B). In this chapter, we are concerned with the regulation of rate by controlling the *activity* of an enzyme. The two principle means of doing this are:

1. Binding of allosteric effectors.
2. Covalent modification by phosphorylation and dephosphorylation of hydroxyl groups on amino acid side chains.

We shall now discuss one example of each process.

G. Phosphofructokinase and control of glycolysis

Glycolysis, the anaerobic degradation of glucose to pyruvate, generates ATP (equation 10·16). The glycolytic pathway is regulated to meet the cellular requirements for this important energy source. The most impor-

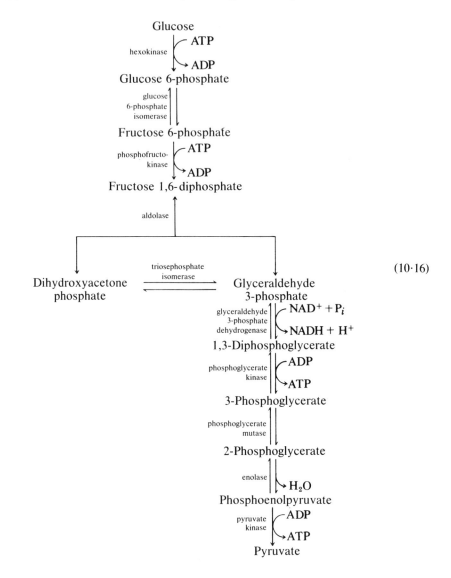

$$(10 \cdot 16)$$

tant control element is the allosteric enzyme phosphofructokinase. It catalyzes the phosphorylation of fructose 6-phosphate to fructose 1,6-diphosphate,

Fructose 6-phosphate + ATP $\rightleftharpoons$

fructose 1,6-diphosphate + ADP $(10 \cdot 17)$

(where both ATP and ADP are bound as their Mg^{2+} salts).

Phosphofructokinase from various organisms is subject to feedback control so that the activity of the enzyme is exquisitely sensitive to the energy level of the cell. Fructose 6-phosphate binds to the enzyme in a positive cooperative manner. The enzymes from eukaryotes are activated by AMP, ADP, and 3',5'-cyclic AMP, and are inhibited by high concentrations of ATP and citrate, a subsequent product of glycolysis. The rate of glycolysis is thus high when the cell requires energy and low when it does not. The enzymes from the prokaryotes *Escherichia coli* and *Bacillus stearothermophilus* are regulated by the same principle but in a slightly less complex manner: they are activated by ADP and GDP, and inhibited by phosphoenolpyruvate only (Figure 10·14). Because protein crystallographic studies of the enzyme from *B. stearothermophilus* are well advanced, and because its kinetics are very similar to those of the enzyme from *E. coli*, which has been studied in even greater depth,[38] the remainder of the discussion is devoted to this pair. (The structural details refer to the thermophilic enzyme, whereas the kinetic data are from the *E. coli* enzyme.)

The enzyme from *B. stearothermophilus* is an α_4 tetramer of subunit M_r 33 900.[39] Early kinetic studies indicated that the enzyme acts in a manner that is qualitatively consistent with an MWC two-state model. The enzyme acts as a K system; i.e., both states have the same value of k_{cat} but different affinities for the principle substrate. In the absence of ligands, the enzyme exists in the T state that binds fructose 6-phosphate more poorly than does the R state. In the absence of ADP, the binding of fructose 6-phosphate is highly cooperative, and $h = 3.8$. The positive homotropic interactions are lowered on the addition of the allosteric effector ADP, with h dropping to 1.4 at 0.8-mM ADP.[40] ADP thus binds preferentially to the R state. The allosteric inhibitor phosphoenolpyruvate binds preferentially to the T state and stabilizes it. The binding of the effector GDP (which binds only to the effector site and not also to the

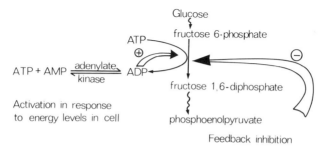

FIGURE 10·14. Phosphofructokinase and the control of glycolysis in *E. coli*. [From P. R. Evans, G. W. Farrants, and P. J. Hudson, *Phil. Trans. R. Soc.* **B293**, 53 (1981).]

active site, as does ADP) has been studied directly by equilibrium dialysis.[41] GDP binds with positive cooperativity to the free enzyme. The cooperativity is increased in the presence of the inhibitor phosphoenolpyruvate. The binding behavior is thus entirely in accord with the predictions of the MWC model (section B1).

1. The structure of the R state

The enzyme has been crystallized in the presence of fructose 6-phosphate. Since the crystals also bind ADP without major structural changes, they are presumably in the R state. The structure has been solved at 2.4-Å resolution (Figure 10·15).[38] The subunits are clearly divided into two domains. Each subunit makes close contacts with only two of the others.

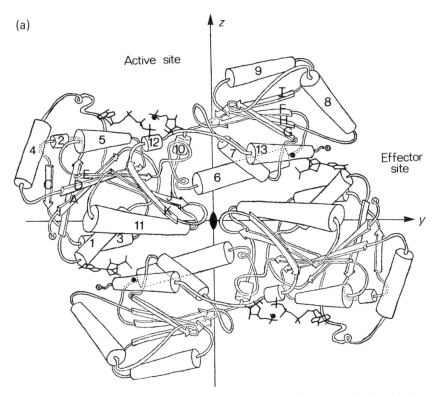

FIGURE 10·15. (a) Schematic representation of two of the four subunits of phosphofructokinase, viewed along the x axis (the other two subunits lie behind). ATP and fructose 6-phosphate (F6P) are shown in the active site. ADP is in the effector site, which lies between two subunits. (b) As above, but viewed along the z axis. From this angle, the binding site for fructose 6-phosphate can just be seen to span two subunits. [From P. R. Evans, G. W. Farrants, and P. J. Hudson, *Phil. Trans. R. Soc.* **B293**, 53 (1981).]

The binding of substrates and effectors has been studied by the difference Fourier method. The binding of fructose 6-phosphate and of ADP was determined directly. Because ATP was unstable under the conditions of the crystallographic studies, the structure of its complex with the enzyme was determined by model building from extrapolation of ADP binding and by using the analogue AMP-PNP (i.e., ATP in which —NH— replaces the —O— between the β- and γ-phosphoryl groups). There are three binding sites per subunit: Sites A and B form the active site, binding the sugar and the nucleotide, respectively; site C is the effector site, binding ADP or phosphoenolpyruvate.

The active site with its catalytic groups is in an extended cleft between the two domains of a subunit (see Figure 10·15). Interestingly, the 6-

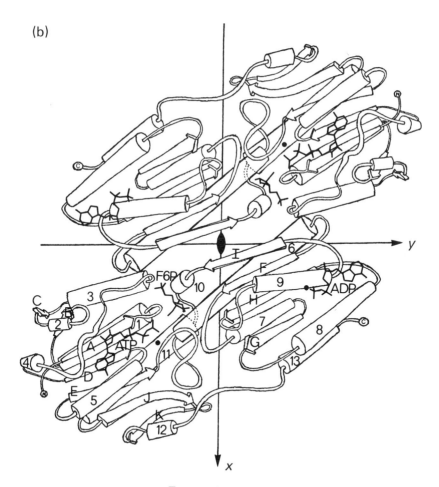

Figure 10·15 (cont'd)

phosphate lies between two subunits, bound by His-249 and Arg-252 from its own subunit and by Arg-162 and Arg-243 from the neighbor. (The 1-hydroxyl is in a suitable position for nucleophilic attack on the γ-phosphoryl of ATP. The carboxylate of Asp-127 is in a likely position to act as a general-base catalyst.)

The effector site lies in a deep cleft between two subunits. Again, the phosphate residues are bound by positively charged side chains from both subunits.

These structural features imply a mechanism for the allosteric changes in a K system, given that a change of quaternary structure means that subunits move relative to each other (Chapter 1). The positioning of effector groups and substrate-binding groups *between* subunits means that binding is poised for regulation by changes of quaternary structure. On the other hand, the location of the catalytic residues of the active site *within* a subunit means that k_{cat} need not be affected by a change in quaternary structure.

The complete elucidation of allosteric stereochemical mechanisms requires the structure of the T state in addition to that of the R state. Fortunately, crystallization of the enzyme from a solution containing phosphoenolpyruvate gives a second form that is presumably in the T state: the crystals crack when the allosteric inhibitor is washed out and fructose 1,6-diphosphate is diffused in. Structural studies are presently in progress(P.R.Evans,personalcommunication).

H. Glycogen phosphorylase and regulation of glycogenolysis[42–44]

Glycogen, a major source of energy for muscle contraction, is the principle storage form of glucose in mammalian cells. Glycogen consists mainly of glucose units linked by $\alpha(1 \rightarrow 4)$ glycosidic bonds, with branches created by residues linked by $\alpha(1 \rightarrow 6)$ glycosidic bonds. The rate-limiting enzyme in glycogenolysis is glycogen phosphorylase, which is frequently called phosphorylase. It catalyzes the sequential phosphorolysis of the $\alpha(1 \rightarrow 4)$-linked units from the nonreducing end to form glucose 1-phosphate (equation 10·18). Phosphorylase exists in two interconvertible forms, a and b. Phosphorylase b, the form in resting muscle, is an α_2 dimer of M_r $2 \times 97\ 333$, each polypeptide chain containing 841 amino acid residues (rabbit muscle enzyme). This form is inactive but it is activated by the allosteric effectors AMP and IMP. ATP and ADP act as allosteric inhibitors. In response to neural or hormonal signals, phosphorylase b is converted to phosphorylase a by the phosphorylation of Ser-14, catalyzed by phosphorylase kinase (equation 10·19). This is accompanied by a further dimerization to give a tetramer of M_r $4 \times 97\ 412$. Phosphorylase a

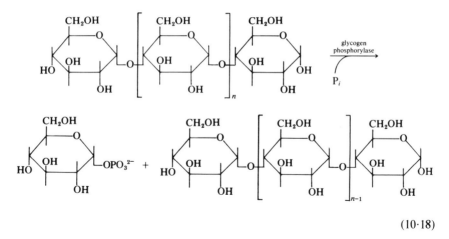

$$(10 \cdot 18)$$

is fully active at saturating substrate concentrations, but at low concentrations of P_i its activity is stimulated by AMP. Glucose is an allosteric inhibitor.

Crucial events in the regulation of the activity of phosphorylase are thus its phosphorylation and dephosphorylation (equation 10·19), but superimposed on these are the allosteric effects. The key factor in the phos-

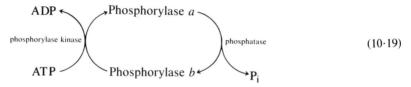

$$(10 \cdot 19)$$

phorylation state of phosphorylase is the activity of the phosphorylase kinase. This enzyme, of complex structure $(\alpha\beta\gamma\delta)_4$, may be activated by two types of stimuli:

1. *Neural control of the activity of phosphorylase kinase.* The electrical stimulation of muscle is mediated by the release of Ca^{2+} ions. These ions also activate phosphorylase kinase, which in turn activates phosphorylase and thus accelerates glycogenolysis to provide the necessary ATP for muscle contraction.

2. *Hormonal control of the activity of phosphorylase kinase.* Just as the activity of phosphorylase is increased by phosphorylation, so is the activity of its phosphorylase kinase (which may be phosphorylated on two serine residues, one in an α subunit and one in a β subunit). Hormonal stimulation (β-adrenergic) leads to the production of 3′,5′-cyclic AMP ("second messenger"), which stimulates the activity of the cyclic-AMP-dependent protein kinase that catalyzes the phosphorylation of phosphorylase kinase.

Four different phosphatases in skeletal muscle catalyze the dephosphorylation of the various enzymes involved in glycogenolysis. Their
activities are controlled by various inhibitors.

Phosphorylase and phosphorylase kinase are but two enzymes in a
complex phosphorylase/glycogen synthase cascade system that regulates
glycogen metabolism.[43,44] In such a cascade, the interconversion of active
and inactive forms of enzymes is not an all-or-none phenomenon, but a
dynamic process that leads to steady state levels of the active form. The
activity of "converter" enzymes such as the kinases and phosphatases
along with multisite interactions involving allosteric effectors provide a
system that can respond to fluctuations in the concentrations of a multitude of cellular metabolites. This allows for a continuous gradation of
enzymatic activity over a wide range of levels to balance the needs of the
cell or system.[44] Protein phosphorylation is a major mechanism by which
external physiological stimuli control intracellular processes in mammalian tissues.[43]

1. The structure of phosphorylase

The crystal structures of rabbit muscle phosphorylase a and b have been
solved at 2.5 and 3 Å, respectively.[45,46] Phosphorylase b crystals were
obtained in the presence of IMP, a weak activator that does not increase
the affinity of the enzyme for the substrate. Crystallographic and solution
studies indicated that the structure is close to that of the T state.[47] The
crystals of phosphorylase a were obtained in the presence of the strong
allosteric inhibitor glucose. The crystals of the two forms are nearly isomorphous, and the a form is also in the T state. The structure of the R
state is not known. The structures of the two crystallized forms are very
similar in terms of the overall folding. Each subunit consists of two major
domains, with the possibility of further subdivision into smaller domains.

There is an important difference between the a and the b forms in
the location of their first 20 amino acids; this region contains the important
residue Ser-14, which is phosphorylated in phosphorylase a. In phosphorylase b the region is disordered, and the first 19 amino acid residues
cannot be located in the electron density map. In phosphorylase a, however, the region is better ordered, although it is not the most clearly
defined one in the map. Ser-14 is found to be at the surface near a subunit
interface. Its phosphoryl group is surrounded by positively charged
groups from both subunits. Thus, phosphorylation causes changes at the
subunit interfaces. The effector molecules AMP and ATP also bind at
sites close to the surface of the subunit interfaces.[48] The 2'-hydroxyl of
the nucleotide interacts with Asn-42 of the related subunit, as well as
with Gln-72 of the subunit to which the rest of the molecule is bound.
There are thus similarities with phosphofructokinase and hemoglobin, in

that effector molecules bind at subunit interfaces and make direct contacts across them.

The catalytic site is some 32 Å from the effector site, between the two major domains of the subunit deep in the center of the molecule. The crystals of phosphorylase b bind the substrate glucose 1-phosphate, whereas the addition of the substrate to crystals of phosphorylase a causes them to crack. This shows that tertiary structural changes are transmitted through the molecule on substrate binding. Some of these changes may be observed by introducing substrate analogues that bind to phosphorylase a and induce a part of the conformational transition leading to activation.[49,50] The situation is similar to that in hemoglobin, in that the active sites are far apart from each other and from the subunit interfaces, yet structural changes may be transmitted across the molecule. Large substrates are involved in the enzymatic reaction, however, so there is more scope for induced conformational changes than is found on the binding of O_2 to hemoglobin.

References

1   G. S. Adair, *J. Biol. Chem.* **63**, 529 (1925).
2   J. Monod, J.-P. Changeux, and F. Jacob, *J. Molec. Biol.* **6**, 306 (1963).
3   M. F. Perutz, *Ann. Rev. Biochem.* **48**, 327 (1979).
4   J. Monod, J. Wyman, and J.-P. Changeux, *J. Molec. Biol.* **12**, 88 (1965).
5   D. E. Koshland, Jr., G. Neméthy, and D. Filmer, *Biochemistry* **5**, 365 (1966).
6   M. Eigen, *Nobel Symp.* **5**, 333 (1967).
7   J. Herzfield and H. E. Stanley, *J. Molec. Biol.* **82**, 231 (1974).
8   A. Levitzki, W. B. Stallcup, and D. E. Koshland, Jr., *Biochemistry* **10**, 3371 (1971).
9   R. A. MacQuarrie and S. A. Bernhard, *Biochemistry* **10**, 2456 (1971).
10 M. J. Irwin, J. Nyborg, B. R. Reid, and D. M. Blow, *J. Molec. Biol.* **105**, 577 (1976).
11 G. Biesecker, J. I. Harris, J. C. Thierry, J. E. Walker, and A. J. Wonacott, *Nature, Lond.* **266**, 328 (1977).
12 A. R. Fersht, R. S. Mulvey, and G. L. E. Koch, *Biochemistry* **14**, 13 (1975).
13 H. R. Bosshard, G. L. E. Koch, and B. S. Hartley, *Eur. J. Biochem.* **53**, 493 (1975).
14 A. Conway and D. E. Koshland, Jr., *Biochemistry* **7**, 4011 (1968).
15 J. Schlessinger and A. Levitzki, *J. Molec. Biol.* **82**, 547 (1974).
16 A. R. Fersht, unpublished data.
17 R. Hill, *Proc. R. Soc.* **B100**, 419 (1925).
18 M. M. Rubin and J.-P. Changeux, *J. Molec. Biol.* **21**, 265 (1966).
19 S. J. Edelstein, *Nature, Lond.* **230**, 224 (1971).
20 K. Kirschner, E. Gallego, I. Schuster, and D. Goodall, *J. Molec. Biol.* **58**, 29 (1971).
21 K. H. Mayo, *J. Molec. Biol.* **146**, 589 (1981).
22 A. Nasuda-Kouyama, H. Tachibana, and A. Wada, *J. Molec. Biol.* **146**, 451 (1983).
23 J. M. Baldwin and C. Chothia, *J. Molec. Biol.* **129**, 175 (1979).

24 M. F. Perutz, *Nature, Lond.* **228**, 734 (1970).
25 A. Arnone, *Nature, Lond.* **237**, 146 (1972).
26 A. Arnone and M. F. Perutz, *Nature, Lond.* **249**, 34 (1974).
27 C. Chothia, S. Wodak, and J. Janin, *Proc. Natn. Acad. Sci. USA* **73**, 3793 (1976).
28 C. Bohr, K. A. Hasselbach, and A. Krogh, *Skand. Arch. Physiol.* **16**, 402 (1904).
29 J. L. Hoard, in *Hemes and hemoproteins* (B. Chance, R. W. Estabrook, and T. Yonetani, Eds.), Academic Press, p. 9 (1966).
30 M. F. Perutz, S. S. Hasnain, P. J. Duke, J. L. Sessler, and J. E. Hahn, *Nature, Lond.* **295**, 535 (1982).
31 A. Warshel, *Proc. Natn. Acad. Sci. USA* **74**, 1789 (1977).
32 B. R. Gelin and M. Karplus, *Proc. Natn. Acad. Sci. USA* **74**, 801 (1977).
33 J. P. Collman, R. R. Gagne, C. A. Reed, W. T. Robinson, and G. A. Rodley, *Proc. Natn. Acad. Sci. USA* **71**, 1326 (1974).
34 G. B. Jameson, F. S. Molinaro, J. A. Ibers, J. P. Collman, J. I. Brauman, B. Rose, and K. S. Suslick, *J. Am. Chem. Soc.* **102**, 3225 (1980).
35 B. Shaanan, *Nature, Lond.* **296**, 5858 (1982).
36 J. P. Collman, J. I. Brauman, I. J. Collins, B. Iverson, and J. L. Sessler, *J. Am. Chem. Soc.* **103**, 2450 (1981).
37 S. E. V. Phillips and B. P. Schoenborn, *Nature, Lond.* **292**, 81 (1981).
38 P. R. Evans, G. W. Farrants, and P. J. Hudson, *Phil. Trans. R. Soc.* **B293**, 53 (1981).
39 E. Kolb, P. J. Hudson, and J. I. Harris, *Eur. J. Biochem.* **108**, 587 (1980).
40 D. Blangy, H. Buc, and J. Monod, *J. Molec. Biol.* **31**, 13 (1968).
41 D. Blangy, *Biochimie* **53**, 135 (1971).
42 R. J. Fletterick and N. B. Madsen, *Ann. Rev. Biochem.* **49**, 31 (1980).
43 P. Cohen, *Nature, Lond.* **296**, 613 (1982).
44 P. B. Chock, S. G. Rhee, and E. R. Stadtman, *Ann. Rev. Biochem.* **49**, 813 (1980).
45 S. Sprang and R. J. Fletterick, *J. Molec. Biol.* **131**, 523 (1979).
46 I. T. Weber, L. N. Johnson, K. S. Wilson, D. G. R. Yeates, D. L. Wild, and J. A. Jenkins, *Nature, Lond.* **274**, 433 (1978).
47 J. A. Jenkins, L. N. Johnson, D. I. Stuart, E. A. Stura, K. S. Wilson, and B. Zanotti, *Phil. Trans. R. Soc.* **B293**, 23 (1981).
48 L. N. Johnson, E. A. Stura, K. S. Wilson, M. S. P. Sansom, and I. T. Weber, *J. Molec. Biol.* **134**, 639 (1979).
49 S. R. Sprang, E. J. Goldsmith, R. J. Fletterick, S. G. Withers, and N. B. Madsen, *Biochemistry* **21**, 5364 (1982).
50 S. G. Withers, N. B. Madsen, S. R. Sprang, and R. J. Fletterick, *Biochemistry* **21**, 5372 (1982).

11

Forces between molecules, and enzyme–substrate binding energies

The distinguishing feature of enzyme catalysis is that the enzyme binds the substrate so that the reactions proceed in the confines of the enzyme–substrate complex. In order to gain insight into the strength and specificity of the binding, we shall discuss in a somewhat empirical and phenomenological manner the interactions between nonbonded atoms. In particular, we shall concentrate on the magnitudes of the energies involved. Besides being responsible for binding, the noncovalent interactions are important in further ways. One important consideration to be discussed in Chapter 12 is that these interactions can be used for lowering the activation energy of a chemical step instead of directly contributing to the binding energy. In addition, these forces play a considerable role in maintaining protein structure.

The obvious interactions between nonbonded atoms are those due to electrostatic forces between charged groups. But even nonpolar molecules have an attraction for one another. This was pointed out by J. D. van der Waals a century ago as one of the reasons for the breakdown of the ideal gas laws. Some of these interactions between neutral gas molecules have been well characterized, and we now know that several "van der Waals" compounds exist. For example, the noble gas dimers Ne_2, A_2, and Xe_2 have bond energies of 0.2, 0.92, and 2.2 kJ/mol (0.05, 0.22, and 0.53 kcal/mol), respectively.[1]

A. Interactions between nonbonded atoms

1. Electrostatic interactions

All forces between atoms and molecules are electrostatic in origin, even those between nonpolar molecules. But we shall reserve the term electrostatic for interactions that occur between charged or dipolar atoms and molecules. Electrostatic interactions are the best-understood type. In practice, however, their quantitative importance is difficult to assess because the interaction energies depend crucially on the dielectric constant D of the surrounding medium. This was discussed in Chapter 2, where an example was given of how a repulsive energy of 530 kJ (130 kcal) between two juxtaposed positively charged nitrogen atoms *in vacuo* is reduced by a factor of 16 when they are surrounded by water. The surrounding medium is polarized by the charges to set up a neutralizing field.

The problem of calculating electrostatic effects in proteins is especially difficult because of their heterogeneity of structure.[2] The dielectric constant is not uniform throughout the protein but depends on the microenvironment, which can range from being very polar near charged groups or dipoles to being nonpolar in regions of hydrocarbon side chains.

The following varieties of electrostatic interaction energies exist:

1. Between ions with net charges; the energy falls off as $1/Dr$.
2. Between randomly oriented permanent dipoles; the energy falls off as $1/Dr^6$.
3. Between an ion and a dipole induced by it; the energy falls off as $1/Dr^4$.
4. Between a permanent dipole and a dipole induced by it; the energy falls off as $1/Dr^6$.

2. Nonpolar interactions (van der Waals or dispersion forces)

Typical potential energy curves for the interaction of two atoms are illustrated in Figure 11·1. There is characteristically a very steeply rising repulsive potential at short interatomic distances as the two atoms approach so closely that there is interpenetration of their electron clouds. This potential approximates to an inverse twelfth-power law. Superimposed upon this is an attractive potential due mainly to the London *dispersion forces*. This follows an inverse sixth-power law. The total potential energy is given by

$$U = \frac{A}{r^{12}} - \frac{B}{r^6} \tag{11·1}$$

The distance dependence of $1/r^6$ for the attractive potential is characteristic of the interaction between dipoles. This is because the attractive dispersion forces result from the mutual induction of electrostatic dipoles. Although a nonpolar molecule has no net dipole averaged over a period of time, at any one instant there will be dipoles due to the local fluctuations of electron density. Because the energies depend on the induction of a dipole, polarizability is an important factor in the strength of the interaction between any two atoms.

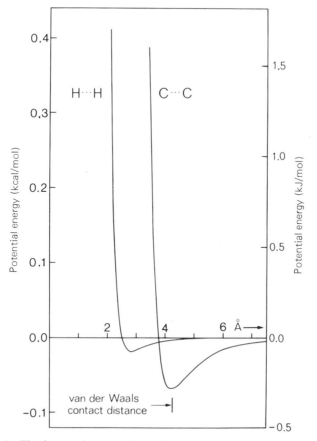

FIGURE 11·1. The interaction energies of two hydrogen atoms and two (tetrahedral) carbon atoms in a protein. (Calculated from the data in Table 11·3.)

The value of B in equation 11·1 for the interaction of two atoms i and j may be calculated from the Slater-Kirkwood equation,

$$B = \frac{\frac{3}{2}e(\hbar/m^{1/2})\alpha_i\alpha_j}{(\alpha_i/N_i)^{1/2} + (\alpha_j/N_j)^{1/2}} \tag{11·2}$$

where α is the polarizability, e is the charge on the electron, m is its mass, $\hbar$ is Dirac's constant, and N is the effective number of outer-sphere electrons. Some values of α and N are listed in Table 11·1. Note the low polarizability of O and the high value for S, indicating the relative importance of dispersion energies in the interactions of these atoms.

The interatomic distance at the bottom of the potential well, the most favorable distance of separation, is known as the *van der Waals contact*

TABLE 11·1. *Polarizabilities and van der Waals attractions of some atoms and groups*

Atom/group	Polarizability[a] (mL $\times$ 10^{24})	N^b	B (for self-interaction)[c] (kcal·$\mathring{A}^6$/mol)
—H	0.40	1	46
—O—	0.63	6	220
—OH	0.99	7	470
—CH₂—	1.80	7	1160
—S—	3.00	16	3760
—SH	3.34	17	4560

[a] From A. R. Fersht and C. Dingwall, *Biochemistry* **18**, 1245 (1979), plus additional calculated values.
[b] The effect number of outer-sphere electrons. [From J. A. McCammon, P. G. Wolynes, and M. Karplus, *Biochemistry* **18**, 927 (1979).]
[c] The value of *B*, the attractive potential, in the "6–12" equation (11·1), as calculated from the Slater-Kirkwood equation for two of the indicated atoms or groups interacting.

distance. A particular atom has a characteristic van der Waals radius. These radii are additive, so that the optimal distance of contact between two atoms may be found by the addition of their two van der Waals radii. The van der Waals radii are not as sharply defined as covalent bond radii. This is because the potential energy wells are so shallow that contact distances may vary by 0.1 $\mathring{A}$ (0.01 nm) or so with little change in energy. Thus, in a crystal structure, it is found that there is a range of values, depending upon the local steric constraints. Several sets of van der Waals radii determined from statistical analyses of contact distances in crystals are in use (Table 11·2). Also listed in Table 11·2 are the minimum values, determined from the potential energy curve and the calculated flexibility of proteins. Some interaction energies for atoms at their optimal separations are given in Table 11·3.

Although the attractive forces are weak and the van der Waals energies low, they are additive and make significant contributions to binding when they are summed over a molecule. For example, it is found experimentally from heat-of-sublimation data that each methylene group in a crystalline hydrocarbon has 8.4 kJ/mol (2 kcal/mol) of van der Waals energy, and that each CH group in benzene crystals has 6.7 kJ/mol (1.6 kcal/mol) (Table 11·4). It has been calculated that the van der Waals energy between the D subsite of lysozyme and the occupying glucopyranose ring is about -58.5 kJ/mol (-14 kcal/mol).[3]

3. The hydrogen bond

A particularly important bond in biological systems is the hydrogen bond. This consists of two electronegative atoms, one of which is usually ox-

TABLE 11·2. *Van der Waals radii*

Atom	Van der Waals radius (Å)			Minimum radius (Å)[d]
	a	*b*	*c*	
H	1.2	1.4	—	—
O (hydroxyl)	1.5	—	1.6	1.3
O (carbonyl)	1.5	1.7	—	1.3
O⁻ (carboxyl)	1.5	1.8ᵢ	1.6	—
N (amide)	1.55	2	1.65	1.3
NH_3^+ (ammonium)	1.65[e]	—	1.75	—
N^+ (imidazolium)	1.55	—	—	—
CH (tetrahedral)	1.85[e]	2.1	1.90	1.5
C (trigonal)	1.7	1.8	1.80	1.45
CH (aromatic)	1.8[e]	—	1.90	—
S	1.8	—	1.90	1.5

[a] From A. Bondi, *J. Phys. Chem.* **68**, 441 (1965); D. A. Brant and P. J. Flory, *J. Am. Chem. Soc.* **87**, 2791 (1965).
[b] From A. Warshel and M. Levitt, *J. Molec. Biol.* **103**, 227 (1976). These values are close to the potential minimum; the values in the preceding column are slightly smaller.
[c] From J. A. McCammon, P. G. Wolynes, and M. Karplus, *Biochemistry* **18**, 927 (1979).
[d] The minimum contact radius in a protein. [From M. Levitt, *J. Molec. Biol.* **82**, 393 (1974).]
[e] The radius includes the bonded hydrogen.

ygen, bound to the same proton. The bonds in the systems of interest are asymmetric, the proton being at its normal covalent bond distance from the atom to which it is formally bonded, and at a distance from the other somewhat shorter than the usual van der Waals contact distance. The optimal configuration is linear, but bending causes only small energy losses. In the hydrogen bond between the =NH group and the carbonyl oxygen in amide crystals, the O···N distance is typically 2.85 to 3.00 Å,

TABLE 11·3. *Dispersion (van der Waals) energies for pairs of atoms at optimal separations*[a]

Interaction	Van der Waals energy	
	kJ/mol	kcal/mol
H···H	0.0778	0.0186
=O···O=	0.217	0.0519
N···N	0.572	0.1366
C···C	0.284	0.0679
C···C[b]	1.891	0.4519

[a] From A. Warshel and M. Levitt, *J. Molec. Biol.* **103**, 227 (1976).
[b] Carbonyl or carboxyl carbon.

TABLE 11·4. *Dispersion and electrostatic energies in some crystals*[a]

| | Calculated energies | | | | | |
| | Dispersion (van der Waals) | | Electrostatic[b] | | Observed heat of sublimation | |
Crystal	kJ/mol	kcal/mol	kJ/mol	kcal/mol	kJ/mol	kcal/mol
Benzene	−42	−9.9	−5.4	−1.3	−44.8	−10.7
n-Hexane	−54	−12.9	0	0	−50.2	−12.0
n-Pentane	−42	−10.0	0	0	−43.9	−10.5
Perylene	−119	−28.4	−2.5	−0.6	−125.5	−30
Phenanthrene	−85	−20.3	−4.2	−1.0	−91.8	−21.7
Adipamide (4 H bonds)	−86	−20.6	−77	−18.4	—	—
Formamide (2 H bonds)	−24	−5.7	−43	−10.3	−73.2	−17.5
Malonamide (4 H bonds)	−52	−12.3	−81	−19.4	−120.5	−28.8
Oxamide (4 H bonds)	−45	−10.8	−60	−14.4	−118.0	−28.2
Succinamide (4 H bonds)	−65	−15.5	−76	−18.2	−135.2	−32.3
Urea (3 H bonds)	−28	−6.7	−65	−15.6	−92.8	−22.2

[a] The energies are calculated from empirical energy functions (see Table 11·3). [From P. Dauber and A. T. Hagler, *Accts. Chem. Res.* **13**, 105 (1980); A. T. Hagler, P. Dauber, and S. Lifson, *J. Am. Chem. Soc.* **101**, 5131 (1979), and references therein.]
[b] The electrostatic contribution is due to the hydrogen bonds.

and the $O \cdots H$ distance 1.85 to 2.00 Å.[4,5] The variation of the potential energy with the $N \cdots O$ distance is similar to that in Figure 11·1, the minimum distance of approach being about 2.4 to 2.5 Å. In the $H_2O \cdots H—OH$ hydrogen bond, the distance between the two oxygens is 2.76 Å and the $O \cdots H$ distance is 1.77 Å.

The energies of hydrogen bonds have been variously estimated to be between 12 and 38 kJ/mol (3 and 9 kcal/mol).[4–6] A reliable value, estimated from the data in Table 11·4, is 21 kJ/mol (5 kcal/mol) for the amide-amide $NH \cdots O$ bond. Bonds of this strength are of particular importance, since they are stable enough to provide significant binding energy, but sufficiently weak to allow rapid dissociation. If the activation energy for the breaking of the bond is the whole of the bond strength, then transition state theory may be used to calculate that bonds of energies 12.5, 25.0 and 37.6 kJ/mol (3, 6, and 9 kcal/mol) dissociate with rate constants of 4 × 10^{10}, 3 × 10^8, and 2 × 10^6 s^{-1}, respectively.

The backbone NH groups of the enzyme are not only used for binding the substrate, but they may also act as the solvation shell for negative charges developing in the transition state. In particular, as was pointed out in Chapter 2, a binding site for the carbonyl oxygen of the substrates of the serine proteases consists of two backbone amido NH groups. When the oxygen becomes negatively charged during the transition state of the reaction, it is stabilized by the dipole moments of the amide groups.

4. The hydrophobic bond[7-9]

The hydrophobic bond is a way of describing the tendency of nonpolar compounds such as hydrocarbons to transfer from an aqueous solution to an organic phase. The classic theory of the hydrophobic bond is that it results not so much from the direct interaction of solvent and solute molecules as from the reorganization of the normal hydrogen-bonding network in water by the presence of a hydrophobic compound. Water consists of a dynamic, loose network of hydrogen bonds. The presence of a nonpolar compound causes a local rearrangement in this network. In order to preserve the number of hydrogen bonds, each one having an energy of about 25 kJ/mol (6 kcal/mol), the water molecules line up around the nonpolar molecule. The hydrophobic solute therefore does not cause large enthalpy changes in the solvent but instead decreases its entropy due to the increase in local order. A hydrophobic molecule is driven into the hydrophobic region of a protein by the regaining of entropy by water.

There is, however, a second component that results from dispersion energies. Dispersion forces are weak in water because of the low polarizability of oxygen (Table 11·1) and because of the low atom density (the dispersion energies are additive). This is an additional factor favoring the self-association of hydrocarbons, as they have a higher atom density and the polarizability of $-CH_2-$ is greater than that of O.

One convenient way of measuring the hydrophobicity of a molecule is to measure its partition between the organic and aqueous phases when it is shaken with an immiscible mixture of an organic solvent, often n-octanol, and water. The distribution of the solute between the two phases depends on the competing tendencies of the hydrophobic regions to be squeezed into the organic phase by the hydrophobic bond, and the polar regions to be solvated and drawn into the aqueous phase.

a. Hydrophobicity of small groups—The Hansch equation[10,11]

From determinations of the partitioning of several series of substituted compounds between n-octanol and water, C. Hansch and coworkers found that many of the substituents make a constant, and additive, contribution to the hydrophobicity of the parent compound. If the ratio of the solubility of the parent compound (H—S) in the organic phase to that in the aqueous phase is P_0, and that of the substituted compound (R—S) is P, then the hydrophobicity constant for R, π, is defined by

$$\pi = \log \frac{P}{P_0} \qquad (11·3)$$

Note that $RT \ln (P/P_0)$, i.e., $2.303RT\pi$, is the incremental Gibbs free energy of transfer of the group R from n-octanol to water (relative to the hydrogen atom H). Some values of π are listed in Table 11·5. Points to be noted are:

1. The values of π for groups that are not strongly electron-donating or -withdrawing are virtually constant and independent of the group to which they are attached. Furthermore, their effects are additive. For example, the methylene group has a π of 0.5, and the addition of each additional methylene adds a further increment of 0.5 to π. The 0.5 log unit is equal to a change of 2.84 kJ/mol (0.68 kcal/mol) in the Gibbs free energy. (In this context, the substitution of a methyl group for a hydrogen is the same as the addition of a methylene group, since it is equivalent to the interposing of a methylene group between the hydrogen and the rest of the molecule.)
2. The behavior of groups that can conjugate with the benzene ring, such as the nitro and amino groups, is variable and depends on the other groups attached to the ring.

b. Hydrophobicity varies as surface area[12–15]

It has been noted that there is an empirical correlation between the surface area of a hydrophobic side chain of an amino acid and its Gibbs energy

TABLE 11·5. *Some values of π^a*

Group[b]	π	Incremental Gibbs energy of transfer from n-octanol to water	
		kJ/mol	kcal/mol
—CH_3	0.5	2.85	0.68
—CH_2CH_3	1.0	5.71	1.36
—$(CH_2)_2CH_3$	1.5	8.56	2.05
—$(CH_2)_3CH_3$	2.0	11.41	2.73
—$(CH_2)_4CH_3$	2.5	14.26	3.41
—CH<$^{CH_3}_{CH_3}$	1.3	7.42	1.77
—CH<$^{CH_2CH_3}_{CH_3}$	1.8	10.27	2.45
—CH_2Ph	2.63	15.00	3.59
—OH^c	−1.16	−6.62	−1.58
—$NHCOCH_3{}^c$	−1.21	−6.90	−1.65
—$OCOCH_3{}^c$	−0.27	−1.54	−0.37

[a] From C. Hansch and E. Coats, *J. Pharm. Sci.* **59**, 731 (1970).
[b] Relative to the hydrogen atom.
[c] Bound to aliphatic compounds.

of transfer from water to an organic phase. A value of 1 $\mathring{A}^2$ of surface area gives a hydrophobic energy of 80 to 100 J/mol (20 to 25 cal/mol). (The surface of a group is defined by rolling a ball with a radius equal to the van der Waals radius of water over the van der Waals surface of the group. The surface is the locus of the center of the ball.)

We can rationalize this correlation between surface area and hydrophobicity in terms of a simple model involving the energy of formation of a cavity in water. The surface tension of water is 72 dyn/cm (0.072 N/m), so that to form a free surface of water of 1 $\mathring{A}^2$ requires 7.2×10^{-22} J (1.72×10^{-22} cal). Multiplying this by Avogadro's number gives a value of 435 J/$\mathring{A}^2$/mol (104 cal/$\mathring{A}^2$/mol). Thus, creating a cavity in water to be occupied by a hydrophobic group costs 435 J/$\mathring{A}^2$/mol of surface area. This will be offset somewhat by a gain in dispersion energy from the interaction of water with the solute when the cavity is filled, and this increment will also vary with surface area.

The energy of formation of the cavity in water is similar to the total energy change in hydrophobic bond formation, and illustrates that removal of the cavity is part of the driving force of the bond in the partitioning experiments.

B. The binding energies of enzymes and substrates

It is difficult to estimate the contribution of hydrogen bonding, electrostatic linkages, and hydrophobic bonding to the overall binding energy of a substrate and enzyme by extrapolation from the simple physical measurements cited in section A. The major cause of the difficulty is that the binding process is an *exchange* reaction: the substrate exchanges its solvation shell of water for the binding site of the enzyme. The net binding energy represents the *differences* between the binding energies of the substrate with water and the substrate with the enzyme.

The evaluation of the net energy of hydrogen bonding is difficult because the substrate is normally hydrogen-bonded to water molecules in aqueous solution, as is the enzyme. The formation of hydrogen bonds in the enzyme–substrate complex involves the displacement of hydrogen-bonded water molecules:

$$\text{E}\cdots\text{H}-\text{O}\Big\backslash_{\text{H}}^{\text{H}} + \text{S}\cdots\text{O}\Big\backslash_{\text{H}}^{\text{H}} \rightleftharpoons \text{E}\cdots\text{S} + \Big/_{\text{H}}^{\text{H}}\text{O}-\text{H}\cdots\text{O}\Big\backslash_{\text{H}}^{\text{H}} \qquad (11\cdot4)$$

There is thus no net gain in the number of hydrogen bonds. But there is an increase in *entropy* on the formation of *intra*molecular bonds in the complex. The energy of an individual hydrogen bond is composed of a favorable attractive energy term and an unfavorable entropy term because two molecules are linked together to form one (see Chapter 2, section

B4, and this chapter, section C). But if the substrate is immobilized in the enzyme–substrate complex anyway, there is no further loss of entropy on formation of intramolecular hydrogen bonds. In other words, the loss of entropy has to be "paid for" only once. Intramolecular hydrogen bonding is favored by the entropy gain from the release of bound water molecules. A rough estimate is that there is a gain of about 40 J/deg (10 cal/deg) per mole of water released.

Evaluation of the energy of salt linkages is difficult because the ions that are involved are solvated in solution. The solvation energies are very high; a $-CO_2{}^-$ ion is estimated to be stabilized by about 270 kJ/mol (65 kcal/mol).[3] Also, the energy of a salt bridge depends strongly on the dielectric constant of the surrounding medium. A further factor is that the formation of a buried salt bridge is favored entropically since water of solvation is released:

$$E-NH_3{}^+(H_2O)_m + (H_2O)_n{}^-O_2C-S \rightleftharpoons$$

$$E-NH_3{}^+\cdots{}^-O_2C-S + (m + n)H_2O \qquad (11\cdot5)$$

The evaluation of hydrophobic bond energies is difficult because there are fundamental differences between hydrophobic bonds of a solute between an organic solvent and water (as measured by partitioning experiments) and bonds formed by the binding of hydrophobic substrates to a protein. The transfer of a solute from the aqueous to the organic phase may be divided into three hypothetical steps: formation of a cavity in the organic phase; transfer of the solute to the cavity; and closing of the cavity left in the aqueous phase. The transfer of a hydrophobic substrate to a hydrophobic region in an enzyme involves the occupation of a *preformed* cavity, and probably the transfer of water from this to the aqueous phase.

As an alternative to using model reactions, the contributions of the various factors to the binding energy of an enzyme and substrate can be determined from direct measurements on enzymatic reactions.

1. Estimation of increments in binding energy from kinetics

One way of measuring the contribution of a substituent R in a substrate R—S to binding, $\Delta\Delta G_b$, is to compare the dissociation constants of R—S and H—S from the enzyme:

$$\Delta G_{b(R-S)} = -RT \ln K_{S(R-S)} \qquad (11\cdot6)$$

$$\Delta G_{b(H-S)} = -RT \ln K_{S(H-S)} \qquad (11\cdot7)$$

so that

$$\Delta\Delta G_b = -RT \ln \frac{K_{S(H-S)}}{K_{S(R-S)}} \qquad (11\cdot8)$$

But, as will be seen in the next chapter, this often underestimates the binding energy of the larger substrate, since enzymes frequently use binding energy to lower the activation energies of reactions rather than using it to give tighter K_M's. A much better method is to compare the values of k_{cat}/K_M for the two substrates. This quantity includes both the activation energies and the binding energies, and avoids the underestimates that result from using dissociation constants alone. It is shown in Chapter 13, equation 13·3, that

$$\Delta\Delta G_b = -RT \ln \frac{(k_{cat}/K_M)_{(R-S)}}{(k_{cat}/K_M)_{(H-S)}} \tag{11·9}$$

($-\Delta\Delta G_b$ is the incremental Gibbs free energy of transfer of R from the enzyme to water, relative to H. This is related to π of section A4a.)

a. Intrinsic binding energies

The intrinsic binding energy of a group R is the maximum binding energy possible when there is perfect complementarity between it and its binding cavity in an enzyme. In practice, enzymes do not necessarily utilize all of the intrinsic binding energy of a group, for various reasons. Consider the utilization of the binding energy of the hydroxyl group of tyrosine with various enzymes. At one extreme, there is the binding to the phenylalanyl-tRNA synthetase. Here, as will be discussed in Chapter 13, the enzyme has evolved to bind tyrosine as weakly as possible, and there is no binding site for the hydroxyl group. In the middle of the range is chymotrypsin. This enzyme has approximately equal specificities for phenylalanine and tyrosine, so it uses only a fraction of the intrinsic binding energy. At the other extreme, the tyrosyl-tRNA synthetase has evolved to bind the —OH of tyrosine as tightly as possible to maximize the specificity against phenylalanine. The binding energy in this example must be the maximum achievable and must therefore tend to the intrinsic value.[16]

b. Estimation of the upper limits of binding energies by measurements on the aminoacyl-tRNA synthetases

The evolutionary pressure on the aminoacyl-tRNA synthetases to bind the distinctive features of their correct substrates as tightly as possible (Chapter 13) affords an experimental method of measuring the maximum possible binding energies. This is done by measuring the values of k_{cat}/K_M for the pyrophosphate exchange reaction (Chapter 7, section D1) and using equation 11·9. Fortunately, one of the assumptions in the derivation of that equation—that the group R has a negligible inductive effect on the reaction—holds well here. The data are listed in Table 11·6. Note the following:

Alkyl groups. A —CH_3 group contributes about 13 kJ/mol (3.2 kcal/mol) of binding energy, as determined by experiments on many different enzymes. This is some 5 times higher than the values for hydrophobic bind-

TABLE 11·6. *Binding energies of aminoacyl-tRNA synthetases and various groups*[a]

Group[b]	Incremental Gibbs energy of transfer from enzyme to water[b]	
	kJ/mol	kcal/mol
—CH_3	14	3.2
—CH_2CH_3	27	6.5
—$CH(CH_3)_2$	40	9.6
—S—	23	5.4
(HS—	38	9.1)
(HO—	29	7.0)

[a] From A. R. Fersht, J. S. Shindler, and W.-C. Tsui, *Biochemistry* **19**, 5520 (1980); W.-C. Tsui and A. R. Fersht, *Nucl. Acids Res.* **9**, 4627 (1981).
[b] Relative to the hydrogen atom.

ing listed in Table 11·5 for transfer from *n*-octanol to water. (The values in Table 11·6 are for experiments equivalent to those in 11·5, except that the transfer is from the enzyme to water.) The Gibbs energy of transfer for a larger side chain, e.g., that of valine, is also 5 times higher from an enzyme to water than from *n*-octanol to water.

Hydroxyl and thiol groups. The Gibbs free energy change for the —OH group listed in the table was measured from the relative activity of tyrosine and phenylalanine with the tyrosyl-tRNA synthetase. The binding site for this is now known and is illustrated in Chapter 12, Figure 12·5. The hydroxyl group forms two hydrogen bonds and undergoes van der Waals interactions with the enzyme. In the absence of substrate, there must be a bound water molecule in the site that is displaced on the binding of tyrosine. It is not known whether the binding of phenylalanine displaces the water molecule or not; it may be energetically more favorable for the enzyme to be distorted on the binding of phenylalanine rather than to be desolvated. The observed value of 29 kJ/mol (7 kcal/mol) is thus a minimum estimate of the incremental Gibbs energy of transfer for the —OH group to water. The same considerations apply to the value for the —SH group.

The *salt linkage* between the α-ammonium ion of tyrosine and the enzyme is illustrated in Figure 12·5. Its energy can be estimated, from the relative binding of tyrosine and its deaminated derivative (structures 11·10) and from equation 11·8, to be 18 kJ/mol (4.3 kcal/mol).[17]

$$\text{HO—}\langle\text{C}_6\text{H}_4\rangle\text{—CH}_2\underset{\overset{|}{NH_3^+}}{\overset{CO_2^-}{CH}} \quad vs. \quad \text{HO—}\langle\text{C}_6\text{H}_4\rangle\text{—CH}_2CH_2CO_2^- \qquad (11\cdot10)$$

c. Typical binding energies from measurements on chymotrypsin
Chymotrypsin, an enzyme of broad specificity, has a hydrophobic binding pocket (Chapter 1). Kinetic measurements on this enzyme afford a general idea of the strength of hydrophobic binding with enzymes.

Alkyl groups. The values of k_{cat}/K_M for the chymotrypsin-catalyzed hydrolysis of a series of esters of the form $R—CH(NHAc)CO_2CH_3$, where R is an unbranched alkyl chain, increase with increasing hydrophobicity of R.[18] The decrease in the activation energy is 2.2 times greater than the free energy of transfer of the alkyl groups from water to *n*-octanol (Figure 11·2).[19] The hydrophobic binding pocket appears to be 2.2 times more hydrophobic than *n*-octanol.

The inhibition constants of a series of substituted formanilides increase with increasing hydrophobicity. A plot of the logarithms of the constants against π yields a straight line of slope -1.5.[20] This again shows that the active site of chymotrypsin is more hydrophobic than *n*-octanol (Figure 11·3).

Salt linkage. The catalytically active conformation of chymotrypsin is

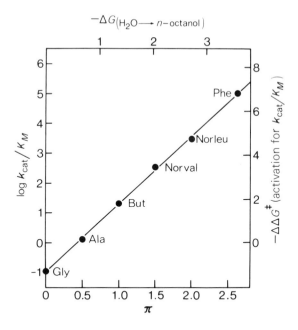

FIGURE 11·2. The relationship between the hydrophobicity of the side chain of the amino acid and k_{cat}/K_M for the hydrolysis of *N*-acetyl-L-amino acid methyl esters by chymotrypsin. Energies are in kcal/mol. [From V. N. Dorovskaya, S. D. Varfolomeyev, N. F. Kazanskaya, A. A. Klyosov, and K. Martinek, *FEBS Lett.* **23**, 122 (1972).]

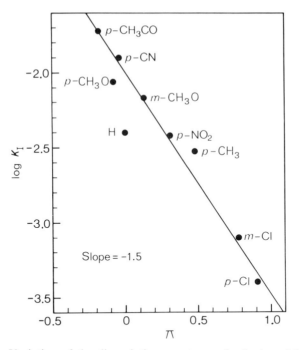

FIGURE 11·3. Variation of the dissociation constants of substituted formanilides with π for the partition between n-octanol and water.

stabilized by a salt bridge between the α-$NH_3{}^+$ group of Ile-16 and the $-CO_2{}^-$ of Asp-194. This conformation is in equilibrium with another in which the α-$NH_3{}^+$ group is free in solution. By measuring the equilibrium constants between the two forms at high pH, where the amino group is unprotonated, and at low pH, where it is in the $-NH_3{}^+$ form, it is found that the salt bridge stabilizes the active conformation by 12.1 kJ/mol (2.9 kcal/mol).[21] This is a buried salt bridge. Those on the surface of hemoglobin have lower stabilization energies.[22]

2. Why are enzymes more hydrophobic than organic solvents?

One reason why chymotrypsin is more hydrophobic than n-octanol is that the binding pocket is present in the absence of substrate, unlike the cavity that has to be made in an organic solvent to accommodate the solute transferred from the aqueous phase. Furthermore, the binding pocket contains some 16 molecules of water. Thus, hydrophobic bond formation with chymotrypsin is like forming *two* "normal" hydrophobic bonds, since there are two energetically unfavorable water–hydrophobic interfaces, one for the substrate and one for the enzyme-bound water.

The tight binding of methylene groups by the aminoacyl-tRNA synthetases has an additional cause. When valine occupies the binding site of the isoleucyl-tRNA synthetase, for example, there is a "hole" in the complex that would be occupied by the additional methylene group in isoleucine if it were bound. It was pointed out earlier that the dispersion energy of a methylene group in a crystalline hydrocarbon is 8.4 kJ/mol (2 kcal/mol). Since proteins are as closely packed as solids,[23] an empty "hole" adds a further 8.4 kJ/mol to the hydrophobic bond energy of the methylene group. The hydrophobic binding in these examples is clearly dominated by the contribution of dispersion energies, as opposed to the classic theory of the hydrophobic bond. Since dispersion energies are additive and so are dependent on the density of atoms, dispersion contributions will be important in enzymatic binding where specific, close-packed interactions are involved.

3. Summary

The dissociation constant of an enzyme and substrate reflects the *relative* stabilities of the substrate when it is bound to the enzyme and when it is free in solution. The constant depends on: the strength of the hydrogen bonding of the enzyme and the substrate to each other, compared with the combined strengths of their separate hydrogen bonding to water molecules; the stability of the salt linkages between the enzyme and the substrate, compared with the tendency of the individual ions to be solvated by water; the dispersion energies, compared with those in water; and also the hydrophobic bonding. These differences in energy, when they are summed over a whole molecule, can be quite considerable. It is of interest that the hydrogen bond and the salt linkage with the enzyme as well as the hydrophobic bond are favored entropically by the release of constrained water molecules.

The *absolute* energy of interaction between the enzyme and the substrate depends on the dispersion forces and the absolute energies of the hydrogen bonds and salt linkages. The hydrogen bonds and the salt linkages provide the strongest attractive forces. Since these are also able to stabilize unfavorable charge formation in the transition state, they are especially important in catalysis.

C. Entropy and binding[24]

The pairing of monomeric complementary bases, such as A (adenosine) and U (uracil), is not detectable in aqueous solution due to the competition of hydrogen bonding with water molecules. However, a triplet of bases binds strongly to a complementary anticodon of a tRNA. The triplet UUC (uracil-uracil-cytosine) binds to tRNA[Phe] with an association constant of

$2 \times 10^3 \ M^{-1}$.[25,26] Two tRNA molecules with complementary anticodons associate even more strongly; the association constant between $tRNA^{Phe}$ and $tRNA_2^{Glu}$ is $2 \times 10^7 \ M^{-1}$.[27] The reason for this increase in the strength of hydrogen bonding is entropy. When a single A and U associate, they gain the energy of the complementary base pairing but lose the energy of hydrogen bonding with water. They also lose their independent entropies of translation and overall rotation, but in turn there is a gain of entropy as the hydrogen-bonded water molecules are released. When a complementary pair of triplets bind, three times as many molecules of water are freed from hydrogen bonding, but there is still the loss of only one set of entropies of translation and overall rotation. The situation is very reminiscent of the advantages of an intramolecular reaction over its intermolecular counterpart (Chapter 2, section B4d). The lesson is that although single hydrogen bonds are weak in solution, multiple hydrogen bonds may be very stable.

A related phenomenon is that the binding of a dimer X—Y to an enzyme may be far greater than expected from the binding of X and Y separately, because the dimer loses only one set of translational and rotational entropies overall.

This phenomenon has been known for many years to inorganic chemists as the *chelate effect*. The magnitude may be illustrated by one of their examples, the replacement by ammonia and polyamines of some or all of the six water molecules that are coordinated to the Ni^{2+} ion. It is seen in Table 11·7 that there are enormous increases in the association constants of the ligands as the number of amino groups increases.

D. Protein-protein interactions

The interfaces of multisubunit proteins, such as hemoglobin, and of protein-protein complexes, such as the trypsin–trypsin inhibitor complex, are close-packed and snug-fitting. All hydrogen bonds and salt linkages are paired. In other words, the two surfaces are complementary to each

TABLE 11·7. *The chelate effect on the binding of amines to Ni^{2+}* [a]

Ligand	Association constants (M^{-1})					
	K_1	K_2	K_3	K_4	K_5	K_6
	468	132	41	12	4	0.8
NH_3	2×10^7	1×10^6	2×10^4	—	—	—
$H_2N(CH_2)_2NH_2$	6×10^{10}	1×10^8	—	—	—	—
$H_2N(CH_2)_2NH(CH_2)_2NH_2$	2×10^{14}	—	—	—	—	—
$H_2N(CH_2)_2NH(CH_2)_2NH(CH_2)_2NH_2$						
$(H_2N(CH_2)_2NH(CH_2)_2)_2NH$	3×10^{17}	—	—	—	—	—

[a] From the Chemical Society (London) Special Publications 17 (1964) and 25 (1971).

TABLE 11·8. *Some limiting values of binding energies*[a]

Binding cavity		Unfavorable binding energy	
Constructed for	Occupied by	kJ/mol	kcal/mol
—CH₃	—OH	14	3.5
—H	—CH₃	>32	>7.6
—H	—OH	15	3.7
—OH	—H	29	7.0
—NH₃⁺	—H	18	4.3

[a] Determined from measurements on aminoacyl-tRNA synthetase (see Table 11·6).

other. This would appear to be the basis of protein-protein recognition.

Although opinions differ on the overall contribution of hydrogen bonding and salt linkage to the total stabilization energy of protein-protein complexes, and it has been suggested that many association constants are entirely accounted for by hydrophobic bonding, there is no doubt that hydrogen bonds and electrostatic interactions are important for specificity.[28-30] Whatever the positive contribution of correctly formed hydrogen bonds and salt bridges in the "correct" complexes, the presence of *unpaired* hydrogen-bond donors/acceptors and ions in "incorrect" complexes provides considerable driving energy for their dissociation.

The binding experiments on the aminoacyl-tRNA synthetases are in many ways a model for protein-protein interactions, since both processes involve the stacking of side chains of amino acids. It is thus expected that specific interactions between alkyl side chains on one protein and a well-defined cavity in another could provide considerable energy (Table 11·6)—far above that expected from classic hydrophobic binding—and hence specificity. Data listed in Table 11·8 indicate the magnitudes of some of the interactions that could give rise to specificity: for example, the salt linkage of an —NH₃⁺ group or the interactions of an —OH group, as discussed earlier. An alternative process for specificity is the recognition of a too-large alkyl group on the "wrong" ligand; where necessary, an additional methyl group may cause sufficient steric hindrance to lower a binding constant by a factor of 10^5 to 10^6. Now that it is possible to change individual amino acid residues in proteins by site-directed mutagenesis (Chapter 14), there will in the near future be direct measurements of the factors that give rise to specificity and binding in protein-protein interactions.

References

1 G. E. Ewing, *Accts. Chem. Res.* **8**, 185 (1975).
2 A. Warshel, *Accts. Chem. Res.* **17**, 284 (1981).
3 A. Warshel and M. Levitt, *J. Molec. Biol.* **103**, 227 (1976).

4   A. T. Hagler, S. Lifson, and E. Huler in *Peptides, polypeptides, and proteins* (E. R. Blout, F. A. Bovey, M. Goodman, and N. Lotan, Eds.), Wiley, p. 35 (1974).
5   P. Dauber and A. T. Hagler, *Accts. Chem. Res.* **13**, 105 (1980).
6   A. T. Hagler, P. Dauber, and S. Lifson, *J. Am. Chem. Soc.* **101**, 5131 (1979).
7   W. Kauzmann, *Adv. Protein Chem.* **14**, 1 (1959).
8   M. H. Klapper, *Progr. Bioorg. Chem.* **2**, 55 (1973).
9   M. H. Abraham, *J. Am. Chem. Soc.* **101**, 5477 (1979).
10 T. Fujita, J. Iwasa, and C. Hansch, *J. Am. Chem. Soc.* **86**, 5175 (1964).
11 A. Leo, C. Hansch, and D. Elkins, *Chem. Rev.* **71**, 525 (1971).
12 R. B. Hermann, *J. Phys. Chem.* **76**, 2754 (1972).
13 M. J. Harris, T. Higuchi, and J. H. Rytting, *J. Phys. Chem.* **77**, 2694 (1973).
14 J. A. Reynolds, D. B. Gilbert, and C. Tanford, *Proc. Natn. Acad. Sci. USA* **71**, 2925 (1974).
15 C. Chothia, *Nature, Lond.* **248**, 338 (1974).
16 A. R. Fersht, J. S. Shindler, and W.-C. Tsui, *Biochemistry* **19**, 5520 (1980).
17 D. V. Santi and V. A. Peña, *J. Med. Chem.* **16**, 273 (1973).
18 J. R. Knowles, *J. Theoret. Biol.* **9**, 213 (1965).
19 V. N. Dorovskaya, S. D. Varfolomeyev, N. F. Kazanskaya, A. A. Klyosov, and K. Martinek, *FEBS Lett.* **23**, 122 (1972).
20 J. Fastrez and A. R. Fersht, *Biochemistry* **12**, 1067 (1973).
21 A. R. Fersht, *J. Molec. Biol.* **64**, 497 (1972).
22 M. F. Perutz, *Nature, Lond.* **228**, 726 (1970).
23 M. I. Page, *Biochem. Biophys. Res. Comm.* **72**, 456 (1976).
24 W. P. Jencks, *Adv. Enzymol.* **43**, 219 (1975).
25 J. Eisinger, B. Feuer, and T. Yamane, *Nature New Biology, Lond.* **231**, 126 (1971).
26 O. Pongs, R. Bald, and E. Reinwald, *Eur. J. Biochem.* **32**, 117 (1973).
27 J. Eisinger and N. Gross, *Biochemistry* **14**, 4031 (1975).
28 W. P. Jencks, *Catalysis in chemistry and enzymology*, McGraw-Hill, pp. 351, 399 (1969).
29 C. Chothia and J. Janin, *Nature, Lond.* **256**, 705 (1975).
30 J. Janin and C. Chothia, *J. Molec. Biol.* **100**, 197 (1976).

Further reading

W. P. Jencks, *Catalysis in chemistry and enzymology*, McGraw-Hill, Chapters 6–9 (1969).
W. P. Jencks, "Binding energy, specificity, and enzymic catalysis—the Circe effect," *Adv. Enzymol.* **43**, 219–410 (1975).

12

Enzyme–substrate complementarity and the use of binding energy in catalysis

A. Utilization of enzyme–substrate binding energy in catalysis

Chapter 2 explained how a combination of entropic factors, acid-base catalysis, and electrostatic effects can account for a large fraction of the magnitude of the catalysis by some enzymes. In addition to these catalytic factors, the one outstanding characteristic of enzymes is that they specifically bind their substrates, and the binding energies involved may be very large. Ever since J. B. S. Haldane suggested in 1930 that these energies may be used to distort the substrate to the structure of the products,[1] theoreticians have explored the various ways in which the binding energy of the enzyme and substrate may be used to lower the activation energy of the chemical steps.

Transition state theory is particularly useful in analyzing theories of enzyme catalysis. We shall now apply this approach to the simple Michaelis-Menten mechanism (where $K_M = K_S$; Chapter 3) to see how the binding energy automatically lowers the activation energy of k_{cat}/K_M and how some of the binding energy may be used to lower the activation energy of k_{cat}.[2]

1. Binding energy lowers the activation energy of k_{cat}/K_M

$$\text{E} + \text{S} \underset{\Delta G_S}{\overset{K_M}{\rightleftharpoons}} \text{ES} \xrightarrow[\Delta G^{\ddagger}]{k_{cat}} \text{products} \tag{12·1}$$

$$\text{E} + \text{S} \underset{\Delta G_T^{\ddagger}}{\overset{k_{cat}/K_M}{\rightleftharpoons}} \text{ES}^{\ddagger} \tag{12·2}$$

Recall from Chapter 3 that the rate constant for free enzyme reacting with free substrate to give products is k_{cat}/K_M (equation 12·1). Expressed in terms of transition state theory (Chapter 2), the equilibrium constant between E + S and the transition state $\text{ES}^{\ddagger}$ is proportional to the activation energy $\Delta G_T^{\ddagger}$ of k_{cat}/K_M (equation 12·2). This activation energy is composed of two terms, an energetically unfavorable term $\Delta G^{\ddagger}$, due to the activation energy of the chemical steps of bond making and breaking, and a compensating energetically favorable term ΔG_S, due to the realization of the binding energy. That is,

$$\Delta G_T^{\ddagger} = \Delta G^{\ddagger} + \Delta G_S \qquad (12\cdot3)$$

(where $\Delta G_T^{\ddagger}$ and $\Delta G^{\ddagger}$ are algebraically positive and ΔG_S is negative). This is illustrated in Figure 12·1 for the simple Michaelis-Menten mechanism.

Substituting equation 12·3 into equation 2·5 to express k_{cat}/K_M in terms of transition state theory gives

$$RT \ln \frac{k_{cat}}{K_M} = RT \ln \frac{kT}{h} - \Delta G^{\ddagger} - \Delta G_S \qquad (12\cdot4)$$

2. Interconversion of binding and chemical activation energies

The maximum binding energy between an enzyme and a substrate occurs when each binding group on the substrate is matched by a binding site on the enzyme. In this case the enzyme is said to be complementary in structure to the substrate. Since the structure of the substrate changes throughout the reaction, becoming first the transition state and then the

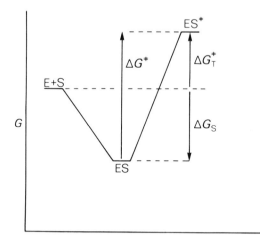

FIGURE 12·1. Gibbs energy changes for the scheme

$$E + S \underset{K_S}{\rightleftharpoons} ES \xrightarrow{k_{cat}} products$$

(ΔG_S is algebraically negative, and $\Delta G^{\ddagger}$ and $\Delta G_T^{\ddagger}$ are positive.)

products, the structure of the *undistorted* enzyme can be complementary to only one form of the substrate. We shall show (section A3b) that it is catalytically advantageous for the enzyme to be complementary to the structure of the transition state of the substrate rather than to the original structure. If this happens, the increase in binding energy as the structure changes to that of the transition state lowers the activation energy of k_{cat}. On the other hand, if the enzyme is complementary to the structure of the unaltered substrate, the decrease in binding energy on the formation of the transition state will increase the activation energy of k_{cat}.

Because these ideas are crucial to the understanding of the role of binding energy in enzyme catalysis, they are amplified in the remainder of this section. Suppose that we are able to increase the binding energy of a particular enzyme with its substrate by adding an extra group to one of the amino acid side chains. For example, we could add a hydroxyl group to an alanine residue, to convert it to a serine that could hydrogen-bond with the substrate and contribute a binding energy of ΔG_R. Now we examine the consequences of controlling at which stage in the reaction the binding actually takes place.

First we consider Figure 12·2, which represents the condition of [S] greater than K_M, so that the enzyme is saturated with substrate and the reaction rate is $v = k_{cat}[E]_0$. There are three extreme cases. The simplest is when the full binding energy of the group is realized equally well in the enzyme–substrate (ES) and the transition state (ES‡) complexes, so that the Gibbs free energies of both are lowered equally. In other words, there is both enzyme–substrate and enzyme–transition state complementarity. As shown in Figure 12·2, this decreases K_M but it does not affect the value of k_{cat}. Thus, when the enzyme is saturated with substrate and the binding energy is realized equally in ES and ES‡, the rate of reaction is not increased and there is no catalytic advantage to adding the group. In the second case, we suppose that the binding energy is realized only in the enzyme–substrate complex. Its energy is lowered whereas that of the transition state remains the same. The activation energy of k_{cat} increases, so the rate of reaction decreases. This example corresponds to enzyme–substrate complementarity. In the third case, which corresponds to enzyme–transition state complementarity, the binding energy is realized only in the transition state complex. Its energy is lowered while that of the enzyme–substrate complex remains the same. This lowers the activation energy of k_{cat} and so increases the rate of reaction.

We consider now the analogous experiments under the condition of [S] less than K_M, as in Figure 12·3. Here the Gibbs energy of ES is greater than that of E + S, and the reaction rate is $v = (k_{cat}/K_M)[E]_0[S]$. Under this condition, both the cases of enzyme–transition state and of combined enzyme–substrate and enzyme–transition state complementarity lower the

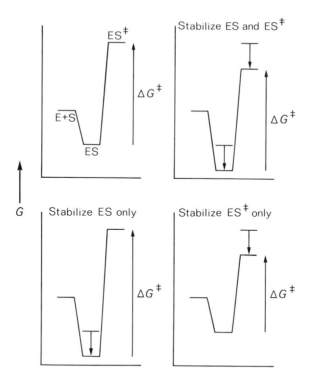

FIGURE 12·2. Where to utilize some extra binding energy ΔG_R when $[S] > K_M$? The Gibbs energy changes are for the reaction under the experimental condition of saturating $[S]$, so that $v = k_{cat}[E]_0$. On stabilization of only $ES^{\ddagger}$, the activation energy is lowered by ΔG_R, whereas stabilization of only ES leads to an increase of activation energy by that amount. Stabilization of ES and $ES^{\ddagger}$ equally has a neutral effect.

activation energy by ΔG_R, and hence increase rate. Where the additional binding energy of the extra group stabilizes only the enzyme–substrate complex—i.e., where there is enzyme–substrate complementarity only —there is no rate advantage from the additional binding energy.

The situation in which the enzyme has a group that can bind to the substrate only in the transition state is an example of *transition state stabilization*. It is important to note that the presence on the enzyme of a group that can bind only to the transition state of the substrate decreases

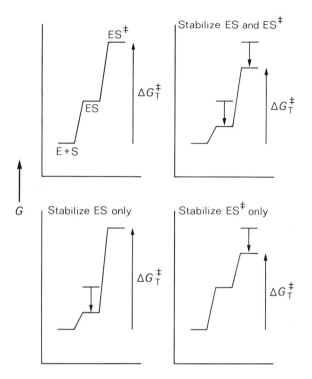

FIGURE 12·3. Where to utilize some extra binding energy ΔG_R when $[S] < K_M$? The Gibbs energy changes are for the reaction condition of subsaturating $[S]$, so that $v = (k_{cat}/K_M)[E]_0[S]$. The activation energy is lowered by ΔG_R on the stabilization of only $ES^{\ddagger}$, or of ES and $ES^{\ddagger}$ (in the latter case, as long as $[S]$ remains below K_M; otherwise, there is a transition to Figure 12·2). Stabilization of ES only does not affect $\Delta G_T^{\ddagger}$.

the activation energy for a chemical step, and, further, does not have to distort the substrate to do so.

   The above experiments are not fanciful. Using the method of site-directed mutagenesis (Chapter 14), it is now possible to alter amino acid residues in a protein as in the Ala → Ser example. The results of one such study are discussed in section B3.

   In the following section, we shall analyze quantitatively by transition state theory the consequences of enzyme–substrate complementarity. To do this we shall divide the components of activation energies into two groups: those arising from chemical terms and those arising from changes in the binding energy as the reaction proceeds.

The idea of complementarity in enzyme–substrate interactions was introduced by E. Fischer with his famous "lock and key" analogy.[3] In modern terminology this would represent enzyme–substrate complementarity. The currently favored concept of enzyme–transition state complementarity was introduced by Haldane[1] and elaborated by L. Pauling.[4]

3. Enzyme complementarity to transition state implies that k_{cat}/K_M is at a maximum

a. Enzyme complementarity to the initial substrate

Suppose that the maximum amount of intrinsic binding energy available is ΔG_b. In this case the full ΔG_b is realized in the initial enzyme–substrate complex, so binding will be good; that is, K_M, the dissociation constant of the enzyme–substrate complex, will be low. But the formation of the transition state will lead to a reduction in binding energy as the substrate geometry changes to give a poorer fit, and so will lower k_{cat} as described above. If the adverse energy change caused by the poorer fit is ΔG_R, and the free energy of activation due to the chemical bond making and breaking involved in k_{cat} is $\Delta G_0^{\ddagger}$ (Figure 12·4), then the observed free energy of activation for k_{cat} is given by

$$\Delta G^{\ddagger} = \Delta G_0^{\ddagger} + \Delta G_R \tag{12·5}$$

and

$$\Delta G_S = \Delta G_b \tag{12·6}$$

The Gibbs energy of activation for k_{cat}/K_M is given by $\Delta G^{\ddagger} + \Delta G_S$, i.e.,

$$\Delta G_T^{\ddagger} = \Delta G_0^{\ddagger} + \Delta G_R + \Delta G_b \tag{12·7}$$

b. Enzyme complementarity to the transition state

Here the full binding energy ΔG_b is realized in the transition state. The adverse energy term ΔG_R in the initial enzyme–substrate complex will increase K_M, but the gain in binding energy as the reaction reaches the transition state will increase k_{cat}. Thus,

$$\Delta G^{\ddagger} = \Delta G_0^{\ddagger} - \Delta G_R \tag{12·8}$$

and

$$\Delta G_S = \Delta G_b + \Delta G_R \tag{12·9}$$

Again, the Gibbs energy of activation for k_{cat}/K_M is given by

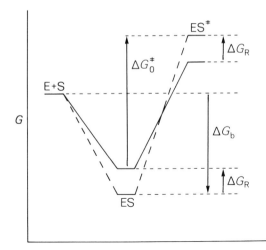

FIGURE 12·4. Gibbs energy changes for the schemes in Figures 12·1 and 12·2, when the enzyme is complementary in structure to (1) the substrate (dashed curve), and (2) the transition state (solid curve). [ΔG_b is algebraically negative, and $\Delta G_0^{\ddagger}$ and ΔG_R are positive.]

$$\Delta G_T^{\ddagger} = \Delta G^{\ddagger} + \Delta G_S$$

i.e.,

$$\Delta G_T^{\ddagger} = \Delta G_0^{\ddagger} + \Delta G_b \qquad (12·10)$$

because ΔG_R cancels out.

Comparison of equations 12·7 and 12·10 shows that k_{cat}/K_M is higher by a factor of $\exp(\Delta G_R/RT)$ when the enzyme is complementary to the transition state rather than to the initial substrate. It should be noted that in enzyme–transition state complementarity, k_{cat}/K_M is independent of the interactions in the initial enzyme–substrate complex, as shown by the term ΔG_R dropping out of the equations.

B. Experimental evidence for the utilization of binding energy in catalysis and enzyme–transition state complementarity

1. Classic experiments: Structure–activity relationships of modified substrates

Some of the most instructive evidence for the utilization of binding energy comes from kinetic experiments on the serine proteases. We recall from Chapter 1 that these enzymes have a series of subsites for binding the amino acid residues of their polypeptide substrates. Table 12·1 shows that as larger groups occupy the leaving group site in chymotrypsin, their binding energy is used to increase k_{cat}/K_M. Similarly, increasing the length of the polypeptide chain of substrates of elastase increases k_{cat}/K_M. Interestingly, in the examples given, the binding energy of the additional groups does not lower K_M; i.e., the binding energy is not used to bind the substrate, but rather to increase k_{cat}.

Similar behavior is observed with pepsin. As shown in Table 12·1, the hydrolyses of a wide range of substrates are catalyzed with K_M's of

TABLE 12·1. *Interconversion of activation and binding energies*

Enzyme and substrate	k_{cat} (s^{-1})	K_M (mM)	k_{cat}/K_M $(s^{-1} M^{-1})$
Chymotrypsin[a]			
Ac-Tyr-NH₂	0.17	32	5
Ac-Tyr-Gly-NH₂	0.64	23	28
Ac-Tyr-Ala-NH₂	7.5	17	440
Ac-Phe-NH₂	0.06	31	2
Ac-Phe-Gly-NH₂	0.14	15	10
Ac-Phe-Ala-NH₂	2.8	25	114
Elastase[b] (cleavage at —NH₂)			
Ac-Ala-Pro-Ala-NH₂	0.09	4.2	21
Ac-Pro-Ala-Pro-Ala-NH₂	8.5	3.9	2.2×10^3
Ac-Gly-Pro-Ala-NH₂	0.02	33	0.5
Ac-Pro-Gly-Pro-Ala-NH₂	2.8	43	64
Pepsin[c] (cleavage of Phe-Phe bond in A-Phe-Phe-OP4P)			
Phe-Gly	0.5	0.3	1.7×10^3
Z-Phe-Gly	25	0.11	2.2×10^5
Z-Ala-Gly	145	0.25	5.8×10^5
Z-Ala-Ala	282	0.04	7×10^6
Z-Gly-Ala	409	0.11	3.7×10^6
Z-Gly-Ile	13	0.07	1.8×10^5
Z-Gly-Leu	134	0.03	4.2×10^6
Phe-Gly-Gly	6	0.6	1×10^4
Z-Phe-Gly-Gly	127	0.13	9.8×10^5
Mns[d]	0.002	0.1	20
Mns-Gly[d]	0.13	0.03	3.7×10^3
Mns-Gly-Gly[d]	16	0.07	2.3×10^5
Mns-Ala-Ala[d]	112	0.06	2×10^6

[a] At 25°C, pH 7.9. [From W. K. Baumann, S. A. Bizzozero, and H. Dutler, *FEBS Lett.* **8**, 257 (1970); *Eur. J. Biochem.* **39**, 381 (1973).]
[b] At 37°C, pH 9. [From R. C. Thompson and E. R. Blout, *Biochemistry* **12**, 51 (1973).]
[c] At 37°C, pH 3.5; OP4P = 3-(4-pyridyl)propyl-1-oxy. The N-terminal portions "A" are listed. [From G. P. Sachdev and J. S. Fruton, *Biochemistry* **9**, 4465 (1970).]
[d] At 25°C, pH 2.4; Mns = mansyl. [From G. P. Sachdev and J. S. Fruton, *Proc. Natn. Acad. Sci. USA* **72**, 3424 (1975).]

Further compilations for chymotrypsin, elastase, and α-lytic protease are given by: C. A. Bauer, *Biochemistry* **17**, 375 (1978); C. A. Bauer, G. D. Brayer, A. R. Sielecki, and M. N. G. James, *Eur. J. Biochem.* **120**, 289 (1981); S. A. Bizzozero, W. K. Baumann, and H. Dutler, *Eur. J. Biochem.* **122**, 251 (1982).

about 0.1 mM. The additional binding energy of the groups on the larger substrates is again used to increase the k_{cat} rather than to decrease the K_M. The k_{cat}/K_M is accordingly higher for the larger substrates.

A striking series of examples occurs with the chymotrypsin-catalyzed hydrolysis of synthetic ester substrates (Table 12·2). As the size of the hydrophobic side chain that fits into the hydrophobic primary binding site of the enzyme is increased, k_{cat}/K_M increases over a range of 10^6. The increase in the binding energy of the larger substrates is distributed between lowering the dissociation constant of the enzyme–substrate complex and increasing the acylation and deacylation rate constants.

Section C explains how this use of binding energy to increase k_{cat} rather than to lower K_M gives higher reaction rates.

2. Transition state analogues: Probes of complementarity[5,6]

Direct evidence for enzyme–transition state complementarity has come from x-ray diffraction experiments on the serine proteases and lysozyme (section D5c of this chapter, and also Chapter 1), and from studies on the binding of transition state analogues. This approach was suggested by Pauling in the 1940s.[4] Chemists, with their knowledge of organic reaction mechanisms, can guess at the structure of the transition state of an enzymatic reaction. Compounds that mimic the transition state may then be synthesized and their binding to the enzyme compared with that of the substrate.

a. Lysozyme and glucosidase

Lysozyme catalyzes the hydrolysis of the polysaccharide component of plant cell walls and synthetic polymers of $\beta(1 \rightarrow 4)$-linked units of N-acetylglucosamine (NAG) (Chapter 1). It is expected from studies on

TABLE 12·2. *Kinetic parameters for the hydrolysis of N-acetyl amino acid methyl esters by chymotrypsin*[a]

$$(E + RCO_2CH_3 \xrightleftharpoons{K_S} E \cdot RCO_2CH_3 \xrightarrow{k_2} RCO—E \xrightarrow{k_3} E + RCO_2H)$$

Amino acid	k_2 (s^{-1})	k_3 (s^{-1})	K_S (mM)	k_{cat}/K_M $(s^{-1} M^{-1})$
Gly	0.49	0.14	3.38×10^3	0.13
But	8.8	1.7	417	21
Norval	35.6	5.93	100	360
Norleu	103	19	34	3×10^3
Phe	796	111	7.6	1×10^5
Tyr[b]	5×10^3	200	17	3×10^5

[a] α-Chymotrypsin at 25°C, pH 7.8. (Data from Table 7·3.)
[b] Ethyl ester.

nonenzymatic reactions that one of the intermediates in the hydrolytic reaction is a carbonium ion in which the conformation of the glucopyranose ring changes from a full-chair to a sofa (half-chair) conformation (Chapter 1). The transition state analogue I, in which the lactone ring mimics the carbonium ion–like transition state II, binds tightly to lysozyme: $K_{diss} = 8.3 \times 10^{-8}\ M$.[7]

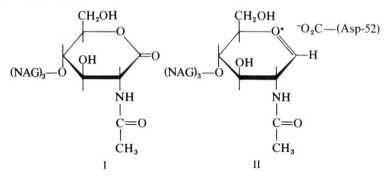

$$\text{I} \qquad\qquad \text{II}$$

This may be compared with the dissociation constants of $10^{-5}\ M$ and $5 \times 10^{-6}\ M$ for $(NAG)_4$ and NAG-NAM-NAG-NAG binding in the A, B, C, and D subsites.[7,8] The 100-fold tighter binding of the transition state analogue may be due in part to the electrostatic interaction of the negatively charged Asp-52 with the partial positive charge on the carbonyl carbon of the lactone.

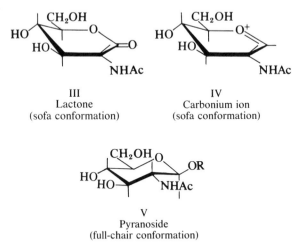

III
Lactone
(sofa conformation)

IV
Carbonium ion
(sofa conformation)

V
Pyranoside
(full-chair conformation)

More striking is the binding of the lactone III to β-N-acetyl-D-glucosaminidase. The dissociation constant of $5 \times 10^{-7}\ M$ is 4000 times smaller than the K_M of $2 \times 10^{-3}\ M$ for the pyranoside substrate V.[9] However, it is possible in this example that the enzyme forms a covalent bond with the analogue so that the tight binding does not result solely from noncovalent binding.[10]

b. Proline racemase

During the racemization of proline (structure VI), the chiral carbon must at some stage become trigonal. In accordance with a trigonal transition state, both structures VII and VIII bind 160 times more tightly than proline.[11,12]

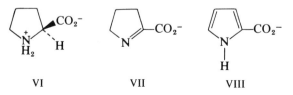

| VI | VII | VIII |

c. Cytidine deaminase

The dissociation constant of tetrahydrouridine (structure IX) is about 10 000 times smaller than the combined constants for the reaction products, uridine (structure X) and ammonia. Tetrahydrouridine presumably resembles the transition state, which is similar to the tetrahedral intermediate XI.[13]

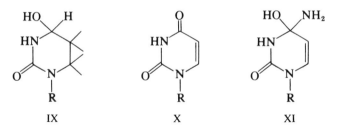

| IX | X | XI |

d. Assessment of the results of binding transition state analogues

The transition state analogues that have been designed so far give a measure of the part of the catalysis that is due to the difference in complementarity of the enzyme for the transition state and the substrate. In the four examples given above, the transition state analogues bind between 10^2 and 10^4 times more tightly than the original substrates. This is good evidence that enzymes have evolved to be complementary in structure to the transition state. Furthermore, it shows that k_{cat} may be increased by a factor of 10^2 to 10^4 at the expense of increasing K_M. Considering that these synthetic analogues might be extremely crude in mimicking the transition state structure, it seems likely that the increase in complementarity between the substrate and the transition state is worth at least 20 kJ/mol (5 kcal/mol). As will be discussed in Chapter 15, the binding site for the carbonyl oxygen of a substrate of a serine protease is deficient by one hydrogen bond. When the transition state for the reaction is formed, the additional hydrogen bond is made. This increase in complementarity must also be worth a factor of about 20 to 25 kJ/mol (5 to 6 kcal/mol).

e. Some possible errors in interpreting transition state analogue data
The effects of enzyme–transition state complementarity on the binding of transition state analogues may be masked by extraneous binding artifacts. Chapter 11 showed that small groups can involve large binding energies when a specific binding site is involved. For example, a methylene group may contribute 12 kJ/mol (3 kcal/mol), and a hydroxyl in a hydrogen bond 30 kJ/mol (7 kcal/mol), to the binding energy. Also, where specific binding sites are not involved, fairly large energies may come from general hydrophobic effects: the substitution of a chloro group for an acetyl on a phenyl ring on an anilide substrate causes it to bind 50 times more tightly to the hydrophobic pocket of chymotrypsin.[14]

Difficulties also arise with analogues for multisubstrate reactions due to the chelate effect. It was pointed out in Chapter 11 that multidentate ligands, such as EDTA, bind tightly to metal ions whereas unidentate ligands do not. The difference is due to entropy; the binding of six unidentate ligands leads to the loss of six sets of translational and rotational entropies. The same applies to the binding of a "multisubstrate" or a "multiproduct" analogue to an enzyme. Assume, for instance, that A and B bind separately and adjacently to an enzyme with Gibbs energies of association x and y kJ, respectively. Then if A–B binds in an identical manner, its free energy of association is given by

$$\Delta G_{ass} = x + y + S \tag{12·11}$$

where S is an energetically favorable term because only one set of entropies is lost on the binding of A–B, compared with two on the binding of A and B separately. The binding of multisubstrate analogues should be very tight, without any effects due to enzyme–transition state complementarity. For example, the binding of structure XII to aspartate trans-

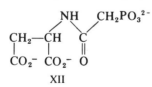

XII

carbamoylase is very tight; $K_{diss} = 2.7 \times 10^{-8}\ M$. However, this value is only equal to the product of the dissociation constants of succinate and carbamoyl phosphate ($9 \times 10^{-4} \times 2.7 \times 10^{-5}\ M^{-2}$), which approximate to the fragments of XII.[15] The transition state structure is derived from XIII.

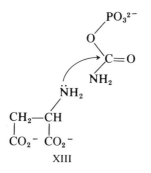

XIII

3. The modern approach: Structure–activity experiments with modified enzymes

As will be discussed in detail in Chapter 14, amino acid residues in an enzyme may be changed in a systematic manner by using site-directed mutagenesis. It is thus possible to prepare mutant enzymes that lack a side chain involved in binding the substrate, and to thereby measure the

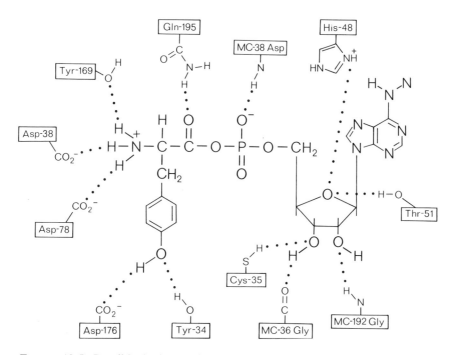

FIGURE 12·5. Possible hydrogen bonds between the tyrosyl-tRNA synthetase (from *B. stearothermophilus*) and tyrosyl adenylate, as indicated by x-ray protein crystallography. MC = the bonding to the polypeptide main chain. (Courtesy of D. M. Blow and P. Brick.)

TABLE 12·3. *Kinetic parameters for the activation of tyrosine by the tyrosyl-tRNA synthetase and its synthetic mutants*[a]

Enzyme	k_{cat} (s^{-1})	K_M (mM)	k_{cat}/K_M (s^{-1} M^{-1})
Wild-type	7.6	0.9	8.4×10^3
His-48 → Gly-48	1.6	1.4	1.14×10^3
Cys-35 → Gly-35	2.8	2.6	1.12×10^3
His-45 → Asn-45	0.003	1.0	3

[a] The enzyme from *B. stearothermophilus*. The reactions are at 25°C, pH 7.8. The constants are for variation of [ATP]. (From text references 16–18.)

effects of the binding energy on catalysis. The first such study has now been performed, using the tyrosyl-tRNA synthetase.[16] It may be recalled from Chapter 7, section D, that this class of enzymes catalyzes the aminoacylation of tRNA in a two-step reaction in which the enzyme first forms an enzyme-bound aminoacyl adenylate complex:

$$E + Tyr + ATP \longrightarrow E \cdot Tyr\text{-}AMP + PP_i \qquad (12 \cdot 12)$$

The complex is relatively stable in the absence of tRNA or pyrophosphate, and this has facilitated the solution of the crystal structure of the complex of the enzyme from *Bacillus stearothermophilus* by x-ray diffraction. The hydrogen-bonding contacts between the enzyme and tyrosyl adenylate are illustrated in Figure 12·5. Mutant enzymes have been prepared in which two of the residues binding the ribose ring, Cys-35 and His-48, have been mutated to glycine residues (which lack side chains).[17,18] The kinetic data for the formation of tyrosyl adenylate by these mutant enzymes (Table 12·3) show that they have lower values of k_{cat} as well as higher values of K_M. We can thus perform the experiment outlined at the beginning of this chapter, and note the effects of adding an extra side chain to the binding site of an enzyme. When the mutant with Gly-48 is converted to the wild-type enzyme containing His-48, most of the intrinsic binding energy of the side chain is used to increase k_{cat}, and only a little is used to decrease K_M. Most striking, however, is the effect of His-45. This residue makes no interactions with the tyrosyl adenylate but is positioned to form a hydrogen bond with the γ-phosphoryl group of ATP as it is attacked by the carboxylate of tyrosine. His-45 lowers the energy of the transition state by 19 kJ/mol but does not affect the binding of ATP.

C. Evolution of the maximum rate: Strong binding of the transition state and weak binding of the substrate

In the last section we showed that enzymes have evolved to bind the transition states of substrates more strongly than the substrates themselves. It will now be seen that it is catalytically advantageous to bind substrates *weakly*.

Although enzyme–transition state complementarity maximizes k_{cat}/K_M, this is not a sufficient criterion for the maximization of the overall reaction rate. The reason is that the maximum reaction rate for a particular concentration of substrate depends on the individual values of k_{cat} and K_M. It can be seen in Table 12·4, where some rates are calculated for various values of k_{cat} and K_M (subject to k_{cat}/K_M being kept constant), that maximum rates are obtained for K_M greater than [S]. The maximization of rate requires *high* values of K_M. That is, enzymes should have evolved to bind substrates weakly.

1. The principle of maximization of K_M at constant $k_{cat}/K_M{}^2$

This principle contradicts the widely held belief that strong binding, or a low K_M, is an important component of enzymatic catalysis. The two additional proofs that follow emphasize the importance of high K_M's, and the graph indicates the physical reason.

a. Graphical illustration of the importance of high K_M's
Suppose that, as in Figure 12·6a, the substrate is at a higher concentration than the K_M for the reaction. The enzyme–substrate complex is at a lower energy than the free enzyme and substrate, and the activation energy is $\Delta G_T^{\ddagger} + \Delta G$. However, if, as in Figure 12·6b, everything is the same except that K_M is now higher than [S], so that the ES complex is at a higher energy than E + S, the activation energy is at the lower value of $\Delta G_T^{\ddagger}$.

TABLE 12·4. *Illustration of the importance of the evolution of k_{cat} and K_M at constant values of k_{cat}/K_M and [S]* [a]

$(k_{cat}/K_M = 10^6 \ M^{-1} \ s^{-1}, \ [S] = 10^{-3} \ M)$

K_M (M)	k_{cat} (s^{-1})	Rate[b] (s^{-1})
10^{-6}	1	1
10^{-5}	10	9
10^{-4}	10^2	90
10^{-3}	10^3	500
10^{-2}	10^4	909
10^{-1}	10^5	990
1	10^6	999

[a] The hypothetical processes are: (1) The enzyme has evolved to be complementary to the transition state of the substrate, so that k_{cat}/K_M is maximized. (2) While maintaining k_{cat}/K_M, the enzyme evolves to *increase* K_M. The values assigned to k_{cat}/K_M and [S] are arbitrary.
[b] Moles of product produced per mole of enzyme per second.

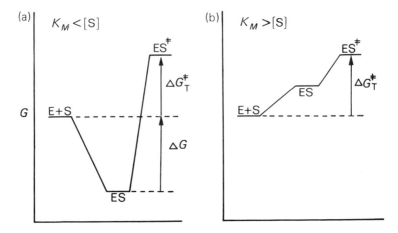

FIGURE 12·6. Two cases of enzyme evolution. In both cases the enzymes bind the transition states equally well, but in (a) the substrate is bound strongly, and in (b) the enzyme has evolved to bind the substrate weakly ([S] is the same in both graphs). The activation energy in (a) is for $ES \rightarrow ES^{\ddagger}$, i.e., $\Delta G_T^{\ddagger} + \Delta G$, whereas in (b) it is for $E + S \rightarrow ES^{\ddagger}$, i.e., $\Delta G_T^{\ddagger}$. (The changes in Gibbs energies are for the concentration of substrate used in the experiment, and not for standard states of 1 M.)

The low K_M leads to a thermodynamic "pit" into which the reaction falls and has to climb out. The high K_M leads to the enzyme–substrate complex being "a step up the thermodynamic ladder."

b. An algebraic illustration
In Chapter 3 it was shown that the Michaelis-Menten equation may be cast into the useful form

$$v = [E][S] \frac{k_{cat}}{K_M} \qquad (12\cdot13)$$

to relate the reaction rate to the concentration of free enzyme, [E].
    The evolution of an enzyme to give maximum rate may be divided into two hypothetical steps based on equation 12·13:

1. The value of k_{cat}/K_M is maximized by having the enzyme be complementary to the transition state of the substrate.
2. The [E] is maximized by having the K_M be high, so that as much of the enzyme as possible is in the unbound form.

    There is an evolutionary pressure on K_M to increase with a consequent increase in k_{cat}. This evolutionary pressure rapidly decreases as K_M becomes greater than [S]. At $K_M = [S]$, half of the enzyme is unbound, so according to equation 12·12 the rate is 50% of the maximum possible. At

$K_M = 5[S]$, five-sixths of the enzyme is unbound, so the rate is 83% of the maximum. Any further increase in K_M gives only a marginal increase in the rate.

How far a K_M evolves relative to the substrate concentration depends on the change in structure when the substrate becomes the transition state. A limit must eventually be reached; at this point, any increase in K_M must be matched by a weakening of transition state binding. The problem will be most severe for large metabolites present at high concentrations.

c. Exceptions to the principle of high K_M's: Control enzymes
Implicit in the preceding arguments for high K_M's is the assumption of the maximization of rate. Although this is valid for most enzymes, there are cases in which rate is subordinate to *control*. Metabolic pathways are characterized by their regulation, which is usually maintained by the activity of certain key enzymes on the pathway (Chapter 10, section F). The activities of these control enzymes are themselves often controlled by variations in the K_M's of critical substrates via allosteric effects. The K_M's for control enzymes have evolved for the purposes of regulation, and are not necessarily subject to the rate arguments of the previous section.

A low K_M could sometimes be advantageous for the first enzyme on a metabolic pathway. This would then control the rate of entry to the pathway and prevent it from being overloaded and accumulating reactive intermediates. For example, hexokinase, the first enzyme in glycolysis, has a K_M for glucose of 0.1 mM, whereas the concentration of glucose in the human erythrocyte is about 5 mM. A tenfold increase or decrease in the glucose concentration will hardly alter the rate of glycolysis.

2. Experimental observations on K_M's

As was shown in Chapter 11, the binding energy of an enzyme and substrate is potentially very high. However, K_M's are usually found to be relatively high. An extreme example of this is NAD$^+$. This large substrate has two ribose moieties, one adenine ring, a nicotinamide residue, and a pyrophosphate linkage. If all the intrinsic binding energy of these groups were realized, a dissociation constant of less than 10^{-20} M could be attained. Indeed, the dissociation constant for the binding of the first NAD$^+$ to the tetrameric glyceraldehyde 3-phosphate dehydrogenase has been found to be immeasurably strong at less than 10^{-11} M.[19] Yet the K_M's and dissociation constants of NAD$^+$ with dehydrogenases are often found to be in the range of 0.1 to 1 mM. Even more striking is the dissociation constant of 10^{-13} M for ATP and myosin,[20] which may be compared with the K_M's of 0.1 to 10 mM that are often found for ATP.

The comparison of K_M's with physiological substrate concentrations

TABLE 12·5. *Metabolite concentrations and K_M's for some glycolytic enzymes*[a]

Enzyme	Source	Substrate	Concentration (μM)	k_M (μM)	K_M/[S]
Glucose	Brain	G6P	130	210	1.6
6-phosphate	Muscle[b]	G6P	450	700	1.6
isomerase		F6P	110	120	1.1
Aldolase	Brain	FDP	200	12	0.06
	Muscle[c]	FDP	32	100	3.1
		G3P	3	1000	333
		DHAP	50	2000	40
Triosephosphate	Erythrocyte[d]	G3P	18	350	19
isomerase	Muscle[e]	G3P	3	460	153
		DHAP	50	870	17
Glyceraldehyde	Brain	G3P	3	44	15
3-phosphate	Muscle[f]	G3P	3	70	23
dehydrogenase		NAD	600	46	0.08
		P_i	2000		>10[g]
Phosphoglycerate	Brain	1,3DPG	<1	9	>9
kinase		ADP	1500	70	0.05
	Erythrocyte[h]	3PG	118	1100	9.3
	Muscle[i]	3PG	60	1200	200
		ADP	600	350	0.6
Phosphoglycerate	Brain	3PG	40	240	6
mutase	Muscle[j]	3PG	60	5000	83
Enolase	Brain	2PG	4.5	33	7
	Muscle[k]	2PG	7	70	10
Pyruvate kinase[l]	Erythrocyte[m]	PEP	23	200	9
		ADP	138	600	4.4
Lactate	Brain	Pyr	116	140	1.2
dehydrogenase	Erythrocyte[n]	Pyr	51	59	1.2
		Lac	2900	8400	2.9
		NADH	0.01[o]	10[p]	100
		NAD	33	150	4.6
Glycerol	Mouse	Gly-P	170	37	0.22
phosphate	Muscle[q]	Gly-P[r]	220	190	0.9
dehydrogenase		DHAP	50	190	3.8

[a] Abbreviations: G6P = glucose 6-phosphate, F6P = fructose 6-phosphate, FDP = fructose 1,6-diphosphate, G3P = glyceraldehyde 3-phosphate, DHAP = dihydroxyacetone phosphate, P_i = orthophosphate, 1,3DPG = 1,3-diphosphoglycerate, 3PG = 3-phosphoglycerate, 2PG = 2-phosphoglycerate, PEP = phosphoenolpyruvate, Pyr = pyruvate, Lac = lactate (all D-sugars); Gly-P = L-glycerol phosphate. Mouse brain enzymes and mouse brain metabolites from O. H. Lowry and J. V. Passonneau, *J. Biol. Chem.* **239**, 31 (1964). Human erythrocyte metabolites from S. Minakami, T. Saito, C. Suzuki, and H. Yoshikawa, *Biochem. Biophys. Res. Commun.* **17**, 748 (1964). Human erythrocyte enzymes: see below.

is difficult in many cases due to a lack of knowledge of the concentrations, but there are some well-characterized examples. One particular case is *carbonic anhydrase*, because the concentrations of carbon dioxide and bicarbonate in the blood are easily measured. Under physiological conditions, the enzyme is only about 6% saturated with each substrate, and the K_M of carbon dioxide is too high to be measured.[21]

a. Substrate concentrations and K_M's in glycolysis
Good data are available for glycolysis. The glycolytic enzymes are particularly well studied and understood, and metabolite concentrations have

Rat diaphragm metabolites from E. A. Newsholme and P. J. Randle, *Biochem. J.* **80**, 655 (1961); H. J. Hohorst, M. Reim, and H. Bartels, *Biochem. Biophys. Res. Commun.* **7**, 137 (1962). Rabbit skeletal muscle enzymes: see below. Metabolite concentrations were calculated on an intramolecular water content of 60% for brain and muscle cells, and 70% for erythrocytes. No allowance has been made for compartmentation in the muscle and brain cells, but gross metabolite concentrations are usually close to those in the cytosol [A. L. Greenbaum, K. A. Gumaa, and P. McLean, *Archs. Biochem. Biophys.* **143**, 617 (1971)]. The values for mouse brain are those immediately on decapitation. The use of peak levels does not cause significant differences.
[b] From J. Zalitis and I. T. Oliver, *Biochem. J.* **102**, 753 (1967).
[c] From W. J. Rutter, *Fedn. Proc.* **23**, 1248 (1964); P. D. Spolter, R. C. Adelman, and S. Weinhouse, *J. Biol. Chem.* **240**, 1327 (1965).
[d] From A. S. Schneider, W. N. Valentine, M. Hattori, and H. L. Heins, *New Engl. J. Med.* **272**, 229 (1965).
[e] From P. M. Burton and S. G. Waley, *Biochim. Biophys. Acta* **151**, 714 (1968).
[f] From M. Oguchi, E. Gerth, B. Fitzgerald, and J. H. Park, *J. Biol. Chem.* **248**, 5571 (1973).
[g] The K_M of ~6 mM for P_i refers to high G3P concentrations where the acylenzyme accumulates. At low concentrations of G3P, the K_M is immeasurably high [P. J. Harrigan and D. R. Trentham, *Biochem. J.* **143**, 353 (1974)]. Note: The *unhydrated* forms of G3P and DHAP are probably the substrates of the reactions. The concentrations tabulated are for both the hydrated and the unhydrated forms, but the values of K_M for the unhydrated forms and their concentrations are overestimated in the same ratio [D. R. Trentham, C. H. McMurray, and C. I. Pogson, *Biochem. J.* **114**, 19 (1969); S. J. Reynolds, D. W. Yates, and C. I. Pogson, *Biochem. J.* **122**, 285 (1971)].
[h] From A. Yoshida and S. Watanabe, *J. Biol. Chem.* **247**, 440 (1972).
[i] From D. R. Rao and P. Oesper, *Biochem. J.* **81**, 405 (1961).
[j] From R. W. Cowgill and L. I. Pizer, *J. Biol. Chem.* **223**, 885 (1956); S. Grisolia and W. W. Cleland, *Biochemistry* **7**, 1115 (1968).
[k] From F. Wold and R. Barker, *Biochim. Biophys. Acta* **85**, 475 (1964).
[l] It is debatable whether or not this is a control enzyme; PEP is certainly well below the K_M in any case. The data quoted are for the presence of 500-μM FDP, in which case Michaelis-Menten kinetics hold. In the absence of FDP, sigmoid kinetics holds with a $K_{0.5}$ of 650 μM.[m]
[m] From S. E. J. Staal, J. F. Koster, H. Kamp, L. van Milligan-Boersma, and C. Veeger, *Biochim. Biophys. Acta* **227**, 86 (1971).
[n] From J. S. Nisselbaum and O. Bodansky, *J. Biol. Chem.* **238**, 969 (1963).
[o] Calculated from the lactate/pyruvate ratio, assuming NAD and NADH at equilibrium, and using an equilibrium constant of 1.11×10^{-4}. [From R. L. Veech, L. V. Eggleston, and H. A. Krebs, *Biochem. J.* **115**, 609 (1969).]
[p] From S. Rapoport, *Essays in Biochemistry* **4**, 69 (1969).
[q] From T. P. Fondy, L. Levin, S. J. Sollohub, and C. R. Ross, *J. Biol. Chem.* **243**, 3148 (1968).
[r] From R. M. Denton, R. E. Yorke, and P. J. Randle, *Biochem. J.* **100**, 407 (1966).

been determined for three diverse types of cells (brain, erythrocyte, and muscle). Data for the nonregulatory glycolytic enzymes are listed in Table 12·5 and illustrated in Figure 12·7. The histogram in the figure shows that the K_M's tend to be in the range of 1 to 10 and 10 to 100 times the substrate concentrations. Notable among these enzymes is triosephosphate isomerase. This well-studied enzyme, which has been described as "evolutionarily perfect"[22] because of its catalytic efficiency, has very high K_M's for both substrates.

It is not known at present whether the examples in which the K_M's are below the substrate concentrations are due to the enzymes' inability to evolve further or to metabolic reasons.

3. The perfectly evolved enzyme for maximum rate

We can use the two hypothetical steps of section C1b; i.e., that k_{cat}/K_M be maximized and that K_M be greater than [S], to set up criteria for judging the state of evolution of an enzyme whose function is to maximize rate. We recall from Chapter 3 that the maximum value of k_{cat}/K_M is the rate constant for the diffusion-controlled encounter of the enzyme and substrate, and from Chapter 4 that this is about 10^8 to 10^9 $s^{-1} M^{-1}$. A perfectly evolved enzyme should have a k_{cat}/K_M in the range of 10^8 to 10^9 $s^{-1} M^{-1}$, and a K_M greater than [S]. Using the data for k_{cat}/K_M listed in Table 4·4 and the substrate concentrations and K_M values mentioned in this chapter, it appears that carbonic anhydrase and triosephosphate isomerase are

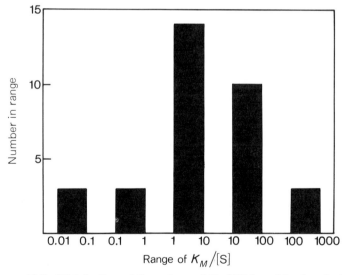

FIGURE 12·7. Distribution of the values of $K_M/[S]$ found in glycolysis.

perfectly evolved for the maximization of rate, which agrees with the conclusions of W. J. Albery and J. R. Knowles on the isomerase.[22]

There is one further consideration. The value of k_{cat}/K_M cannot be at the diffusion-controlled limit for a reaction that is thermodynamically unfavorable. This point stems from the Haldane equation (Chapter 3, section H), which states that the equilibrium constant for a reaction in solution is given by the ratio of the values of k_{cat}/K_M for the forward and reverse reactions. Clearly, k_{cat}/K_M for an unfavorable reaction cannot be at the diffusion-controlled limit, since k_{cat}/K_M for the favorable reverse reaction would have to be greater than the diffusion-controlled limit to balance the Haldane equation. The value of k_{cat}/K_M for an unfavorable reaction is limited by the diffusion-controlled limit multiplied by the unfavorable equilibrium constant for the reaction.

It was pointed out in Chapter 3, section A3, that when k_{cat}/K_M is at the diffusion-controlled limit, Briggs-Haldane rather than Michaelis-Menten kinetics are obeyed. Thus, the more advanced an enzyme is toward the evolution of maximum rate, the more important are Briggs-Haldane kinetics.

D. Molecular mechanisms for the utilization of binding energy

We have discussed in general terms the catalytic advantages of enzyme–transition state complementarity combined with high K_M's, and have seen that this combination tends to occur in practice. We shall now deal with the specific mechanisms that are used to achieve it.

1. Strain

Strain is the classic concept of Haldane and Pauling.[1,4] The enzyme has an active site whose structure is complementary to the structure of the transition state of the substrate rather than to the substrate itself. On binding, the substrate is strained or distorted. In Haldane's words: "Using Fischer's lock and key simile, the key does not fit the lock perfectly but exercises a certain strain on it" (Figure 12·8). Nowadays, the modified concept of *transition state stabilization* is gaining favor. According to this idea, it is not that the substrate is distorted but rather that the transition state makes better contacts with the enzyme than the substrate does, so that the full binding energy is not realized until the transition state is reached. Nevertheless, we shall use the term strain to cover the general situation of enzyme–transition state complementarity.

2. Induced fit

Chapter 10 showed how the induced-fit theory nicely describes some of the phenomena associated with allosteric enzymes. This theory had been

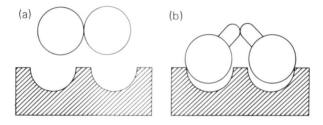

FIGURE 12·8. Haldane's picture of strain. The binding site on the enzyme (a) stretches the substrate toward the structures of the products, and (b) compresses the products toward the structure of the substrate.

introduced earlier to account for specificity in simple enzymes. It includes the notion that in the absence of substrate, the enzyme does not have a structure complementary to that of the transition state. However, the enzyme is floppy and the substrate rigid, so that when the enzyme–substrate complex is formed, the catalytic groups on the enzyme are aligned in their optimal orientations for catalysis: that is, the structure of the enzyme is complementary to the transition state only after binding has occurred. In the classic strain mechanism, the K_M is increased by binding energy being used to distort the substrate; in induced fit, it is increased by binding energy being used to distort the enzyme.

a. Disadvantages of the induced-fit mechanism in nonallosteric enzymes

In the strain mechanism, the value of k_{cat}/K_M is at a maximum, since the undistorted enzyme is complementary to the undistorted transition state. Induced fit lowers the k_{cat}/K_M because it is only the *distorted* enzyme that is complementary to the undistorted transition state: the k_{cat}/K_M is decreased by the energy required to distort the enzyme.

An alternative way of viewing the induced-fit process is to divide it into hypothetical steps. We can suppose that there is an equilibrium between the inactive form of the enzyme, E_{in}, which is the major species ($[E_{in}] \approx [E]_0$), and a small fraction of active enzyme, E_{act}, in which the catalytic groups are correctly aligned:

$$E_{act} \underset{S}{\overset{K_M}{\rightleftharpoons}} E_{act}S \xrightarrow{k_{cat}}$$
$$K \updownarrow \qquad\qquad \updownarrow K'$$
$$E_{in} \underset{S}{\overset{K_M}{\rightleftharpoons}} E_{in}S$$

Scheme I

By treating the K_M's in Scheme I as simple dissociation constants and defining $K = [E_{act}]/[E_{in}]$, it may be shown that

$$v = \frac{[E]_0[S]k_{cat}K'/(1 + K')}{[S] + (K_MK'/K)(1 + K)/(1 + K')} \tag{12·14}$$

and

$$\left(\frac{k_{cat}}{K_M}\right)_{obs} = \frac{k_{cat}}{K_M}\left(\frac{K}{1 + K}\right) \tag{12·15}$$

Since $K \ll 1$,

$$\left(\frac{k_{cat}}{K_M}\right)_{obs} = K\frac{k_{cat}}{K_M} \tag{12·16}$$

The observed value of k_{cat}/K_M is much less than if all the enzyme is in the active conformation.

Equation 12·14 may be further simplified if $K' \gg 1$, i.e., if virtually all of the enzyme is in the active form when it is bound with substrate. Under these conditions,

$$(k_{cat})_{obs} = k_{cat} \tag{12·17}$$

and

$$(K_M)_{obs} = \frac{K_M}{K} \tag{12·18}$$

The value of k_{cat} is the same as if all the enzyme were in the active conformation, but the value of K_M is far higher.

Induced fit thus increases K_M without increasing k_{cat}, and so decreases k_{cat}/K_M. It mediates against catalysis.

Equation 12·16 shows that as far as k_{cat}/K_M is concerned, it is as if a small fraction of the enzyme is permanently in the active conformation. This is the same for all substrates since K is substrate-independent.

3. Nonproductive binding

Although nonproductive binding is not a mechanism for increasing K_M, it is appropriately discussed here since it gives rise to effects that are qualitatively similar to those of strain and induced fit. This theory was originally invoked to account for specificity in the relative reactivities of larger, specific substrates compared with smaller, nonspecific substrates. It is assumed that as well as the productive binding mode at the active site, there are alternative, nonproductive modes in which the smaller substrates may bind and not react.

An example is the binding of polysaccharide substrates to lysozyme. In order for reaction to occur, the substrate must bind across sites D and

E of the six subsites A, B, C, D, E, and F. There is some strain associated with binding in subsite D, and occupying it does not increase the overall binding energy. Trimers and tetramers bind nonproductively in A, B, and C (Figure 12·9) and in A, B, C, and D. However, the favorable binding energy of occupying sites E and F causes hexamers to bind productively in subsites A through F.

As well as this "gross" nonproductive binding, there is an alternative example in which the substrate is bound at the active site but in the wrong orientation (Figure 12·10). This mechanism has been proposed to account for the low deacylation rates of the nonspecific acylenzymes of chymo-trypsin.[23]

4. The unimportance of strain, induced fit, and nonproductive binding in specificity

Specificity, in the sense of discrimination between competing substrates, is independent of the above three effects. The reasons are discussed in detail in Chapter 13. The basic reason is that specificity depends on k_{cat}/K_M, and strain and nonproductive binding do not affect this value (equations 12·11 and 3·36). Equation 12·16 shows that induced fit does alter k_{cat}/K_M for the active conformation, but equally for all substrates (i.e., by a factor of K).

5. Experimental evidence concerning the existence and the nature of strain and induced-fit processes

a. Steady state kinetics
Although the results of steady state kinetic measurements are often interpreted as supporting one of the three mechanisms, with the exact one depending on the whim of the experimentalist, the evidence is usually ambiguous. The approach generally used is to compare the k_{cat} and K_M values for a series of substrates, as in Tables 12·1 and 12·2, to see if the specificity is manifested in increasing k_{cat} rather than decreasing K_M. If this is found, it is good evidence that one of the processes is occurring, but it does not indicate which one. The problem is that all three mech-

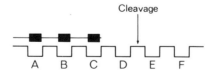

FIGURE 12·9. Nonproductive binding with substrates and lysozyme. Small substrates may bind at alternative sites along the extended active site of lysozyme, avoiding the cleavage site, which has a lower affinity.

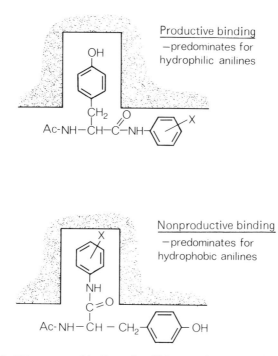

FIGURE 12·10. Wrong-way binding of anilides to chymotrypsin. Synthetic substrates with extraneous hydrophobic residues bind the wrong way in the hydrophobic binding site of chymotrypsin.

anisms predict the same result: that binding energy is converted into chemical activation energy. The following arguments may be made.

Strain. The additional groups on the larger, specific substrates are used to strain the substrate rather than to provide binding. (In the transition state stabilization model this is modified to: "The additional binding energy is not realized until the transition state is reached.")

Induced fit. The additional groups on the larger, specific substrates are used to provide energy for the distortion of the enzyme. The smaller, nonspecific substrates bind predominantly to the inactive form of the enzyme. The k_{cat} is lower for these since less is productively bound, but the K_M is correspondingly lower since the binding energy is not used to convert the inactive conformation to the active one.

Nonproductive binding. The larger substrate binds in the productive mode only, but the smaller one, in addition to binding more weakly in the productive mode, binds in nonproductive modes, lowering the K_M. The k_{cat} is correspondingly lower.

Although it is not possible to distinguish between strain and induced fit by a comparison of k_{cat} and K_M values from steady state kinetics, it is sometimes possible to bring in additional evidence to rule out or to favor nonproductive binding. One of the clearest examples of this is a study on elastase (Figure 12·11).[24] The k_{cat} for the hydrolysis of Ac-Pro-Ala-Pro-Ala-NH$_2$ is ~100 times higher than that for Ac-Ala-Pro-Ala-NH$_2$ (Table 12·1). This difference is unlikely to be caused by nonproductive binding, since the presence of the Ala-Pro-Ala should be sufficient to ensure productive binding. Comparison of the binding of substrates and transition state analogues suggests that the residues in S_{5-4} destabilize the hydrolyzed —CONH$_2$ group by about 8 kJ/mol (2 kcal/mol) relative to the transition state. Transition state analogues that have the —CONH$_2$ replaced by —CHO bind as a tetrahedral adduct to Ser-195, mimicking the tetrahedral intermediate–like transition state of the hydrolytic reaction. Their binding is enhanced by residues in S_{5-4}. It is not possible to say *a priori* whether the mutual destabilization between the S_{5-4} subsites and the substrate amide group is due to a conformational change transmitted through the protein or to an unfavorable interaction that is relieved on formation of the transition state.

b. Pre–steady state kinetics and direct binding measurements
Many well-documented cases of changes of fluorescence and circular dichroism on the addition of substrates and inhibitors to enzymes are consistent with induced conformational changes.[25] Some of these changes have been measured by rapid reaction techniques. However, the results are generally not interpretable in terms of specific mechanisms.

c. X-ray diffraction studies
The best information available comes from x-ray diffraction studies on crystalline enzymes.

1. *Lysozyme: A strain mechanism?* During the hydrolysis of polysaccharide substrates by lysozyme, the sugar ring bound in site D becomes a carbonium ion and its conformation changes from a full chair to a sofa (half-chair) (Chapter 1, section E of Chapter 15, and section B2a of this chapter). The original model-building studies on the enzyme suggested that the substrate was bound in such a way that the residue in the D site

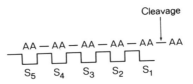

FIGURE 12·11. The binding subsites in elastase.

was forced into the sofa conformation due to unfavorable interactions with the enzyme; i.e., a classic strain mechanism.[26] More recently, M. Levitt has re-examined the binding of $(NAG)_6$ to lysozyme, using accurately refined coordinates of the enzyme and sophisticated calculations of the interactions with the substrate rather than examination of wire models.[8,27] The coordinates of both the enzyme and the substrate were optimized by expressing all the bond lengths, bond angles, torsion angles, and nonbonded interatomic distances as empirical energy functions and minimizing the energy by computation. His results for the D sugar ring and D subsite are given in Figure 12·12, where the torsion and angle strain in the ring, as well as the nonbonded interactions between the enzyme and the substrate, are plotted. The three approaches used for each meas-

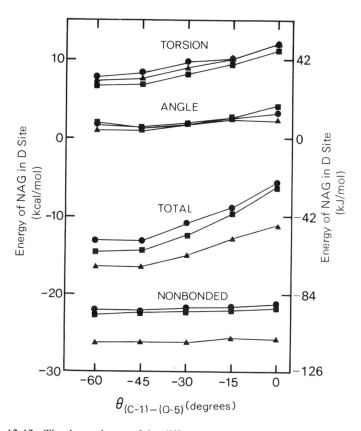

FIGURE 12·12. The dependence of the different energy contributions of the D sugar ring on the torsion angle $\theta_{(C-1)-(O-5)}$ for various artificially locked $(NAG)_6$ substrates bound to lysozyme (see text). The value of $\theta_{(C-1)-(O-5)}$ is $-60°$ for a full chair and $0°$ for a sofa. [From M. Levitt, in *Peptides, polypeptides and proteins* (E. R. Blout, F. A. Bovey, M. Goodman, and N. Lotan, Eds.), Wiley, p. 99 (1974).]

urement are illustrated. First (●), the substrate is assumed to be flexible and it is fitted into a rigid enzyme; second (▲), both the enzyme and the substrate are assumed to be flexible, and the pair are convergently energy-refined to give a mutual best fit; and third (■), the enzyme is assumed to be partially flexible. In all cases the most favorable situation is on the left-hand side of the figure, where the substrate is bound as the full chair, rather than on the right-hand side, where its conformation is the sofa. This suggests, then, that the substrate binds in its usual full-chair conformation in subsite D and does not take up the sofa conformation until the carbonium ion (or transition state) is formed. The classic strain mechanism of substrate distortion seems unlikely, although the transition state analogue studies suggest that there could be transition state stabilization.

2. *Hexokinase: Induced fit.* There is good evidence that yeast hexokinase, which phosphorylates glucose with the terminal phosphate of ATP, involves induced fit. X-ray diffraction studies show that the binding of both the sugar and the nucleotide causes extensive changes in tertiary structure. Also, the binding of the nucleotide to one form of the enzyme is promoted by sugar binding.[28,29] These results are in accord with earlier solution studies on the ATPase activity of the enzyme in the absence of glucose. Hexokinase phosphorylates glucose at the 6 position with a V_{max} of 800 μmol/min/mg of protein, and a K_M for ATP of 0.1 mM. In the absence of glucose, ATP is hydrolyzed with a V_{max} of 0.02 μmol/min/mg and a K_M of 4.0 mM. The addition of lyxose, which cannot be phosphorylated because it lacks the hydroxymethyl group at position 6, increases

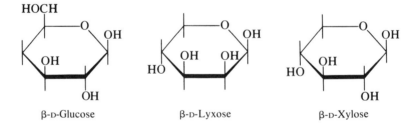

β-D-Glucose   β-D-Lyxose   β-D-Xylose

the V_{max} for the ATPase reaction by a factor of 18, and decreases the K_M of ATP by a factor of 40, to the value for the phosphorylation of glucose.[30,31] The changes induced by xylose in the crystal structure of the enzyme are similar to those induced by glucose (Figure 12·13).

3. *The serine proteases: Transition state stabilization.* Chymotrypsin and the serine proteases are among the best-characterized systems because of the high-resolution x-ray diffraction studies that have been done on the enzymes, and on their complexes with certain naturally occurring polypeptide inhibitors that mimic substrate binding (Chapter 1).[32–35] The catalytic activity of the enzyme may be destroyed by chemically con-

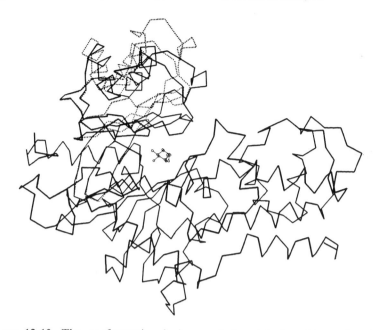

FIGURE 12·13. The conformational changes induced in hexokinase by glucose binding. The solid lines show the α-carbon backbone of the A isozyme crystallized in the presence of glucose. The dotted lines show the backbone of the part of the B isozyme that has a different structure when crystallization occurs in the absence of glucose. [From W. S. Bennett and T. A. Steitz, *Fedn. Proc. Abstr.* (1977).]

verting the nucleophilic Ser-195 to dehydroalanine by removing the hydroxyl group and a hydrogen atom. The structure of the anhydroenzyme is otherwise left unaltered so that it may be used for studies that separate binding from catalysis.

There is definite evidence for a strain process in peptide hydrolysis.[36] The crystallographic studies indicate a binding site for the leaving group of the peptide substrate. The value of k_{cat}/K_M for the hydrolysis of Ac-Phe-NH$_2$, Ac-Phe-Gly-NH$_2$, and Ac-Phe-Ala-NH$_2$ increases considerably as the size of the leaving group increases to fill the binding site (Table 12·1), but this is due entirely to increases in k_{cat}, with K_M remaining constant. In the reverse reaction—the attack of the leaving group amine on Ac-Phe-chymotrypsin—there is no evidence for Ala-NH$_2$ or Gly-NH$_2$ binding to the enzyme, since saturation kinetics are not observed. Instead, these compounds react far more rapidly than ammonia: the k_{cat}/K_M of NH$_3$ is 8 s^{-1} M^{-1}, of Gly-NH$_2$ is 2000 s^{-1} M^{-1}, and of Ala-NH$_2$ is 6000 s^{-1} M^{-1}.[36] The binding energy of the Ala and Gly residues is used to lower the chemical activation energy. This is not due to an induced-fit effect, since the structure of crystalline trypsin is identical to that of trypsin (and anhydrotrypsin, apart from the hydroxyl of Ser-195) in the crystalline complex with the pancreatic trypsin inhibitor at 1.4- to 1.5-Å (0.14- to 0.15-nm) resolution.

There also appears to be no distortion of the substrate on binding. X-ray diffraction studies on the complex of trypsin and the inhibitor are complicated by the inhibitor being distorted in the absence of enzyme,[37] but NMR studies on the binding of small substrates to chymotrypsin indicate that the substrates are unstrained.[38]

The utilization of the binding energy of the leaving group appears to be an example of transition state stabilization. There is little, if any, distortion of the enzyme or the substrate on binding, but the leaving group does not fit snugly into its binding site until the transition state is reached.

4. *Specific solvation of the transition state: A backbone contribution.* One feature to emerge from the x-ray diffraction work on the serine proteases is that the amido NH groups of the protein act as the solvation shell for the transition state of the reaction. This leads to a fundamental difference between simple chemical reactions in solution and enzyme-catalyzed reactions. In solution, the only orientation effects that have to be considered are those between the reagents, as the solvent is free to solvate any charges that develop. In the enzyme reaction, on the other hand, there is a precise stereochemical relationship between the reacting groups and the effective solvating groups that are part of the enzyme, as illustrated in Figure 12·14. Hydrogen bonds are particularly suitable for strain and specificity in this sense, since their potential varies strongly with distance (Chapter 11). The backbone is also an appropriate structure, since the rigid positioning of the groups admits the possibility of a strain contribution to catalysis by allowing only weak hydrogen bonds with the substrate but stronger ones with the transition state. It has been proposed that for chymotrypsin and trypsin, the carbonyl oxygen of the reacting amide or ester group of the substrate sits between the backbone NH groups of Ser-195 and Gly-193, but forms a good hydrogen bond with Ser-195 (Chapter 1, and Chapter 2, Figure 2·6). The bond with Gly-193 is long and weak, but as the reaction proceeds and the carbonyl oxygen double bond lengthens to become single, the oxygen moves closer to form a strong bond.[36] A similar but somewhat different process has been proposed from model-building studies on subtilisin.[39] The carbonyl oxygen does not sit between the two NH groups in the enzyme–substrate complex, but swings into position as the tetrahedral intermediate is formed. It is also suggested that another hydrogen bond, that between the *N*-acylamino group of the substrate and Ser-124, is made only in the tetrahedral intermediate.

A point worth noting is that specificity in the hydrogen bonding between an enzyme and a substrate is more likely to be caused by the *length* of the bond than by the *angle*, since, as discussed in Chapter 11, the

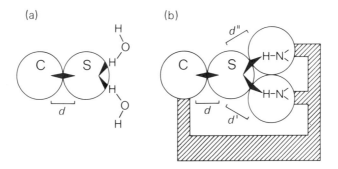

FIGURE 12·14. Illustration of the geometric constraints in enzymatic catalysis, due to the catalyst (C) and the solvating groups (H—N≡) being part of the enzyme structure. The shaded cones represent the permissible angles for orbital overlap or hydrogen-bond formation; d is the interatomic distance. (a) The reaction between the catalyst molecule and the substrate causes development of a negative charge that is solvated by water. (b) The distance between the catalyst and the hydrogen-bond donors is constrained because these structures are part of the enzyme. [From A. R. Fersht, *Proc. R. Soc.* **B187**, 397 (1974).]

energy of hydrogen bonds varies strongly with distance but only weakly with bending from linearity.[2]

The mechanism of the serine proteases is a good example of what is meant by transition state stabilization. Further examples of strain and induced fit are discussed in Chapter 15 (carboxypeptidase—induced fit; papain—strain, induced fit).

6. Conclusions about the nature of strain: Strain or stress?

Although strain may be manifested in some cases by a genuine distortion of the substrate, it is likely that the strain will generally be distortionless. This could be due either to the substrate and the enzyme having unfavorable interactions that are relieved in the transition state, or to the transition state having additional binding interactions that are not realized in the enzyme–substrate complex. In both cases there would be forces that *tend* to distort the substrate toward the transition state. Since non-bonded interactions have weak force constants (apart from van der Waals repulsion) and enzymes and substrates are flexible, it is improbable that the substrate would be distorted by its interactions with the enzyme. Rotation about single bonds, as in conformational changes in lysozyme substrates, is possible, but the stretching of single bonds or the twisting of double bonds appears to be less likely, as these require strong forces. There is ample binding energy available, but for it to be used to distort a substrate the energy would have to change greatly in magnitude over a short distance, i.e., provide a strong *force*. On the basis of energy calculations, Levitt has suggested the following tentative rule: "Small distortions of a substrate conformation that cause large increases in strain energy cannot be caused by binding to the enzyme."[8,40] He suggests also that the largest forces that can be exerted are less than 12 kJ/mol/Å (3 kcal/mol/Å or 120 kJ/mol/nm), so that to strain a substrate by about 12 kJ (3 kcal), atoms must be moved by 1 Å (0.1 nm). In extreme cases genuine distortion might occur; but in general, strain will involve the subtle interplay of favorable and unfavorable interactions. It is more likely that the enzyme rather than the substrate is distorted, because the enzyme is less rigid. Indeed, as we have seen, there are many examples of the distortion of an enzyme on the binding of a substrate. Perhaps some of these distortions are an unavoidable consequence of the flexibility of proteins, and of the impossibility of constructing an active site that is precisely complementary to the substrate. Also, a low-energy conformational transition in which part of the active site closes over the substrate (as with peptides and carboxypeptidase, and with NAD^+ and dehydrogenases—Chapter 15) may be a small price for the enzyme to pay to provide easy access to its active site.

The term strain has a specific meaning in physics and engineering: it implies that an object is physically distorted. Its companion term, stress, means that an object being subjected to forces is not distorted by them. *Using these precise physical terms, it is probably apt to say that in the enzyme–substrate complex, the enzyme is often strained, whereas the substrate is often stressed.*

Strain and stress in enzymes arise from several different causes. We have seen in this chapter, and we shall see further in Chapter 15, that stress and strain may be divided into two processes, substrate destabilization and transition state stabilization. Substrate destabilization may consist of: steric strain, where there are unfavorable interactions between the enzyme and the substrate (e.g., with proline racemase, lysozyme); desolvation of the enzyme (e.g., by displacement of two bound water molecules from the carboxylate of Asp-52 of lysozyme); and desolvation of the substrate (e.g., by displacement of any bound water molecules of a peptide[41]). Transition state stabilization may consist of: the presence of transition state binding modes that are not available for the substrate (e.g., with serine proteases, cytidine deaminase); relief of steric strain; and reestablishment of the solvation of the enzyme or substrate (or the formation of electrostatic bonds) in the transition state.

7. Strain vs. induced fit vs. nonproductive binding

In addition to providing enzyme–transition state complementarity, the strain mechanism makes a positive contribution to catalysis by increasing k_{cat} and K_M to give increased rates. The induced-fit mechanism mediates against catalysis by increasing K_M without a corresponding increase in k_{cat}, compared with all the enzyme being in the active form in the absence of substrate. The nonproductive binding mechanism does not affect the catalysis of the specific substrates, but just provides additional binding sites for competitive, nonspecific substrates. These could be deleterious to catalysis if they became effective inhibitors of the binding of the specific substrates. Since none of the mechanisms gives any additional specificity, it would seem that in nonregulatory enzymes strain should be the most important of the three processes. Nonproductive binding could just be an artifact of systems *in vitro* with no biological importance.[2]

E. Effects of rate optimization on accumulation of intermediates and internal equilibria in enzymes

1. Accumulation of intermediates

We saw in Chapter 7 that much effort has been put into the detection of chemical intermediates in enzymatic reactions. It has been found, though, that these do not accumulate in the reactions of many of the most common hydrolytic enzymes with their natural substrates under physiological conditions (Table 12·6). For this reason, the mechanisms of pepsin and car-

TABLE 12·6. *Enzymes and intermediates*

Enzyme (class)	Substrate	Intermediate	Accumulation[a]
Chymotrypsin (serine proteases)	Peptides	Acylenzyme	—
Pepsin (acid proteases)	Peptides	Acylenzyme(?) Aminoenzyme(?)	— —
Carboxypeptidase	Peptides	?	—
Papain (thiol proteases)	Amides	Acylenzyme	—
Pig liver esterase (liver esterases)	Aliphatic esters	Acylenzyme	—
Acetylcholine esterase (choline esterases)	Acetylcholine	Acylenzyme	+
Acid phosphatase	Phosphate monoesters	Phosphorylenzyme	+
Lysozyme (glycosidases)	Polysaccharides	Carbonium ion (ester)	—[b]

[a] Whether or not the physiologically relevant substrates involve the accumulation of an enzyme-bound intermediate at saturating substrate concentrations. It should be noted that if the physiological concentration of the substrate is below its K_M value, an intermediate does not accumulate, even if it would at saturating concentrations.
[b] An intermediate accumulates for an artificial leaving group (*p*-nitrophenol) with glycosidase.

boxypeptidase are still unresolved. In cases in which intermediates do accumulate, it is often through the use of synthetic, highly reactive substrates, such as esters with chymotrypsin, or through the effects of pH with alkaline phosphatase. There is a good theoretical explanation. It is a corollary of the principle of maximization of rate by the mutual increasing of k_{cat} and K_M that the accumulation of intermediates lowers the reaction rate.[2] Any intermediate that does accumulate lowers the K_M for the reaction, causing saturation at lower substrate concentrations. In other words, the accumulation of an intermediate means that the reaction has fallen into a "thermodynamic pit."

The "accumulation problem" is most severe for enzymes such as the digestive enzymes, which have to cope with pulses of high substrate concentrations. If the concentration of the substrate is below the K_M for the reaction under physiological concentrations, no intermediate accumulates *in vivo* in any case, since the enzyme is unbound. But in a test tube experiment in which the experimenter can use artificially high concentrations of substrate, an intermediate can sometimes be made to accumulate. An example occurs with glyceraldehyde 3-phosphate dehydrogenase. As Table 12·5 shows, the concentration of the aldehyde is below the K_M *in vivo*. But in the laboratory, the acylenzyme accumulates at saturating substrate concentrations.

2. Balanced internal equilibria

Although an enzyme cannot alter the equilibrium constant for a reaction in solution, there are no similar constraints on the equilibrium constant for the reagents when they are bound to the enzyme; i.e., there are no constraints on the internal equilibrium constant (Chapter 3, section H). There is, in fact, an inherent tendency for the internal equilibrium constant for the formation of an unstable intermediate or product to be more favorable than that in solution.[22] This stems from the Hammond postulate (Chapter 2, section A2) and the principle of enzyme–transition state complementarity: the transition state resembles the unstable intermediate or product, so the enzyme is closer in complementarity to the unstable species than to the substrate. The enzyme thus binds the unstable species more tightly than the substrate, and consequently increases the internal equilibrium constant in favor of the intermediate or product. However, the unstable species must not be bound too tightly or the reaction may fall into a "thermodynamic pit." For reversible reactions, the optimal situation is when all the intermediates have the same Gibbs free energy, so that the internal equilibria are balanced with each equilibrium constant being 1.[22] A compilation of internal equilibrium constants for a range of phosphotransferases reveals that most are indeed close to 1.[42]

References

1 J. B. S. Haldane, *Enzymes*, Longmans, Green and Co., p. 182 (1930). M.I.T. Press (1965).
2 A. R. Fersht, *Proc. R. Soc.* **B187**, 397 (1974).
3 E. Fischer, *Ber. Dt. Chem. Ges.* **27**, 2985 (1894).
4 L. Pauling, *Chem. Engng. News* **24**, 1375 (1946); *Am. Scient.* **36**, 51 (1948).
5 R. Wolfenden, *Accts. Chem. Res.* **5**, 10 (1972).
6 G. E. Lienhard, *Science, N.Y.* **180**, 149 (1973).
7 I. I. Secemski and G. E. Lienhard, *J. Biol. Chem.* **249**, 2932 (1974).
8 M. Levitt, in *Peptides, polypeptides and proteins* (E. R. Blout, F. A. Bovey, M. Goodman, and N. Lotan, Eds.), Wiley, p. 99 (1974).
9 D. H. Leaback, *Biochem. Biophys. Res. Comm.* **32**, 1025 (1968).
10 G. Legler, M. L. Sinnott, and S. G. Withers, *J. Chem. Soc. Perk. II*, 1374 (1980).
11 G. J. Cardinale and R. H. Abeles, *Biochemistry* **7**, 3970 (1968).
12 M. V. Keenan and W. L. Alworth, *Biochem. Biophys. Res. Comm.* **57**, 500 (1974).
13 R. M. Cohen and R. Wolfenden, *J. Biol. Chem.* **246**, 7561 (1971).
14 J. Fastrez and A. R. Fersht, *Biochemistry* **12**, 1067 (1973).
15 K. D. Collins and G. R. Stark, *J. Biol. Chem.* **246**, 6599 (1971).
16 G. Winter, A. R. Fersht, A. J. Wilkinson, M. Zoller, and M. Smith, *Nature* **299**, 756 (1982).
17 A. J. Wilkinson, A. R. Fersht, D. M. Blow, and G. Winter, *Biochemistry* **22**, 3581 (1983).
18 A. R. Fersht, J. -P. Shi, A. J. Wilkinson, D. M. Blow, P. Carter, M. M. Waye, and G. Winter, *Angewandte Chemie* (1984, in press).
19 J. Schlessinger and A. Levitzki, *J. Molec. Biol.* **82**, 547 (1974).
20 H. G. Mannherz, H. Schenck, and R. S. Goody, *Eur. J. Biochem.* **48**, 287 (1974).
21 J. C. Kernohan, W. W. Forrest, and F. J. W. Roughton, *Biochim. Biophys. Acta* **67**, 31 (1963).
22 W. J. Albery and J. R. Knowles, *Biochemistry* **15**, 5627, 5631 (1976); *Angewandte Chemie* **16**, 285 (1977).
23 R. H. Henderson, *J. Molec. Biol.* **54**, 341 (1970).
24 R. C. Thompson, *Biochemistry* **13**, 5495 (1974).
25 N. Citri, *Adv. Enzymol.* **37**, 397 (1973).
26 C. C. F. Blake, L. N. Johnson, G. A. Mair, A. C. T. North, D. C. Phillips, and V. R. Sarma, *Proc. R. Soc.* **B167**, 378 (1967).
27 A. Warshel and M. Levitt, *J. Molec. Biol.* **103**, 227 (1976).
28 W. F. Anderson and T. A. Steitz, *J. Molec. Biol.* **92**, 279 (1975).
29 S. P. Colowick, *The Enzymes* **9**, 1 (1973).
30 G. DelaFuente, R. Lagunas, and A. Sols, *Eur. J. Biochem.* **16**, 226 (1970).
31 G. DelaFuente and A. Sols, *Eur. J. Biochem.* **16**, 234 (1970).
32 A. Rühlmann, D. Kukla, P. Schwager, K. Bartels, and R. Huber, *J. Molec. Biol.* **77**, 417 (1974).
33 R. Huber, D. Kukla, W. Bode, P. Schwager, K. Bartels, J. Deisenhofer, and W. Steigemann, *J. Molec. Biol.* **89**, 73 (1974).
34 R. M. Sweet, H. T. Wright, J. Janin, C. H. Chothia, and D. M. Blow, *Biochemistry* **13**, 4212 (1974).
35 W. Bode, P. Schwager, and R. Huber, *FEBS Lett., Tenth Anniversary Issue* **40** (suppl.), 3 (1975).

36 A. R. Fersht, D. M. Blow, and J. Fastrez, *Biochemistry* **12**, 2035 (1973).
37 J. Deisenhofer and W. Steigemann, *Acta Cryst.* **B31**, 238 (1975).
38 G. Robillard, E. Shaw, and R. G. Shulman, *Proc. Natn. Acad. Sci. USA* **71**, 2623 (1974).
39 J. D. Robertus, J. Kraut, R. A. Alden, and J. J. Birktoft, *Biochemistry* **11**, 4293 (1972).
40 M. Levitt, Ph.D. Thesis, University of Cambridge (England), p. 270 (1972).
41 R. Wolfenden, *Biochemistry* **17**, 201 (1978).
42 A. Hassett, W. Blättler, and J. R. Knowles, *Biochemistry* **21**, 6335 (1982).

Further reading

W. P. Jencks, "Binding energy, specificity, and enzymic catalysis—the Circe effect," *Adv. Enzymol.* **43**, 219–410 (1975).

13

Specificity and editing mechanisms

Specificity is a grossly overworked and often misused word. The most important meaning for the enzymologist refers to an enzyme's *discrimination* between several substrates competing for an active site: for example, the specificity of a particular aminoacyl-tRNA synthetase for a particular amino acid and a particular tRNA in a mixture of all the amino acids and all the tRNAs. This is the definition of specificity that is relevant to biological systems. It concerns the situation in which a desired and an undesired substrate are competing for an enzyme, and deals with the problem of the relative rate of reaction of the undesired substrate and desired substrate in a mixture of the two. Specificity in this sense is a function of both substrate binding and catalytic rate: if the undesired substrate and the enzyme have a k_{cat} that is 1000 times lower than the k_{cat} for the desired substrate, but the undesired substrate binds 1000 times more tightly, the preferential binding will compensate for the lower rate. For this reason, as discussed below, the k_{cat}/K_M is the important kinetic constant in determining specificity, since it combines both the rate and the binding terms.

A meaning of specificity that is really a misuse of the term refers to the activity of an enzyme toward an alternative substrate in the *absence* of a specific substrate, as can happen in an experiment *in vitro*. In such a test tube experiment, a substrate is often described as "poor" because it involves either a high value of K_M or a low value of k_{cat}. In biological systems both k_{cat} *and* K_M are important.

The difference between the two meanings is crucial to the status of strain, induced fit, and nonproductive binding in catalysis. As we discussed in Chapter 12 and as we shall amplify below, these do not affect biological specificity, since they alter k_{cat} and K_M in a mutually compensating manner without altering k_{cat}/K_M.

A. Limits on specificity

The basic problem in specificity is: How does an enzyme discriminate against a substrate that is smaller than, or the same size as (isosteric with), the specific substrate? There is no difficulty in discrimination against a substrate that is larger than the specific substrate, since the binding cavity at the active site may be constructed to be too small to fit the larger competitor. But a smaller competitor must always be able to bind, and it cannot be excluded by steric hindrance.[1] There will just be less binding energy available to be used for catalysis. Crude examples of this have been discussed in regard to the serine proteases. The larger aromatic amino acid derivatives cannot bind in the small binding pocket of elastase, but the smaller amino acid derivatives can bind to and react with chymotrypsin (Chapter 1). However, as discussed at the beginning of Chapter 12, the reactions of the smaller substrates involve much lower values of k_{cat} and k_{cat}/K_M. There is also no difficulty in discrimination against substrates with the wrong stereochemistry. As was pointed out in Chapter 8, the substitution of an L-amino acid by a D-amino acid leads to an interchange of two groups around the chiral carbon, so that the substrate cannot be bound productively because of steric effects.

The areas in which high specificity is most essential are DNA replication and protein biosynthesis, because of the necessity of maintaining the genetic information encoded in DNA and faithfully translating it into protein structure. Because of the evolutionary pressure on the enzymes involved in these processes to be as accurate as possible, measurements on the key polymerases and synthetases delineate the maximum possible practical limits on specificity. We used measurements on these enzymes in Chapter 11, section B1, to tabulate maximum values for the binding energies of small groups to proteins.

Good examples of discrimination occur in the reactions of the aminoacyl-tRNA synthetases (see Chapter 7, section D, for their mechanism). These enzymes are responsible for the selection of amino acids during protein synthesis, and so have to discriminate among a multitude of often very similar substrates with high precision. For example, the isoleucyl-tRNA synthetase has to discriminate between isoleucine and valine, and the valyl-tRNA synthetase between valine and threonine (Figure 13·1). Valine, being shorter by one methylene group than isoleucine, binds to the isoleucyl-tRNA synthetase, but 150 times more weakly.[2] Threonine, although it is isosteric with valine, binds 100 to 200 times more weakly to the valyl-tRNA synthetase, because of the burying of the hydroxyl group in the hydrophobic pocket normally occupied by a methyl group of valine.[3] Similarly, the alanyl-tRNA synthetase discriminates against glycine by a factor of 250.[4] Thus, a precisely tailored active site can recognize the absence of a methylene group on a substrate by a relative

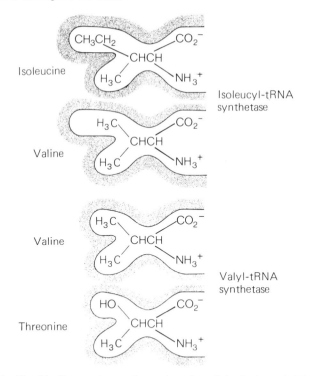

FIGURE 13·1. The binding cavity at the active site of the isoleucyl-tRNA synthetase must be able to bind valine as it binds the larger isoleucine. The active site of the valyl-tRNA synthetase cannot exclude threonine, because it is isosteric with valine.

rate factor of up to 250, though a factor of 10 or so is more usual. The limits on steric exclusion of a larger substrate from a smaller binding site can be estimated from the relative rates of activation of isoleucine and valine by the valyl-tRNA synthetase. Isoleucine, which again is too large by one methylene group to fit the site tailored for valine, reacts at a rate that is up to 2×10^5 times lower.[5]

The specificity against binding a small substrate to an active site constructed for a larger one may be analyzed by transition state theory in the same way as enzyme–substrate complementarity was in Chapter 12. The activation energy of the reaction is divided into contributions from the chemical activation energy and from the enzyme–substrate binding energy. We shall see that if the smaller or isosteric substrate differs from the specific substrate by lacking an element that has a structure R and a potential binding energy of $\Delta\Delta G_b$, the maximum possible discrimination due to this difference in stereochemistry is $\exp(-\Delta\Delta G_b/RT)$, and this cannot be amplified by strain, by induced fit, by a series of conformational changes, by an additional series of chemical steps, or by two (or more) sites functioning simultaneously. We shall first demonstrate this limit for specific examples in Michaelis-Menten kinetics, and then generalize for any mechanism. It will be seen that specificity is due only to transition state binding.

1. Michaelis-Menten kinetics

It was shown in Chapter 3, section G2, that specificity for competing substrates is controlled by k_{cat}/K_M. If the rate of reaction of the specific substrate A is v_A, and that of the competitor B is v_B, then

$$\frac{v_A}{v_B} = \frac{[A](k_{cat}/K_M)_A}{[B](k_{cat}/K_M)_B} \tag{13·1}$$

This is translated into terms of binding energy by using equations 12·3, 12·4, and 12·10; i.e.,

$$\ln\frac{k_{cat}}{K_M} = \ln\frac{kT}{h} - (\Delta G_0^{\ddagger} + \Delta G_b) \tag{13·2}$$

where $\Delta G_0^{\ddagger}$ is the chemical activation energy and ΔG_b is the binding energy of the enzyme and transition state. If the additional group R on A is not directly chemically involved in the reaction, $\Delta G_0^{\ddagger}$ will be the same for both A and B, apart from inductive effects. Ignoring these for convenience and setting the difference in binding energy at $\Delta\Delta G_b$ gives, from equation 13·2,

$$\frac{(k_{cat}/K_M)_A}{(k_{cat}/K_M)_B} = \exp\left(-\frac{\Delta\Delta G_b}{RT}\right) \tag{13·3}$$

(where $\Delta\Delta G_b$ is algebraically negative).

Equation 13·3 quantifies the *maximum* effect the additional binding energy can have. If the rate of the slow step in the reaction of A is lowered to such an extent in B that another step becomes rate-determining, the activation energy has not been lowered by the full amount, $\Delta\Delta G_b$. Also, if B binds and reacts in alternative modes, these will be in addition to the mode of A and they will add to the overall rate of reaction of B.

One example in which specificity may be lost is when Briggs-Haldane kinetics are occurring (Chapter 3, section A3a). Under these conditions, k_{cat}/K_M is equal to the rate constant for the association of the enzyme and the substrate. Since it is usually found that the higher dissociation constants for smaller substrates arise from a higher rate of dissociation rather than from a lower rate of association, there will be a partial or complete loss of specificity.

The following mechanisms have been suggested as causes of specificity, but we will see why they cannot be so.

1. *Strain.* As was explained in Chapter 12, strain does not affect k_{cat}/K_M but just causes compensating changes in k_{cat} and K_M without altering their ratio.

2. *Induced fit.* Chapter 12 also pointed out that k_{cat}/K_M for enzymes involving induced fit is just the value of k_{cat}/K_M for the active conformation scaled down by a constant factor for all substrates (equations 12·15 and 12·18). Induced fit does not alter the relative values of k_{cat}/K_M from what they would be if all of the enzyme were in the active conformation, and thus does not affect specificity.

3. *Nonproductive binding.* It was shown in Chapter 3, section E, that nonproductive binding does not alter k_{cat}/K_M but decreases both k_{cat} and K_M while maintaining their ratio. Specificity is unaffected.

4. *A series of sequential reactions.* Specificity cannot be amplified by there being a series of steps in which $\Delta\Delta G_b$ is utilized at each one. A simple way of seeing this is to recall from Chapter 3, section F, that in the following equation k_{cat}/K_M is always equal to k_2/K_S, irrespective of the number of additional intermediates:

$$E + S \xrightleftharpoons{K_S} ES \xrightarrow{k_2} ES' \xrightarrow{k_3} ES'' \xrightarrow{k_4} ES''' \xrightarrow{k_5} \text{etc.} \qquad (13\cdot4)$$

2. The general case

The following formal thermodynamic approach can be used quite generally for analyzing binding energy contributions.

Consider any series of reactions

$$E + S \rightleftharpoons ES \rightleftharpoons E'S' \rightleftharpoons E''S'' \rightleftharpoons \rightleftharpoons \rightleftharpoons E^{\ddagger}S^{\ddagger} \xrightarrow{\text{rate-determining}} \qquad (13\cdot5)$$

where E, E′, E″, etc. are different states of the enzyme, and S, S′, S″, etc. are different states of the substrates (covalently altered, etc.). The rate of the reaction may be calculated from transition state theory by ignoring all the intermediate steps and just considering the energetics of the process $E + S \rightarrow E^{\ddagger}S^{\ddagger}$. The Gibbs free energy of activation may be considered to be composed of three terms: $\Delta G_E^{\ddagger}$, a free energy change representing the energy difference between E and $E^{\ddagger}$; $\Delta G_S^{\ddagger}$, representing the energy difference between S and $S^{\ddagger}$; and ΔG_b, the binding energy of $E^{\ddagger}$ and $S^{\ddagger}$:

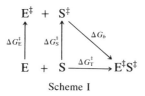

Scheme I

The thermodynamic cycle in Scheme I shows that the activation energy $\Delta G_T^\ddagger$ is given by

$$\Delta G_T^\ddagger = \Delta G_E^\ddagger + \Delta G_S^\ddagger + \Delta G_b \tag{13.6}$$

The rate of reaction is given from transition state theory:

$$v = \frac{kT}{h} \text{[E][S]} \exp\left(-\frac{\Delta G_E^\ddagger + \Delta G_S^\ddagger + \Delta G_b}{RT}\right) \tag{13.7}$$

The relative reaction rates of the two substrates A and B are given by substituting the values of the Gibbs energies for A and B into equation 13.7 and taking the ratio:

$$\frac{v_A}{v_B} = \frac{[A]}{[B]} \exp\left(-\frac{\Delta\Delta G_b + \Delta G_A^\ddagger - \Delta G_B^\ddagger}{RT}\right) \tag{13.8}$$

The difference in the binding energy of A and B, $\Delta\Delta G_b$, comes into the equations only once. As mentioned in the preceding section, $\Delta\Delta G_b$ cannot be used in a cumulative manner over a series of steps to amplify the differences. A reaction might involve several steps, but the specificity due to $\Delta\Delta G_b$ will just be spread out over them.

For similar substrates, $\Delta G_A^\ddagger$ will be similar to $\Delta G_B^\ddagger$, so that

$$\frac{v_A}{v_B} = \frac{[A]}{[B]} \exp\left(-\frac{\Delta\Delta G_b}{RT}\right) \tag{13.9}$$

This procedure may be extended to include common cosubstrates, e.g., NAD^+ or ATP. The Gibbs energy changes involving these cancel out, as does $\Delta G_E^\ddagger$, when the ratios of rates are found from equation 13.7 or its equivalent.

3. Interacting active sites

Can specificity be increased by more than one molecule of substrate binding to an enzyme with multiple binding sites? Such an increase could appear to happen in a test tube experiment with only one substrate present, since absolute rate and not discrimination would be measured. But in a biological experiment with both the specific and the competitive substrates present, it would become clear that specificity cannot be increased in this way, because of the reaction of mixed complexes containing the enzyme and both types of substrates. This may be proved by the formal thermodynamic approach. In the following example we consider the case of "half-of-the-sites reactivity," in which one molecule of substrate S* binds without giving products during the turnover of the enzyme, but the second molecule S goes on to react. By comparing the reaction rates of pairs of complexes, e.g., E·A·A* with E·B·A*, it will be seen that the additional binding energy of the larger substrate, $\Delta\Delta G_b$, can be used only "once":

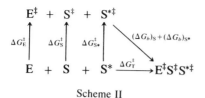

Scheme II

According to transition state theory, the rate is given by

$$v = \frac{kT}{h}[E][S][S^*]\exp\left[-\frac{\Delta G_E^\ddagger + \Delta G_S^\ddagger + \Delta G_{S*}^\ddagger + (\Delta G_b)_S + (\Delta G_b)_{S*}}{RT}\right]$$

(13·10)

If A can bind more strongly than B because the difference in structure contributes a binding energy of $\Delta\Delta G_b$, and if v_{AB^*} is the reaction rate when A is bound at the chemically reacting site and B is bound at the other, etc., then substituting the Gibbs energies into equation 13·7 and assuming that $\Delta G_A^\ddagger = \Delta G_B^\ddagger$ gives

$$\frac{v_{AA^*}}{v_{BA^*}} = \frac{[A]}{[B]}\exp\left(-\frac{\Delta\Delta G_b}{RT}\right)$$

(13·11)

and

$$\frac{v_{AB^*}}{v_{BB^*}} = \frac{[A]}{[B]}\exp\left(-\frac{\Delta\Delta G_b}{RT}\right)$$

(13·12)

so that

$$\frac{v_{AA^*} + v_{AB^*}}{v_{BA^*} + v_{BB^*}} = \frac{[A]}{[B]}\exp\left(-\frac{\Delta\Delta G_b}{RT}\right)$$

(13·13)

Owing to the mixed complexes of A and B that bind to the enzyme, the specificity cannot be enhanced by the binding of two molecules of substrate simultaneously. Substrate A can enhance activity more than B does by binding at a noncatalytic site, but it enhances activity with B as well as with itself.

4. The stereochemical origin of specificity

Specificity between competing substrates depends on the relative binding of their transition states to the enzyme. Enzyme–transition state complementarity maximizes specificity because it ensures the optimal binding of the desired transition state. This is also the criterion for the optimal value of k_{cat}/K_M, which is not surprising, since specificity is determined by k_{cat}/K_M. Maximization of rate parallels maximization of specificity, as long as there is no changeover to Briggs-Haldane kinetics.

The reason why hexokinase phosphorylates glucose in preference to water is that glucose binds well in the transition state, whereas water does not. Whatever the mechanism of the reaction, be it strain or induced fit, the competition between glucose and water is the same. If, for the sake of argument, the V_{max} for the phosphorylation of glucose were the same as that for water, and the binding energy of the glucose were used to give a very low K_M, the glucose would be preferentially phosphorylated due to its preferential binding to the active site. If, on the other hand, all the binding energy of the glucose were used in lowering the activation energy of V_{max}, this would also lead to its preferential phosphorylation due to its greater reactivity when bound.

There is one case in which strain or induced fit could be useful in a type of specificity. These mechanisms are unimportant where competition between substrates is concerned. But given a situation in which there is *no* specific substrate present, these mechanisms could be of use in providing a low absolute activity of the enzyme toward, say, water. For example, induced fit could prevent hexokinase from being a rampant ATPase in the *absence* of glucose (although its absence is extremely unlikely).

B. Editing or proofreading mechanisms

There are two fundamental types of interactions that limit the accuracy of DNA replication and protein biosynthesis. The first and more general type is the complementary base pairing that controls specificity in DNA replication and transcription (Chapter 14). Theory suggests[6] and experiment shows[7] that base pairing is accurate to about 1 part in 10^4 to 10^5. The second type of interaction is the intrinsic binding energy of amino acid side chains to proteins, because this limits the accuracy of amino acid selection. As was discussed at length at the beginning of this chapter and in Chapter 11, amino acid selection is accurate only to about 1 part in 10^2 because of the enzyme's difficulties in using steric exclusion to reject substrates that are isosteric with or slightly smaller than the correct substrate. Yet the overall error rate in the replication of DNA in *Escherichia coli* is only 1 mistake per 10^8 to 10^{10} nucleotides polymerized, and the overall error rate in transcription of the DNA and translation of the message into protein is in general only about 1 per 10^3 to 10^4 amino acid residues incorporated. Given the limits on the accuracy of base pairing and amino acid recognition, this specificity is beyond the theoretical thermodynamic limits for simple enzymes. It is possible only because of the evolution of *editing* or *proofreading* mechanisms. Certain key enzymes involved in polymerization have evolved, in addition to their active site for synthesis, a second, hydrolytic active site which is used to destroy incorrect intermediates or products as they are formed. The synthesis is thus double-checked at each step so that errors may be removed before they are permanently incorporated.

The crux of an editing mechanism is the formation of a high-energy intermediate that is unstable with respect to hydrolysis (equation 13·14). This allows an element of *kinetic* control of product formation to be introduced: the intermediate is at a branch point in the reaction pathway so that it may be channeled either to further synthesis or to hydrolytic products.

$$E + S \rightleftharpoons ES \longrightarrow EI \begin{array}{c} \nearrow \text{further synthesis} \\ \searrow \text{hydrolytic destruction} \end{array} \qquad (13\cdot14)$$

Without these editing mechanisms, the errors occurring during the replication of the genetic material and the synthesis of proteins would be at an intolerably high level.

1. Editing in protein synthesis

In the absence of editing mechanisms, the least accurate component of protein synthesis would be amino acid selection. For example, the extra methylene group of isoleucine causes the isoleucyl-tRNA synthetase to favor the activation of isoleucine over valine by a factor of only 100 to 200. This, combined with the 5-fold higher concentration of valine over isoleucine *in vivo*, would give an error rate of 1 in 20 to 40. Yet the overall error rate found for the mistaken incorporation of valine in positions normally occupied by isoleucine is only 1 in 3000.[8] The phenomenon of editing was first discovered in relation to this enzyme.[9] The isoleucyl-tRNA synthetase was found to form a stable enzyme-bound valyl adenylate complex in the absence of tRNA. But, whereas the addition of tRNA^{Ile} to the correct complex gives Ile-tRNA^{Ile} (equation 13·15), the addition of tRNA^{Ile} to the incorrect complex of valyl adenylate leads to quantitative hydrolysis to valine and AMP and gives no Val-tRNA^{Ile} (equation 13·16). In the presence of valine and tRNA^{Ile}, the isoleucyl-tRNA synthetase is an ATP pyrophosphatase that wastefully, but necessarily, catalyzes the hydrolysis of ATP to AMP via the activation reaction.

$$E \xrightarrow{\text{Ile, ATP}} E \cdot \text{Ile-AMP} \xrightarrow{\text{tRNA}^{\text{Ile}}} \text{Ile-tRNA}^{\text{Ile}} + \text{AMP} + E \qquad (13\cdot15)$$

$$E \xrightarrow{\text{Val, ATP}} E \cdot \text{Val-AMP} \xrightarrow{\text{tRNA}^{\text{Ile}}} \text{Val} + \text{tRNA}^{\text{Ile}} + \text{AMP} + E \qquad (13\cdot16)$$

There are two high-energy intermediates on the reaction pathway that could be edited by hydrolysis: the enzyme-bound aminoacyl adenylate and the aminoacyl-tRNA. A mechanistic study must distinguish between the two. A pathway involving the mischarged tRNA involves the formation of a covalent intermediate—the aminoacylated tRNA—so the three rules of proof may be considered (Chapter 7, section A1). These criteria have been rigorously applied to the rejection of threonine and other amino acids by the valyl-tRNA synthetase (Chapter 7, section D2).[3,10] The enzyme has a distinct and separate hydrolytic site for the deacylation of Thr-tRNA^{Val}; the k_{cat}/K_M for formation of a threonyl adenylate complex is some 600 times lower than that for the activation of

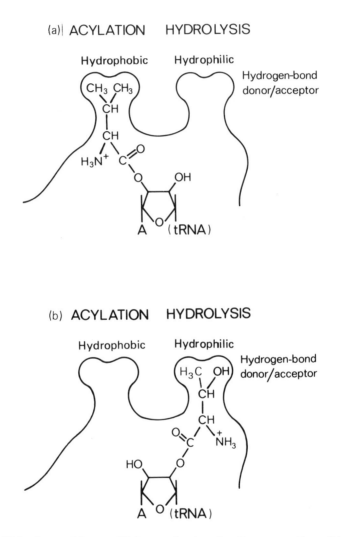

FIGURE 13·2. A possible specificity mechanism for the prevention of the misacylation of tRNA^{Val} with threonine. (a) The hydrophobic acylation site discriminates against threonine. (b) The hydrolytic site specifically uses the binding energy of the hydroxyl of threonine for a binding or catalytic effect. The translocation may occur as illustrated via a $2'- \rightarrow 3'$-hydroxyl acyl transfer. [From A. R. Fersht and M. Kaethner, *Biochemistry* **15**, 3342 (1976).

valine. Thr-tRNA^{Val} is formed but it is rapidly deacylated with a rate constant of 40 s^{-1} before it can dissociate from the enzyme. The deacylation site must presumably have a hydrogen-bond donor and acceptor that binds the hydroxyl group of threonine (Figure 13·2). The correctly charged Val-tRNA^{Val} is deacylated ~3000 times more slowly, presumably because of the unfavorable energetics associated with the wrong amino acid occupying the hydrophilic region tailored for threonine. Nature has thus evolved a specificity mechanism that uses the structural differences twice, but in different ways: threonine binds less well to the activation site but better to the deacylation site than does valine.

The failure to trap mischarged tRNA during editing by other aminoacyl-tRNA synthetases leaves open the possibility that editing may also occur at the level of the aminoacyl adenylate. The isoleucyl-tRNA synthetase provides an example. The decomposition of the E·Val-AMP complex (equation 13·16) requires the addition of tRNA^{Ile}; further, Val-tRNA^{Ile} that is prepared synthetically is rapidly deacylated by the isoleucyl-tRNA synthetase. But, in contrast to the mischarged tRNA formed by the valyl-tRNA synthetase, the Val-tRNA^{Ile} could not be detected during the course of the editing reaction.[11] It is possible that there is a double-check mechanism in which most of the valyl adenylate is destroyed prior to the transfer to tRNA, and in which most of the quantity that *is* transferred is mopped up by the esterase activity of the enzyme (Figure 13·3).

The methionyl-tRNA synthetase edits misactivated homocysteine before the transfer to tRNA^{Met}. The homocysteinyl adenylate complex is rapidly decomposed in the absence of tRNA. The reaction is, however, idiosyncratic, in that it invokes the cyclization of the intermediate to give the thiolactone.[12]

An analogy may be made between the amino acid selection process and a "double sieve" (Figure 13·4).[5] This crudely illustrates both the basic principles of selection for editing and how just two active sites can sort

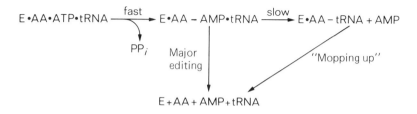

FIGURE 13·3. A possible double-check mechanism, with the major editing step occurring before the transfer of the amino acid to tRNA. [From A. R. Fersht, *Biochemistry* **16**, 1025 (1977); see also W. Freist and F. Cramer, *Eur. J. Biochem.* **131**, 65 (1983).]

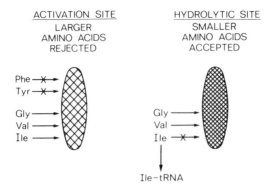

```
     ACTIVATION SITE              HYDROLYTIC SITE
        LARGER                       SMALLER
      AMINO ACIDS                  AMINO ACIDS
       REJECTED                     ACCEPTED
```

Ile–tRNA

FIGURE 13·4. The "double sieve" analogy for the editing mechanism of the isoleucyl-tRNA synthetase. The active site for the formation of the aminoacyl adenylate can exclude amino acids that are larger than isoleucine but not those that are smaller. On the other hand, a hydrolytic site that is just large enough to bind valine can exclude isoleucine while accepting valine and all the smaller amino acids. (On some enzymes, the hydrolytic site offers specific chemical interactions that enable it to bind isosteres of the correct amino acid as well as smaller amino acids.)

the whole range of amino acids by invoking first size and then specific chemical characteristics. Amino acids larger than the correct one are rejected by the first ("coarse") sieve by steric exclusion. All smaller or isosteric amino acids are activated, but at reduced rates because of poorer binding. Finally, the second ("fine") sieve accepts the products of activation or of transfer of the smaller or isosteric amino acids and excludes the products of the correct one, either because it is too large or because it lacks the specific interactions that allow an isostere to bind in the hydrolytic active site.

Not all aminoacyl-tRNA synthetases have editing sites. The cysteinyl- and tyrosyl-tRNA synthetases bind the correct substrates so much more tightly than their competitors that they do not need to edit.[13,14] Similarly, since the accuracy of transcription of DNA by RNA polymerase is better than the overall observed error rate in protein synthesis at about 1 part in 10^4, RNA polymerases do not need to edit.[15] The same should be true for codon–anticodon interactions on the ribosome. However, it is possible that accuracy has been sacrificed to achieve higher rates in this case, which is analogous to a change from Michaelis-Menten to Briggs-Haldane kinetics, and so an editing step is required.[16]

2. Editing in DNA replication

[Chapter 14, sections A1 and A2, may be read at this stage for an introduction to DNA replication and the associated enzymes.]

There are two fundamental differences between selection of amino acids during protein synthesis and the matching of base pairs during the replication of DNA.

The first is that whereas each amino acid has its own activating enzyme precisely tailored to it, a single DNA polymerase with just one active site for synthesis has to cope with all four correct base-pair combinations. The specificity of the reaction is largely delegated to the specificity of the base pairing itself, with its inherent error rate of 1 in 10^4 to 10^5. An even higher accuracy would be obtained if there were four separate enzymes (or just four active sites), each precisely tailored for its own base pair of AT, TA, GC, or CG. But it has been found expedient in evolution for errors to be corrected by editing mechanisms and for a common polymerase to be used.

The second difference enables errors in DNA replication to be corrected with relative ease. During protein synthesis, the growing end of the polypeptide chain is activated and transferred to the next amino acid in the sequence (Figure 13·5). There is no means of removing an incorrectly added residue and reactivating the polypeptide. Error correction has to be made before polymerization. But in the synthesis of DNA, the monomeric nucleotide is activated and added to the unactivated growing chain. This has enabled the evolution of a mechanism for the editing of errors after polymerization has occurred.

DNA synthesis proceeds in the $5' \rightarrow 3'$ direction, with the nucleotides being added to the 3'-hydroxyl of the polynucleotide. At the same time, all prokaryotic DNA polymerases have a $3' \rightarrow 5'$ exonuclease activity that works in the opposite direction (Figure 13·6). There is strong evidence that this is an editing function for the excision of incorrect, mismatched bases.[17] First, there is the evidence that the exonuclease activity is greatest for mismatched bases or single-stranded DNA.[17,18] Second, the mutation frequency in the T4 bacteriophage correlates with the measured rate of the exonuclease activity catalyzed by the DNA polymerase. "Mutator" mutants have a very high mutation rate and some code for a DNA polymerase with a low exonuclease activity[19,20] (Table 13·1). "Antimutators" have a high resistance to mutation and a high exonuclease activity. As Table 13·1 shows, the antimutators are very wasteful in the high amount of deoxynucleoside triphosphate hydrolyzed to incorporate a single base. Third, whereas the accuracy of DNA replication *in vitro* catalyzed by prokaryotic DNA polymerases is much higher than would be expected from the known frequencies of base mispairing, eukaryotic DNA polymerases that lack the $3' \rightarrow 5'$ exonuclease activity are less accurate and make errors at the expected rate.[21,22] Fourth, as described in the next section, the kinetics of error induction by prokaryotic DNA polymerases are consistent with the active participation of an editing mechanism.[23]

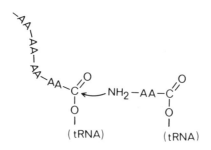

FIGURE 13·5. Protein synthesis involves the transfer of the activated polypeptide chain to the next amino acid residue.

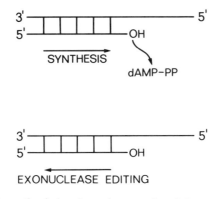

FIGURE 13·6. DNA synthesis involves the transfer of the activated deoxynucleotide monophosphate from its triphosphate to the 3′-hydroxyl of the growing chain. Editing takes place in the opposite direction.

TABLE 13·1. *Correlation of mutation and exonuclease activity of DNA polymerases of T4 phage[a]*

Strain of phage	Phenotype	dTTP wastefully hydrolyzed ÷ dTMP incorporated
L56	Mutator	0.005
L98	Mutator	0.01
74D	Wild type	0.04
L42	Antimutator	1.6
L141	Antimutator	13

[a] From N. Muzyczka, R. L. Poland, and M. J. Bessman, *J. Biol. Chem.* **247**, 7116 (1972).

a. Kinetics of polymerase accuracy: "Kinetic genetics"

Prokaryotic DNA polymerases are so accurate that special kinetic assays have had to be introduced to detect errors *in vitro*. These depend on replicating under controlled conditions the circular DNA of a small bacteriophage that contains a single-point mutation.[23-25] The accuracy of copying can be measured by producing viable phage from the synthetic DNA and then scoring by classic plaque-counting assays the proportion of revertant phage produced: each revertant corresponds to an error in replication. This is exactly analogous to the geneticists' method of measuring spontaneous mutation rates *in vivo*. The error rate of replication of bacteriophage ϕX174 *in vitro* was found to be very similar to the rate of spontaneous mutation *in vivo*: about 1 in 10^6 to 10^7. This shows that errors in base substitution during DNA replication are largely rate-determining in spontaneous mutation.[23,26]

The rate law for incorporation of incorrect nucleotides during DNA replication catalyzed by a polymerase that has an exonuclease activity is more complicated than that for the simple case of two substrates competing for an active site.[23,27,28] There has to be an additional term that allows for the partitioning of the newly added mismatched nucleotide between hydrolytic excision and permanent incorporation via elongation. For example, suppose that an incorrect deoxynucleoside triphosphate, $dNTP_i$, and a correct one, $dNTP_c$, compete for pairing at a site. Then the relative rate of insertion (i.e., of adding to the 3' terminus) is given by

$$R = \frac{[dNTP]_i (k_{cat}/K_M)_i}{[dNTP]_c (k_{cat}/K_M)_c} \tag{13·17}$$

But if in addition there is partitioning, as in equation 13·18, then the concentration of the nucleotide that next follows in the sequence to be

$$E\cdot DNA\cdot dNTP_i \rightarrow E\cdot DNA\text{-}N_i \overset{dNTP_f}{\rightleftharpoons} E\cdot DNA\text{-}N_i\cdot dNTP_f \rightarrow E\cdot DNA\text{-}N_i\text{-}N_f \rightarrow$$

$$\downarrow \text{editing}$$

$$E\cdot DNA + dNMP_i \tag{13·18}$$

incorporated, $dNTP_f$, comes into the calculations. In general, the concentration of $dNTP_f$ follows Michaelis-Menten-type kinetics. If the partitioning ratio equals the fraction F of inserted intermediate giving products, then F is of the form

$$F = \frac{[dNTP]_f}{K_M + \alpha[dNTP]_f} \tag{13·19}$$

where K_M has the dimensions of a Michaelis constant.[23,28] The overall misincorporation frequency is given by RF_i/F_c. In general, F_c is close to 1, since excision of the correct nucleotide is slow. Thus, the misincorporation frequency is close to RF_i. The accuracy consequently depends on the concentration of $dNTP_f$ in a predictable manner. At low concentrations, i.e., where $[dNTP]_f$ is less than K_M, partitioning favors editing. Hence the accuracy is high and the error rate is proportional to $[dNTP]_i[dNTP]_f/[dNTP]_c$. At high $[dNTP]_f$, F_i saturates, so the error rate is proportional simply to $[dNTP]_i/[dNTP]_c$. The accuracy is lower under these conditions, and when $\alpha = 1$ there will be no editing at all.

The dependence of overall misincorporation on the concentration of the next nucleotide to be incorporated provides a diagnostic test for the active participation of an editing mechanism.

b. DNA repair mechanisms

It is possible for base mispairing in duplex DNA to be corrected by repair mechanisms because the information content is duplicated in the two complementary strands. For example, aberrant base pairing arising from environmental damage such as x-rays, ultraviolet radiation, oxidation, and chemical modification may be repaired as follows.

The damaged strand is cut by an endonuclease, bases are removed and then replaced by the polymerase, and the join is sealed by DNA ligase.[29] In these cases the incorrect base is recognized by its chemical differences from the four naturally occurring ones in DNA. There is also a system that can remove deoxyuridine residues that have been mistakenly incorporated instead of thymidine.[30] This situation is similar to the isoleucine/valine case in protein synthesis, with uracil being smaller than

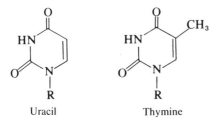

Uracil   Thymine

thymine by one methylene group. Uracil is removed from the DNA by a uracil glycosidase which excises the base from the sugar ring. This activity is analogous to the hydrolytic activity of the isoleucyl-tRNA synthetase toward Val-tRNA^{Ile}. In both cases the hydrolytic site is too small by the size of one methylene group to accommodate the substrate that is to be left intact. In DNA synthesis, the editing is performed by a separate enzyme, since the editing can wait until after polymerization. As this luxury is not permitted in protein synthesis, the hydrolytic function is on the synthetase, so that correction can occur before the misacylated tRNA leaves the enzyme.

Postreplicational mismatch repair has been found to correct errors in base substitution occurring during DNA replication in prokaryotes.[31] This lowers the error rate for the polymerase from 1 in 10^6 to 10^7 to the observed range of values of 1 in 10^8 to 10^{10} in *E. coli*. How does the repair system know in this case which base in a mispair is the incorrect one? The answer appears to be that the parent strand is *tagged* by methylation. A small proportion, some 0.2%, of the cytosine residues are methylated at the 5 position, and a similar proportion of the adenine residues are methylated at the 6 position. As methylation is a postreplicative event, the daughter strand is temporarily undermethylated after replication.

Why is the editing mechanism of the polymerase unable to achieve the necessary accuracy by itself: why does it need the postreplicative mismatch repair mechanism? The answer comes from analyzing the *cost* of editing.

C. The cost of accuracy

It is clear from the data in Table 13·1 on the hydrolysis of substrates accompanying DNA replication that editing costs energy.[20] Not only are the products of incorrect insertions removed, but also some of the correct substrate is wastefully hydrolyzed through insertion followed by exonuclease editing. The fraction of correct substrate wastefully hydrolyzed is defined as the cost. It is further seen that the more efficient the editing, the more the cost. The relationship between cost and accuracy has been analyzed in depth and at various levels of sophistication.[32-36] The following treatment[35,36] gives a very simple equation that provides answers to the questions: *What are the limits of editing? What does it cost? How do these depend on mechanism of selection?*

1. The cost-selectivity equation for editing mechanisms

Most editing mechanisms can be reduced to the sequence outlined in equation 13·20. The substrate first reacts to form the high-energy inter-

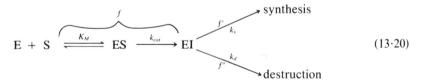

$$\text{E} + \text{S} \underset{K_M}{\overset{f}{\rightleftharpoons}} \text{ES} \overset{k_{\text{cat}}}{\longrightarrow} \text{EI} \begin{array}{c} \overset{f'}{\underset{k_s}{\nearrow}} \text{synthesis} \\ \underset{f''}{\overset{k_d}{\searrow}} \text{destruction} \end{array} \qquad (13·20)$$

mediate EI, which is then partitioned between further synthesis (with a rate constant k_s) and hydrolytic editing (with a rate constant k_d). (Further synthesis could be elongation in DNA replication, escape of the charged tRNA from the aminoacyl-tRNA synthetase in protein synthesis, etc.) The overall accuracy of the enzyme can be expressed in terms of the cost C and three discrimination factors, f, f', and f'', which are defined in the equations that follow. Kinetic quantities for the correct substrate are labeled with the subscript c, and those for the incorrect substrate are labeled i. The discrimination factor f represents the preferential rate of formation of the correct intermediate EI; f is defined by

$$f = \frac{(k_{cat}/K_M)_c}{(k_{cat}/K_M)_i} \tag{13·21}$$

The f' factor represents the higher rate of further synthesis for the correct substrate; f' is defined by

$$f' = \frac{(k_s)_c}{(k_s)_i} \tag{13·22}$$

The f'' factor represents the higher rate of destruction of the incorrectly formed intermediate EI; f'' is defined by

$$f'' = \frac{(k_d)_i}{(k_d)_c} \tag{13·23}$$

The cost is seen from equation 13·20 to be

$$C = \frac{(k_d)_c}{(k_d + k_s)_c} \tag{13·24}$$

whereas the partitioning to give products is given by

$$F_c = \frac{(k_s)_c}{(k_d + k_s)_c} \tag{13·25}$$

$$F_i = \frac{(k_s)_i}{(k_d + k_s)_i} \tag{13·26}$$

Substituting equations 13·22 and 13·23 into 13·26 gives

$$F_i = \frac{(k_s)_c}{(f' f'' k_d + k_s)_c} \tag{13·27}$$

Dividing equation 13·25 by 13·27 and substituting 13·24 gives the relative partitioning ratio for the correct and incorrect substrates:

$$\boxed{\frac{F_c}{F_i} = 1 + (f' f'' - 1)C} \tag{13·28}$$

The relative partitioning ratio is the factor by which specificity is increased by editing. In the absence of editing, v_c/v_i would be equal to $f[S]_c/[S]_i$. In the presence of editing, the overall ratio of the rates of incorporation of correct and incorrect substrates is equal to the ratio of the rates of formation of the intermediate EI multiplied by the relative partitioning ratio; that is,

$$\frac{v_c}{v_i} = \frac{f[S]_c}{[S]_i} \left(\frac{F_c}{F_i} \right) \tag{13·29}$$

We now introduce a new specificity term called the *selectivity*, S, which describes the overall specificity for competing substrates when editing is taking place. S is defined by

$$\frac{v_c}{v_i} = S\frac{[S]_c}{[S]_i} \tag{13·30}$$

Comparing equation 13·30 with 13·29 and substituting 13·28 for the partitioning ratio gives the *cost-selectivity* equation:[35]

$$S = f[1 + (f'f'' - 1)C] \tag{13·31}$$

When there is significant editing, the relative partitioning ratio of equation 13·28 is much greater than 1 and reduces to $f'f''C$. The additional specificity caused by editing is thus proportional both to the cost and to the product $f'f''$ of the two discrimination factors. This pair of discrimination factors may be considered as a single value, say f_{ed}, which represents the overall discrimination in the editing step. The efficiency of editing depends crucially on the magnitude of this factor. Mechanisms may be classified according to its value relative to f.

2. Single-feature recognition: $f = f'f''$

Certain editing processes, such as nucleotide selection in DNA replication, appear to use the same structural feature in selection for editing as is used in the initial selection for synthesis. For example, the incoming dNTP is checked in the insertion reaction by testing its base pairing with that of the template. Selection for editing repeats this process, and the same base pairing is checked again. In these cases, $f'f''$ is unlikely to be greater than f, and in many cases the two discrimination factors will be equal. If so, the partitioning ratio F_c/F_i is equal to $1 + (f - 1)C$. This means that the increase in specificity by editing is limited to a further factor of f, since $C \leqslant 1$. In addition, the factor of f is attained only in the ridiculous situation of $C = 1$, when all of the correct substrate is edited! Similarly, S is limited to f^2, which is attained only when $C = 1$. This significantly limits the usefulness of editing.

As an example, the cost-selectivity equation is plotted in Figure 13·7 for $f = f'f'' = 10^4$ or 10^5, the measured discrimination factors for base pairing. Significant increases in specificity are reached only when the cost becomes appreciable. The measured cost for the replicational DNA polymerase of *E. coli in vitro* is 6 to 13%, depending on the dNTP concerned.[36] This must be at the limits of the tolerable. Under these conditions, $S \approx 10^7$ to 10^9; these values are too low to achieve the error frequency of 1 in 10^8 to 10^{10} observed in *E. coli in vivo*. Hence the necessity of postreplicative mismatch repair mechanisms for attaining the

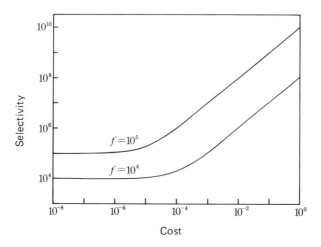

FIGURE 13·7. Plot of selectivity against cost for single-feature recognition, where $f = f'f'' = 10^4$ or 10^5.

desired error rate. In general, higher selectivity at lower cost can be achieved by multistage editing.[32,34]

J. J. Hopfield has suggested a general mechanism called "kinetic proofreading," in which there is no hydrolytic site on the enzyme; instead, the desired intermediates diffuse into solution, where they hydrolyze nonenzymatically.[37] An example is in the selection of amino acids by the aminoacyl-tRNA synthetases (equation 13·32). In this case, $f = f'f''$. For

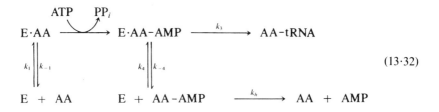

$$(13·32)$$

the isoleucyl-tRNA synthetase, $f \approx 150$, so any subsequent increase in specificity is limited to a factor of less than 15 if the cost is to be kept tolerable. In fact, the measured cost is less than 0.05, and it may be predicted from the cost-selectivity equation that if the Hopfield mechanism were operating it would be increasing specificity by a factor of less than 8.5.[38] We show next that "double sieving" is not limited in the same way.

3. Double-feature recognition: $f'f'' > f$

The double-sieve mechanism is an example of a double-feature selection process whereby different structural features of the substrates are used for f and for $f'f''$. To return to the example of the binding of valine and isoleucine to the isoleucyl-tRNA synthetase, $f \approx 150$ because only the weak forces of binding are being invoked. But in the checking step, if the editing site is tailored for the smaller valine, then the strong forces of steric exclusion can be invoked to prevent the binding of isoleucine. The value of $f'f''$ could be as high as 10^5 or so. This very great discrimination at the editing step means that high specificity can be obtained at low cost. Double-feature recognition is thus inherently far more cost-effective than single-feature.

References

1   L. Pauling, *Festschrift Arthur Stoll*, Birkhäuser Verlag AG, Basel, p. 597 (1958).
2   R. B. Loftfield and E. A. Eigner, *Biochim. Biophys. Acta* **130**, 426 (1966).
3   A. R. Fersht and M. Kaethner, *Biochemistry* **15**, 3342 (1976).
4   W.-C. Tsui and A. R. Fersht, *Nucl. Acids Res.* **9**, 4627 (1981).
5   A. R. Fersht and C. Dingwall, *Biochemistry* **18**, 2627 (1979).
6   M. D. Topal and J. R. Fresco, *Nature* **263**, 235 (1976).
7   A. R. Fersht, J.-P. Shi, and W.-C. Tsui, *J. Molec. Biol.* **165**, 655 (1983).
8   R. B. Loftfield and M. A. Vanderjagt, *Biochem. J.* **128**, 1353 (1972).
9   A. N. Baldwin and P. Berg, *J. Biol. Chem.* **241**, 831 (1966).
10 A. R. Fersht and C. Dingwall, *Biochemistry* **18**, 1238 (1979).
11 A. R. Fersht, *Biochemistry* **16**, 1025 (1977).
12 H. Jakubowski and A. R. Fersht, *Nucl. Acids Res.* **9**, 3105 (1981).
13 A. R. Fersht and C. Dingwall, *Biochemistry* **18**, 1245 (1979).
14 A. R. Fersht, J. S. Shindler, and W.-C. Tsui, *Biochemistry* **19**, 5520 (1980).
15 C. F. Springgate and L. A. Loeb, *J. Molec. Biol.* **97**, 577 (1975).
16 R. C. Thompson and A. M. Karim, *Proc. Natn. Acad. Sci. USA* **79**, 4922 (1982).
17 D. Brutlag and A. Kornberg, *J. Biol. Chem.* **247**, 241 (1972).
18 H. Koessel and R. Roychoudhury, *J. Biol. Chem.* **249**, 4094 (1974).
19 Z. W. Hall and I. R. Lehman, *J. Molec. Biol.* **36**, 321 (1968).
20 N. Muzyczka, R. L. Poland, and M. J. Bessman, *J. Biol. Chem.* **247**, 7116 (1972).
21 G. Seal, C. W. Shearman, and L. A. Loeb, *J. Biol. Chem.* **254**, 5229 (1979).
22 F. Grosse, G. Krauss, J. W. Knill-Jones, and A. R. Fersht, *EMBO J.* **2**, 1515 (1983).
23 A. R. Fersht, *Proc. Natn. Acad. Sci. USA* **76**, 4946 (1979).
24 L. A. Weymouth and L. A. Loeb, *Proc. Natn. Acad. Sci. USA* **75**, 1924 (1978).
25 C. C. Liu, R. L. Burke, U. Hibner, J. Barry, and B. Alberts, *Cold Spring Harbor Symp. Quant. Biol.* **43**, 469 (1979).
26 A. R. Fersht and J. W. Knill-Jones, *Proc. Natn. Acad. Sci. USA* **78**, 4251 (1981).

27 F. Bernardi and J. Ninio, *Biochimie* **60**, 1083 (1978).
28 A. R. Fersht and J. W. Knill-Jones, *J. Molec. Biol.* **165**, 633 (1983).
29 A. Kornberg, *DNA replication*, W. H. Freeman and Company (1980).
30 T. Lindahl, *Proc. Natn. Acad. Sci. USA* **71**, 3649 (1974).
31 B. W. Glickman and M. Radman, *Proc. Natn. Acad. Sci. USA* **77**, 1063 (1980).
32 M. A. Savageau and R. Freter, *Biochemistry* **18**, 3486 (1979).
33 D. J. Galas and E. W. Branscomb, *J. Molec. Biol.* **124**, 653 (1978).
34 M. Ehrenberg and C. Blomberg, *Biophys. J.* **31**, 333 (1980).
35 A. R. Fersht, *Proc. R. Soc.* **B212**, 351 (1981).
36 A. R. Fersht, J. W. Knill-Jones, and W.-C. Tsui, *J. Molec. Biol.* **156**, 37 (1982).
37 J. J. Hopfield, *Proc. Natn. Acad. Sci. USA* **71**, 4135 (1974).
38 R. S. Mulvey and A. R. Fersht, *Biochemistry* **16**, 4731 (1977).

14

Genetic engineering and enzymology: Protein engineering

Many questions in enzymology could be answered if we could synthesize enzymes or just simply change individual amino acid residues at will. For example, the contribution to catalysis of the buried aspartate residue (Asp-102) in chymotrypsin (Chapter 1, section C1a; Chapter 15, section B1a) could be measured by synthesizing the enzyme with an asparagine residue replacing the aspartate. Similarly, the importance of Tyr-148 at the active site of carboxypeptidase (Chapter 15, section B3a) could be assessed by substitution with phenylalanine. In addition to these and other problems of academic interest, it would be of great commercial value to be able to engineer enzymes used in industrial processes so as to give them improved properties, such as better rate and binding constants, changed reaction and substrate specificities, and increased thermostability and stability in organic solvents. This has not been feasible in the past with classic methods of peptide synthesis. But engineering of enzymes is now becoming a reality through the use of recombinant DNA technology and nucleic acid chemistry. These methods should eventually lead to the ultimate achievement: the design and synthesis of novel enzymes.

The techniques of genetic engineering have revolutionized enzymology by allowing: (1) the production of large quantities of known and previously inaccessible enzymes by the cloning and expression of their genes; (2) the rapid determination of their primary structures, since sequencing the DNA of the cloned genes is far faster than sequencing the proteins; and (3) the facile modification of protein structures by mutation of cloned genes. It is now possible to synthesize new enzymes *de novo* via the synthesis of their genes, and to tailor and redesign existing enzymes by mutating their genes. This new technology is called *protein engineering*.

Protein engineering is in its infancy, but it has begun to produce results. In this chapter we outline the basic principles involved, beginning with the properties of DNA that enable it to be so readily manipulated by enzymes. Some of the essential enzymatic reactions are briefly discussed. A comprehensive but lucid account of DNA enzymology is given by A. Kornberg.[1]

A. The structure and properties of DNA

DNA, the genetic material, is a long, unbranched polymer containing the four deoxynucleoside monophosphates: deoxyadenosine monophosphate (dAMP), deoxythymidine monophosphate (dTMP), deoxyguanosine monophosphate (dGMP), and deoxycytidine monophosphate (dCMP) (Figure 14·1). These are linked in the polymer by an ester bond between the 5'-phosphate of the nucleotide and the 3'-hydroxyl of the sugar of the next (Figure 14·2). An essential feature of the bases is that the purine A pairs with the pyrimidine T by hydrogen bonding, as does the purine G with the pyrimidine C (*Watson-Crick pairing rules*, Figure 14·3). Interestingly and importantly, the pairs AT, TA, GC, and CG are almost identical in overall size and shape. This base pairing enables two complementary strands of DNA to form a duplex, as, for example, in the following structure:

```
-T-A-T-G-C-A-C-G-
 | | | | | | | |
-A-T-A-C-G-T-G-C-
```

The duplex is a right-handed double helix with 10 bases per turn. The diameter of the helix is 20 Å (2 nm) and the pitch is 34 Å (3.4 nm). The sugar–phosphate backbone is on the outside of the helix, and the two antiparallel chains are connected by the hydrogen-bonded bases. The DNA in prokaryotes and eukaryotes is generally found in the duplex form, although there are some single-stranded DNA viruses. DNA is a very robust molecule in comparison with many proteins. The simple double-

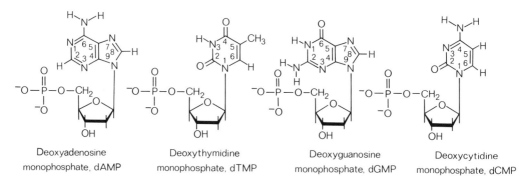

| Deoxyadenosine monophosphate, dAMP | Deoxythymidine monophosphate, dTMP | Deoxyguanosine monophosphate, dGMP | Deoxycytidine monophosphate, dCMP |

FIGURE 14·1. The four deoxynucleoside monophosphates that constitute DNA.

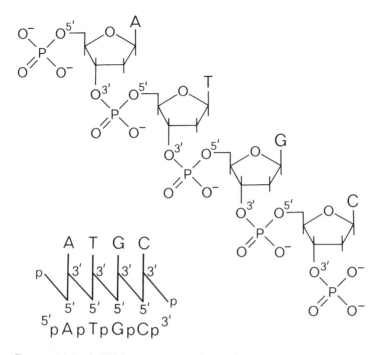

FIGURE 14·2. A DNA sequence and two shorthand notations for it.

helical secondary structure is readily reassembled after denaturation, unlike the complex tertiary protein structures that denature irreversibly. This feature and the duplex nature of DNA give it the properties that are so important to its manipulation in the laboratory: the information in one strand is duplicated in the other, and the two strands are complementary, so that they spontaneously anneal.

As any elementary textbook on molecular biology will relate, the sequences of proteins are stored in DNA in the form of a triplet code. Each amino acid is encoded by one or more triplet combinations of the four bases A, T, G, and C. For example, tryptophan is coded by the sequence TGG. The sequence of triplets is converted into a protein by a process in which DNA is first transcribed into mRNA. This message is then translated into protein on the ribosomes in conjunction with tRNA and the aminoacyl-tRNA synthetases. In prokaryotes, there is a one-to-one relationship between the sequence of triplets in the DNA and the sequence of amino acids in the protein. In eukaryotes, the DNA often contains stretches of intervening sequences or *introns* which are excised from the mRNA after transcription.

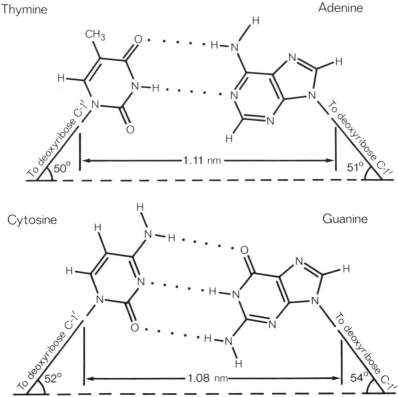

FIGURE 14·3. Complementary base pairing in the DNA double helix or at the active site of a DNA polymerase.

1. DNA may be replicated: DNA polymerases

DNA is replicated by DNA polymerases. The characteristics of this reaction are:

1. There is an absolute requirement for a template strand of DNA.
2. There is an absolute requirement for a DNA or RNA primer, which is annealed to the template and which possesses a 3′-hydroxyl group on the (deoxy)ribose. (Some very large eukaryotic enzymes have been found to have a "primase" activity by which they synthesize their own RNA primers.)
3. Replication proceeds exclusively in the 5′ → 3′ direction, with the four 5′-deoxynucleoside triphosphates serving as the source of monomers.

4. The replication is directed by the template according to the Watson-Crick pairing rules.

   DNA polymerases have just one binding site for all four combinations of base pairing—AT, TA, GC, and CG. The specificity of these sites is dictated by the Watson-Crick pairing rules, in that the sites themselves appear to recognize just the overall shape of a correct purine-pyrimidine pair, with the precise specificity resulting from the complementary nature of the base pairing. The polymerase catalyzes the transfer of a complementary deoxynucleoside monophosphate from its triphosphate to the 3'-hydroxyl of the primer terminus (equation 14·1). As is generally true of

$$\begin{array}{ccc}
\overset{\displaystyle\text{dTMP-PP}}{\underset{\displaystyle\text{HO}\nearrow}{}} & & \text{HO}\diagup \\[2pt]
\begin{array}{l} 5' \text{-T-A-C-G} \\ \;\;\;|\;\;|\;\;|\;\;| \\ 3' \text{-A-T-G-C-A-T-G-C-} \end{array} & \xrightarrow{\text{polymerase}} & \begin{array}{l} 5' \text{-T-A-C-G-T} \\ \;\;\;|\;\;|\;\;|\;\;|\;\;| \\ 3' \text{-A-T-G-C-A-T-G-C-} \end{array} + \text{PP}_i \quad (14\cdot1)
\end{array}$$

nucleotide transfer reactions that release pyrophosphate, there is an absolute requirement for Mg^{2+}. The true substrate in the reaction is the complex between the deoxynucleoside triphosphate and Mg^{2+}. The Mg^{2+} probably complexes with the β- and γ-phosphate groups (Chapter 2, equation 2·70).[2] It was once thought that there was a catalytically essential Zn^{2+} bound to the active site of DNA polymerases. However, measurements on large quantities of DNA polymerase obtained by genetic manipulation (section B2) show that Zn^{2+} is not required.[3]

   There are many different types of DNA polymerases, and they vary greatly in their activities and in the nature of the reactions they catalyze. Some polymerases are involved mainly in the replication of DNA. Others are used for the repair of damaged DNA. There is also an important difference between the enzymes isolated from eukaryotes and those isolated from prokaryotes. Most of the eukaryotic DNA polymerases that have been isolated so far have just the simple $5' \rightarrow 3'$ polymerization activity shown in equation 14·1. All known prokaryotic DNA polymerases, however, are multifunctional. In addition to their $5' \rightarrow 3'$ polymerase activity, they possess a $3' \rightarrow 5'$ exonuclease activity that can excise incorporated deoxynucleoside monophosphates. As was explained in Chapter 13, this is a proofreading or editing activity that enhances specificity by the removal of mismatches. The best known polymerase, DNA polymerase I (Pol I) of *Escherichia coli*, has a further exonuclease activity in the $5' \rightarrow 3'$ direction. This enzyme is a single polypeptide chain of relative molecular mass (M_r) 109 000. The $5' \rightarrow 3'$ polymerase and $3' \rightarrow 5'$ exonuclease activities may be separated by proteolysis. For example, the chain may be cleaved by subtilisin to give a large fragment of M_r 76 000 (the Klenow enzyme) containing the $3' \rightarrow 5'$ exonuclease and polymerase activities, and a small fragment of M_r 36 000 containing the 5′ → 3′ exonuclease activity.

2. Gaps in DNA may be sealed: DNA ligases[4]

Certain nicks in duplex DNA may be resealed by a DNA ligase. These enzymes will form a phosphodiester bond between a 5′-phosphoryl group and a directly adjacent 3′-hydroxyl, using either ATP or NAD^+ as an external energy source (Figure 14·4). The ligase from *E. coli* is well known. It is a single polypeptide chain of M_r 77 000 and it uses NAD^+ as the energy source. *E. coli* infected with phage T4 provides a useful ligase that has an M_r of ~65 000 and that uses ATP. An extended gap in duplex DNA may thus be repaired by first filling in the gap from the 3′-hydroxyl up to the 5′-phosphoryl group with a DNA polymerase, and then resealing the strand with a DNA ligase.

a. The mechanism of DNA ligase
The ligase reaction proceeds via the two covalent intermediates illustrated in Figure 14·5. A ligase adenylate with a phosphoamide bond is formed by the nucleophilic attack of a lysine side chain on either the ATP or the NAD^+, generating either PP_i or NMN (nicotinamide mononucleotide) with the phage T4 or *E. coli* enzyme, respectively. The adenylate is then transferred to the 5′-phosphoryl terminus of the DNA by nucleophilic attack on the α-phosphate of the AMP moiety to give the second intermediate, in which the 5′-phosphoryl group is activated. The ligase catalyzes the nucleophilic attack of the 3′-hydroxyl group on the activated phosphate to seal the nick, releasing AMP. The ligase adenylate may be readily isolated in 1:1 stoichiometry in the absence of DNA, but the DNA adenylate may be isolated only in small amounts from the complete reaction mixture, where it is formed at a steady state concentration.

3. Duplex DNA may be cleaved at specific sequences: Restriction endonucleases[5,6]

Just as proteins may be cleaved by specific proteases at defined residues, so duplex DNA may be cleaved in regions of defined sequence by a class

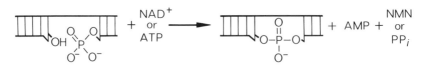

FIGURE 14·4. The DNA ligase reaction. NMN = nicotinamide mononucleotide.

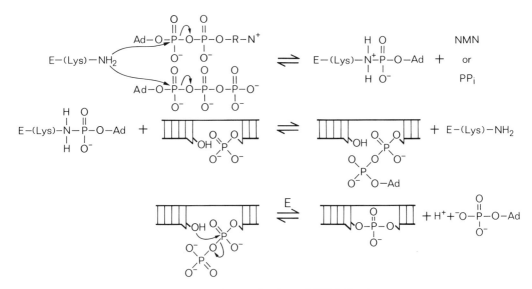

FIGURE 14·5. The mechanism of DNA ligation.

of endonucleases known, for historical reasons, as restriction endonucleases. The class of restriction endonucleases we are interested in is the Type II. These enzymes require only Mg^{2+} as a cofactor, and have M_r's of 20 000 to 100 000. They recognize specific DNA sequences that are generally several nucleotides long and that are symmetrical in a rotational sense, as shown in Table 14·1. Both strands of the DNA are cleaved in a symmetrical manner to give nicks with 3'-hydroxyl and 5'-phosphoryl termini. The cuts may be staggered; for example, *Eco*RI (Table 14·1) generates a protruding 5' terminus (equation 14·2), and *Pst*I generates a

$$5' \quad \begin{array}{c} \text{-G-A-A-T-T-C-} \\ | \ | \ | \ | \ | \ | \\ \text{-C-T-T-A-A-G-} \end{array} \xrightarrow{\textit{EcoRI}} \quad 5' \quad \begin{array}{c} \text{-G} \\ | \\ \text{-C-T-T-A-A} \end{array} \quad + \quad \begin{array}{c} \text{A-A-T-T-C-} \\ | \\ \text{G-} \end{array} \qquad (14\cdot2)$$

protruding 3' terminus (equation 14·3). Other enzymes, such as *Hae*III, give flush (or "blunt") ends because they cut at the center of symmetry.

$$5' \quad \begin{array}{c} \text{-C-T-G-C-A-G-} \\ | \ | \ | \ | \ | \ | \\ \text{-G-A-C-G-T-C-} \end{array} \xrightarrow{\textit{PstI}} \quad 5' \quad \begin{array}{c} \text{-C-T-G-C-A} \\ | \\ \text{-G} \end{array} \quad + \quad \begin{array}{c} \text{G-} \\ | \\ \text{A-C-G-T-C-} \end{array} \qquad (14\cdot3)$$

4. DNA fragments may be joined by using enzymes

An essential process in recombinant DNA technology is the joining of a fragment of DNA from one genome with that from another. For example, the staggered cleavages generated by restriction enzymes in equations

TABLE 14·1. *Specificity of Type II restriction endonucleases*

Enzyme	DNA sequence cleaved

Axis of symmetry

*Eco*RI   5′ —G$\overset{\downarrow}{-}$A—A⊢T—T—C— 3′
—C—T—T⊣A—A$\overset{}{\underset{\uparrow}{-}}$G—

*Bam*I   —G$\overset{\downarrow}{-}$G—A⊢T—C—C—
—C—C—T⊣A—G$\overset{}{\underset{\uparrow}{-}}$G—

*Hae*III   —C—C$\overset{\downarrow}{-}$⊢G—G—
—G—G-$\overset{}{\underset{\uparrow}{|}}$C—C—

*Pst*I   —C—T—G⊣C—A$\overset{\downarrow}{-}$G—
—G$\overset{}{\underset{\uparrow}{-}}$A—C⊢G—T—C—

14·2 and 14·3 give ends that are complementary and thus mutually co-hesive. This means that restriction fragments (i.e., lengths of DNA pro-duced by cutting with restriction enzymes) generated from different pieces of DNA by the same restriction enzyme may be annealed and sealed with DNA ligase, since they contain the requisite 3′-hydroxyl and 5′-phos-phoryl ends.

An important extension of this method is the procedure of *blunt-end ligation*: high concentrations of the T4 ligase will catalyze the ligation of DNA fragments containing flush ends (equation 14·4). Flush ends may

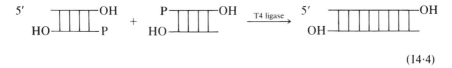

$$(14·4)$$

be generated by a number of enzymatic processes in addition to cleavage by enzymes such as *Hae*III. For example, the protruding 3′ ends gen-erated by *Pst*I in equation 14·3 may be excised by using the 3′ → 5′ exonuclease activity of the T4 DNA polymerase. The protruding 5′ ends generated by *Eco*RI (equation 14·2) may be removed by using a nuclease such as SI, which is specific for the cleavage of single-stranded DNA. Alternatively, the 3′ ends in the *Eco*RI fragments may be extended by "filling in" with the large subfragment of Pol I so that they become flush with the 5′ ends. In combination with these reactions, blunt-end ligation becomes a general means of joining a restriction fragment generated by one restriction enzyme from one piece of DNA with a fragment generated by any other enzyme from another piece.

5. Joining DNA by complementary homopolymeric tails: Terminal transferase[7]

There is an alternative method of generating cohesive tails. The enzyme calf thymus terminal (deoxynucleotidyl) transferase adds deoxynucleoside monophosphate residues from 5'-deoxynucleoside triphosphates to protruding 3'-hydroxyl termini in the absence of a template. For example, as shown in equations 14·5 to 14·7, complementary ends can be generated

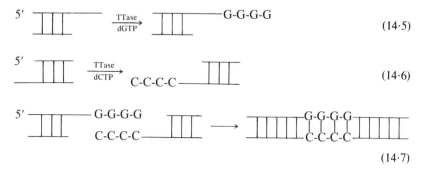

$$(14\cdot5)$$

$$(14\cdot6)$$

$$(14\cdot7)$$

and can then be annealed, with the resultant gap filled in by a DNA polymerase and sealed by a ligase.

6. Processive vs. distributive polymerization

Enzyme-catalyzed polymerization reactions have an important characteristic that is not found elsewhere. Once the enzyme has added a monomeric unit to the growing chain, it can either dissociate and recombine at random with other growing termini, or it can remain attached to the same chain and add further residues. Enzymes that dissociate between each addition and distribute themselves among all the termini are termed *distributive*. Those that process along the same chain without dissociating are termed *processive*. These terms apply also to degradative enzymes such as exonucleases.

A typical processive enzyme is terminal transferase. It adds on deoxynucleoside monophosphates randomly to exposed 3'-hydroxyl termini so that the final products are formed in a statistical distribution. The distribution follows *Poisson's law*.[8] Suppose that the enzyme adds on an average of x residues per chain. Then the probability of a particular chain having k residues added [i.e., $p(k)$] is given by

$$p(k) = \frac{x^k}{k!} e^{-x} \tag{14·8}$$

(where $k!$ = factorial $k = 1 \times 2 \times 3 \times \ldots \times k$). For example, if an average of 1 residue has been added per chain, then it may be calculated from equation 14·8 that $p(0) = 0.368$ ($0! = 1$), $p(1) = 0.368$, $p(2) = 0.184$, $p(3) = 0.061$, $p(4) = 0.015$, etc. That is, 36.8% of the chains have no residues added, 36.8% have 1, etc. Similarly, for $x = 2$, $p(0) = 0.135$, $p(1) = 0.271$, $p(2) = 0.271$, $p(3) = 0.180$, $p(4) = 0.090$, $p(5) = 0.036$, etc.

The balance between processivity and distributivity clearly depends on the ratio of rate constants for polymerization (or degradation) and dissociation. DNA polymerases *in vitro* vary from being distributive to having some degree of processivity. *In vivo*, however, the polymerase is part of a multienzyme DNA polymerizing complex (a "replisome") that is almost certainly highly processive.

B. Cloning enzyme genes for overproduction

The biosynthesis of enzymes is under strict control so that they are produced in the correct quantities for the optimal viability of cells. The "housekeeping" enzymes—those involved in high-volume metabolic routes such as glycolysis—are produced in large amounts, whereas others are produced only in small quantities. Consequently, much of enzymology in the past has concentrated upon the easily accessible metabolic enzymes. Now, however, because of the advent of gene cloning, it is possible to obtain large quantities of proteins that were previously rare.

The gene coding for a particular protein from an organism such as *E. coli* may be cloned by the following strategy: the genomic DNA of that organism is cut into small fragments; these are inserted into a *vector*, a double-stranded DNA molecule that can replicate after the foreign DNA has been inserted into it; the vector is allowed to replicate in a host, such as *E. coli*, and is screened for production of the desired protein (Figure 14·6).[9–11]

The basic enzymology for insertion of fragments of DNA into a vector was covered in section A. The genomic DNA is usually fragmented by partly digesting it with a restriction enzyme. This produces a series of restriction fragments, some of which may contain the desired gene. If the vector DNA may be cut at a single site with the same restriction enzyme, then the restriction fragments may be inserted into the cleaved vector by annealing its "sticky ends" and sealing with ligase as described in section A2. This gives a *gene library* or a *gene bank*. An alternative method of producing DNA fragments is to shear the DNA by physical forces and use the terminal transferase procedure of section A5. Cloning genes or enzymes from eukaryotes is more complicated partly because of problems caused by the introns. These difficulties can be avoided by copying the mRNA (from which the introns have been excised) by using a polymerase

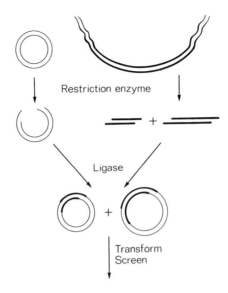

FIGURE 14·6. A scheme for gene cloning.

to give the so-called cDNA, and then cloning this. However, there are often problems in obtaining a high level of expression of eukaryotic genes in prokaryotes.

Advanced methodologies for these processes are now available, and are discussed in detail elsewhere (see Further Reading). Some of the basic principles concerning vectors and screening are discussed in the next section.

1. Vectors

Perhaps the most common vector is the plasmid pBR322. Plasmids are genetic elements that are not part of the major bacterial chromosome, and that replicate autonomously in the cytoplasm. The plasmid pBR322 is a small, covalently closed circle of double-stranded DNA containing 4362 base pairs. It carries genes for resistance to the antibiotics ampicillin and tetracycline, and replicates autonomously in *E. coli* by using the host replication proteins. Many copies of pBR322 may exist in each cell, and it is thus termed a multicopy plasmid. Partly because of the high copy number, a gene cloned into such a plasmid may produce large amounts of enzyme. The precise degree of overproduction depends also on the promoter for the gene, since the DNA from the cloned protein is transcribed from its own promoter. (The DNA sequence of the promoter, the stretch of DNA to which RNA polymerase binds to initiate transcription, controls the efficiency of transcription: a strong promoter gives high production of mRNA, with the possibility of high levels of protein; a weak promoter gives low production.) Another factor is whether or not the organism can tolerate the overproduction. In extreme cases, *E. coli* is found to tolerate up to 50% of its soluble protein being in the form of the cloned protein (e.g., the recA protein discussed in section B3). In other cases, overproduction is lethal. An example of the latter occurs with the overproduction of Pol I.

Pol I has been cloned for overproduction, however, by using a different vector, bacteriophage λ.[12] Phage λ can exist as a lysogen integrated into the genome of *E. coli*. The DNA is replicated along with that of *E. coli*, but the major proteins are not expressed and the phage is not produced until an event occurs that induces the phage. There is a mutant of phage λ that may be induced by simply raising the temperature to 42°C. The gene coding for Pol I has been cloned into this mutant and the enzyme produced as follows. *E. coli* is grown at 30°C until a high cell density is obtained. The temperature is rapidly raised to 42°C for a few minutes. This induces the phage, and the Pol I is produced in significant quantities (under the control of the λ promoters). Although the cells are now doomed, they produce the desired protein on further incubation at 37°C.

2. Screening

A gene library consists of thousands of different restriction fragments cloned into a vector. Only a few of these will contain the desired gene, so screening methods are necessary to find the needles in the haystack. The simplest way of screening the gene library for the gene of the desired protein is by means of a *complementation assay*. This can be performed if there is a mutant of the host that lacks the desired enzyme activity. A vector that contains the missing gene may allow the mutant host to grow under nonpermissive conditions if the necessary protein is expressed from the vector. For example, an available mutant of *E. coli* is temperature-sensitive in its tyrosyl-tRNA synthetase: the bacterium grows at 30°C but not at 42°C. The mutant can be used to screen a gene library of DNA from *E. coli* or *Bacillus stearothermophilus* in pBR322 as follows.[13] The plasmid, which contains the gene for ampicillin resistance, is introduced into the mutant bacterium (i.e., the cells are *transformed*) by mixing the plasmid and bacterium in the presence of $CaCl_2$ under certain conditions. The cells are then grown in the presence of ampicillin at 30°C. This selects the cells that have been transformed and that have the antibiotic resistance coded by the plasmid, and also allows the copy number of the plasmid to increase. The cells are then incubated at 42°C, so that only those that have the plasmids containing the restriction fragments expressing the tyrosyl-tRNA synthetase grow.

If a complementation assay cannot be performed, brute force must be applied to locate the desired gene or its product. One of the better procedures is a *DNA hybridization assay*. If the sequence of the protein is available, a short length of DNA corresponding to the sequence may be synthesized. This may then be labeled with ^{32}P by using a phosphokinase and [γ-^{32}P]ATP. The vector DNA from individual colonies of transformed cells is baked onto nitrocellulose sheets, and the denatured DNA is bathed in a solution of the radioactive probe. The probe will hybridize to any DNA containing the complementary sequence. After the surplus probe is washed off, an autoradiograph of the nitrocellulose sheet locates the colonies containing the desired DNA.[14] In practice, because of the degenerate nature of the genetic code, a mixture of probes must be used. There is an illustration of an autoradiograph at the end of the next section.

Another important procedure is to screen for production of the desired protein by using immunological methods with antibodies raised against it.

3. Examples

Overproduction of a desired enzyme not only affords large amounts of protein, but greatly eases purification: a protein that is overproduced by a factor of 100 needs fewer purification steps to achieve the same quality as preparation without overproduction. An example of a protein that may be overproduced in large amounts and that is readily purified is the recA protein from *E. coli*.[15-20] This fascinating enzyme—which is an ATPase, a protease, and catalyzes the annealing of single-stranded DNA to double-stranded DNA, all contained within M_r 37 800—is a prime example of a product of genetic engineering. Prior to 1976, it was just an unknown protein, protein X, postulated to be involved in certain genetic processes. Between 1977 and 1980, the recA gene was isolated and cloned; the protein was purified in large quantities and its amino acid sequence was determined from its DNA sequence simultaneously by two groups; and the first x-ray diffraction patterns from its crystals were published. Because of the high level of its overproduction, it can be purified to homogeneity with just a few precipitation steps and one gel filtration.

Literally hundreds of enzymes have now been cloned, from DNA ligase, polymerases, and aminoacyl-tRNA synthetases to old favorites such as glycolytic enzymes. The purpose of cloning these latter enzymes is not so much for overproduction but for the introduction of mutations into their genes. Mutations make it possible to modify the structure and activity of the enzyme.

C. Site-directed mutagenesis

Given the gene of a protein, it is possible to mutate it by a variety of methods. Most of these are relatively nonspecific and nonsystematic, although they are still valuable. For example, one group of methods involves opening a stretch of single-stranded DNA in the double-stranded vector by using an endonuclease (e.g., a restriction endonuclease that, in the presence of ethidium bromide, just makes a single cut in one strand), followed by partial digestion by an exonuclease (e.g., by Pol I). The gap may be repaired by a polymerase or ligase under mutagenic conditions to generate a range of substitution or deletion mutants: this is termed gap misrepair.[21] Alternatively, the single-stranded gap may be treated with sodium bisulfite, which catalyzes the deamination of C (cytosine) residues to U (uracil). U codes the same as T (thymine), so this treatment leads to a change of coding. Additions and deletions may be made by using some of the enzymology discussed earlier in this chapter. For example, after both strands have been cut at a single site with a restriction enzyme, the protruding ends can be digested by nucleases and the resultant flush ends can be ligated to generate a truncated genome. Or, where a 5' overhang is generated by a restriction endonuclease, the 3' end may be extended by a polymerase until it is flush. Then, upon blunt-end ligation, an addition has been made to the genome. The drawback of these methods is that they cannot generally be targeted against particular codons in the genomes. Further, there are difficulties in screening for the desired mutants.

1. Oligodeoxynucleotide-directed mutagenesis[22–25]

Oligodeoxynucleotide-directed mutagenesis allows the systematic replacement of every single amino acid residue in a protein in turn. The method affords a simple screening procedure for the detection of a desired mutant. A successful experimental approach is illustrated in Figure 14·7. The gene is cloned into a double-stranded vector, and one of the constituent circles of single-stranded DNA is isolated. A short oligodeoxynucleotide has already been synthesized to be complementary to the region of the gene to be mutated, except for a single-base (or double-base) mismatch. The mismatch is designed to change the codon for the target amino acid residue into the codon for the desired mutant residue. The oligodeoxynucleotide is annealed to the gene in the single-stranded vector and becomes a primer for Pol I (the Klenow fragment) to use in replicating the rest of the genome. The replicated strand is ligated, and the result is a heteroduplex containing one strand of mutant and one strand of wild-type (unmutated) DNA. The heteroduplex is used to transform a host and

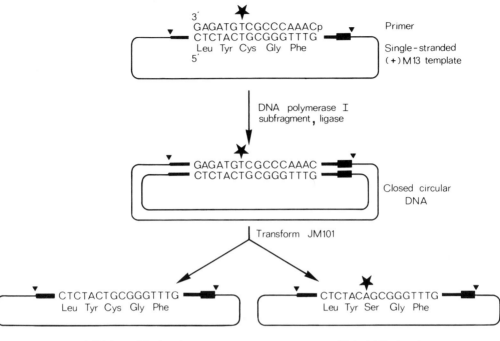

FIGURE 14·7. A scheme for oligodeoxynucleotide-directed mutagenesis. The mismatched primer is designed to mutate the codon for cysteine (TGC) to that for serine (AGC) (see text).

The gene selected for mutation must be obtained in the form of single-stranded DNA before the primer can be annealed. This is done by subcloning the gene from the double-stranded pBR322 into a new vector, bacteriophage M13. The genome of this filamentous phage is a covalent circle of single-stranded DNA that goes through a covalent circular duplex termed RF (= replicative form) during replication in its host, *E. coli*. The subcloning is performed on the RF DNA, which then yields single-stranded viral DNA. Note that the viral DNA is termed the (+) strand as it has the same sense as the message. The complementary strand is termed (−). JM101 is a strain of *E. coli* that is a host for M13. JM101 may be transformed by M13 DNA, as described in section B2, to produce mature phage. [From G. Winter, A. R. Fersht, A. J. Wilkinson, M. Zoller, and M. Smith, *Nature, Lond.* **299**, 756 (1982).]

produce colonies of cells that each contain either the vector with the mutant or the vector with the wild-type gene. The colonies may be screened by introducing the original oligodeoxynucleotide primer, now radioactively labeled as a probe in a DNA hybridization assay[26] (section B2). The probe anneals preferentially to the mutant DNA, to which it is fully complementary, rather than to the wild-type DNA, with which it has a mismatch. This technique is called dot-blot hybridization (Figure 14·8).

This method may be extended to make addition or deletion mutants. For example, by using a longer primer (a 24-base polymer) in which one half is complementary to the residues in one part of the genome, and the other half to the residues in a distant region, the intervening sequence may be deleted.

The example illustrated in Figures 14·7 and 14·8 was the first reported case in which a residue in a protein of known three-dimensional structure was modified in this way. The mutation was in the ATP-binding site of the tyrosyl-tRNA synthetase from *B. stearothermophilus*.[27] The gene was obtained in the form of single-stranded DNA by subcloning it from pBR322 into a new vector, bacteriophage M13. The genome of this filamentous phage is a covalent circle of single-stranded DNA that during replication in its host, *E. coli*, goes through a covalent circular duplex of

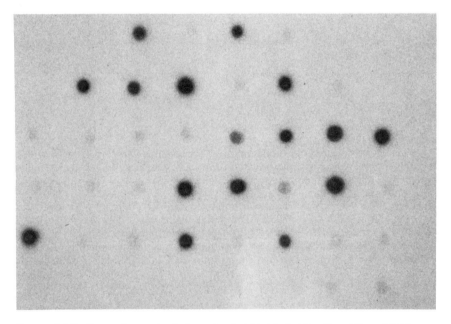

FIGURE 14·8. Screening mutant phage by dot-blot hybridization. Phage DNA was baked on the nitrocellulose filter and was annealed to the mutagenic primer of Figure 14·7 that had been labeled with ^{32}P-phosphate. The filter was washed at successively higher temperatures until one was reached at which the primer was washed from the wild-type DNA but remained bound to the mutant. The autoradiograph shows the 30 percent of clones that contain the mutant. In general, some 1 to 40 percent of the clones are found to be mutant. [From G. Winter, A. R. Fersht, A. J. Wilkinson, M. Zoller, and M. Smith, *Nature, Lond.* **299**, 756 (1982).]

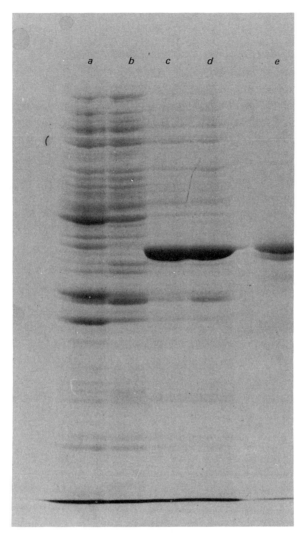

FIGURE 14·9. Expression of tyrosyl-tRNA synthetase from M13 in *E. coli*, detected by polyacrylamide gel electrophoresis. Tracks *a* through *d* are from total cell extracts, whereas track *e* is a marker of authentic tyrosyl-tRNA synthetase. Track *a* is from uninfected *E. coli*; *b* is from cells infected with M13; *c* is from cells infected with M13 containing the wild-type gene; and *d* is from cells infected with M13 containing the mutated gene. [From G. Winter, A. R. Fersht, A. J. Wilkinson, M. Zoller, and M. Smith, *Nature, Lond.* **299**, 756 (1982).]

385

DNA termed RF (= replicative form). All the cloning procedures were performed on the RF DNA, and the mutagenesis was performed on the naturally produced single-stranded viral DNA. The mutant enzyme was produced in high yield in crude cell extracts (Figure 14·9).

As mentioned earlier, the massive overproduction of mutant enzymes facilitated their rapid purification. Subsequent production of mutants in which the residues that hydrogen-bond with the substrate were systematically altered allowed the structure–activity analysis described in Chapter 12, section B3.[28] Interestingly, one mutant was found to have a dramatically lowered K_M for ATP and a higher activity for tyrosine activation (Table 14·2), indicating that this is a promising approach for improving the catalytic activity of industrial enzymes.[29]

It is clearly possible to tour the architecture of the enzyme to assess the role of each residue in catalysis. Additional possibilities are to explore the folding of proteins and to experiment with changing their stabilities and solubilities.

This systematic approach requires knowledge of the three-dimensional structure of a protein from x-ray crystallography. In the absence of such information, experiments may still be performed to detect residues that are involved in catalysis. For example, the β-lactamase that is encoded in pBR322 and that confers the resistance to ampicillin has also been modified by the oligodeoxynucleotide-directed procedure. Two residues at the active site were mutated simultaneously to give complete loss of activity.[30] The method differed from that above in that the protein was not subcloned into a single-stranded vector; instead, a short section of the DNA of pBR322 was exposed by a partial enzymatic digestion prior to the annealing of the mismatched oligodeoxynucleotide primer.

TABLE 14·2. *Protein engineering an increased activity of the tyrosyl-tRNA synthetase in the activation of tyrosine*[a]

Enzyme	k_{cat} (s^{-1})	K_M (mM)	k_{cat}/K_M (s^{-1} M^{-1})
Wild-type	7.6	0.9	8 400
Thr-51 → Ala-51	8.6	0.54	15 900
Thr-51 → Pro-51	12.0	0.058	208 000

[a] See Table 12·3 for conditions and more data. The active site of the enzyme is illustrated in Figure 12·5. The constants are for the variation of [ATP]. Note that in the overall reaction, the aminoacylation of tRNA, the mutant enzyme containing Pro-51 has a 100 fold lower K_M for ATP but also a two fold lower K_{cat} than the wild type. At the high concentrations of ATP in vivo, the wild-type enzyme is more active than the mutant. This is a good illustration of the principles of Chapter 12, that tight binding of the substrate is to be avoided and values of K_M should match the concentration of metabolites *in vivo*. [From A. J. Wilkinson, A. R. Fersht, D. M. Blow, P. Carter, and G. Winter, *Nature, Lond.* **307**, 187 (1984).]

The above experiments were made possible by the rapid solid-phase procedures that have been developed in recent years for synthesizing oligodeoxynucleotides.[31] These methods have improved so fast that inexperienced biochemists can manually synthesize a 12-base polymer in a day, while machines can generate much longer polymers. The first fully synthetic genes coding for and expressing a protein, human leukocyte interferon, have already been made by assembling smaller, overlapping synthetic oligodeoxynucleotides through use of a polymerase and ligase (equation 14·9).[32] For genes synthesized in this way, it is not necessary to use the methods of site-directed mutagenesis to change individual residues, since it is simpler to assemble the mutant gene directly from mutated fragments.

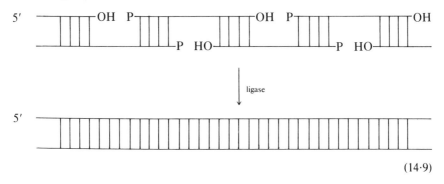

$$(14·9)$$

References

1   A. Kornberg, *DNA replication*, W. H. Freeman and Company (1980) (and 1982 supplement).
2   P. M. J. Burgers and F. Eckstein, *J. Biol. Chem.* **254**, 6889 (1979).
3   K. E. Walton, P. C. Fitzgerald, M. S. Herrman, and W. D. Behnke, *Biochem. Biophys. Res. Commun.* **108**, 1353 (1982).
4   I. R. Lehman, *Science* **186**, 790 (1974).
5   D. Nathans and H. O. Smith, *Ann. Rev. Biochem.* **44**, 273 (1975).
6   R. Yuan, *Ann. Rev. Biochem.* **50**, 285 (1981).
7   F. J. Bollum, *Adv. Enzymol.* **47**, 347 (1978).
8   F. N. Hayes, V. E. Mitchell, R. L. Ratliffe, and D. L. Williams, *Biochemistry* **6**, 2488 (1967).
9   D. A. Jackson, R. H. Symons, and P. Berg, *Proc. Natn. Acad. Sci. USA* **67**, 2904 (1972).
10   P. E. Lobban and A. D. Kaiser, *J. Molec. Biol.* **78**, 453 (1973).
11   S. N. Cohen, *Science* **195**, 654 (1977).
12   N. E. Murray and W. S. Kelley, *Molec. Gen. Genet.* **175**, 77 (1979).
13   D. Barker, *Eur. J. Biochem.* **125**, 357 (1982).
14   M. Grunstein and D. S. Hogness, *Proc. Natn. Acad. Sci. USA* **72**, 3961 (1975).
15   K. McEntee, *Proc. Natn. Acad. Sci. USA* **74**, 5275 (1977).
16   L. J. Gudan and D. W. Mount, *Proc. Natn. Acad. Sci. USA* **74**, 5280 (1977).

17   T. Horii, T. Ogawa, and H. Ogawa, *Proc. Natn. Acad. Sci. USA* **77**, 313 (1980).
18   A. Sancar, C. Stachelek, W. Konigsberg, and W. D. Rupp, *Proc. Natn. Acad. Sci. USA* **77**, 2611 (1980).
19   D. B. Mackay, T. A. Steitz, I. T. Weber, S. C. West, and P. Howard-Flanders, *J. Biol. Chem.* **255**, 6662 (1980).
20   S. M. Cotterill, A. Satterthwait, and A. R. Fersht, *Biochemistry* **21**, 4332 (1982).
21   D. Shortle, P. Grisafi, S. J. Benkovic, and D. Botstein, *Proc. Natn. Acad. Sci. USA* **79**, 1588 (1982).
22   H. Schott and H. Kössel, *J. Am. Chem. Soc.* **95**, 3778 (1973).
23   C. A. Hutchinson III, S. Phillips, M. H. Edgell, S. Gillam, P. Jahnke, and M. Smith, *J. Biol. Chem.* **253**, 6551 (1978).
24   R. B. Wallace, P. F. Johnson, S. Tanaka, M. Schold, K. Itakawa, and J. Abelson, *Science* **209**, 1396 (1980).
25   A. Razin, T. Hirose, K. Itakura, and A. D. Riggs, *Proc. Natn. Acad. Sci. USA* **75**, 4269 (1978).
26   R. Wallace, M. Schold, M. J. Johnson, P. Dembek, and K. Itakura, *Nucl. Acids Res.* **9**, 3647 (1981).
27   G. Winter, A. R. Fersht, A. J. Wilkinson, M. Zoller, and M. Smith, *Nature, Lond.* **299**, 756 (1982).
28   A. J. Wilkinson, A. R. Fersht, D. M. Blow, and G. Winter, *Biochemistry* **22**, 3581 (1983).
29   A. J. Wilkinson, A. R. Fersht, D. M. Blow, P. Carter, and G. Winter, *Nature, Lond.* **307**, 187 (1984).
30   G. Dalbadie-McFarland, L. W. Cohen, A. D. Riggs, C. Morin, K. Itakura, and J. H. Richards, *Proc. Natn. Acad. Sci. USA* **79**, 6409 (1982).
31   K. Itakura, *Trends Biochem. Sci.* **7**, 442 (1982).
32   M. D. Edge, A. R. Greene, G. R. Heathcliffe, P. A. Meacock, W. Schuck, D. B. Scanlon, T. C. Atkinson, C. R. Newton, and A. R. Markham, *Nature, Lond.* **292**, 756 (1981).

Further reading

D. M. Glover, *Gene cloning: the mechanics of DNA manipulation*, Chapman and Hall (1984).
A. Kornberg, *DNA replication*, W. H. Freeman and Company (1980) (and 1982 supplement).
T. Maniatis, E. F. Fritsch, and J. Sambrook, *Molecular cloning*, Cold Spring Harbor Laboratory, Cold Spring Harbor, N.Y. (1982). (An indispensable manual of laboratory techniques.)
J. D. Watson, J. Tooze, and D. T. Kurtz, *Recombinant DNA: A Short Course*, W. H. Freeman and Company (1983).
R. Wu, L. Grossman, and K. Moldave, eds., *Methods in enzymology*, **100, 101**, Academic Press (1983). (Detailed reference volumes on enzymes and methods in recombinant DNA research.).

15

The structures and mechanisms
of selected enzymes

In this chapter we shall discuss the mechanisms of some enzymes whose crystal structures have been solved at high resolution. The emphasis is on how kinetic and structural work have been combined to produce satisfactory descriptions of the reaction mechanisms, and what general lessons have been learned about enzyme catalysis. The chapter also provides a nonsystematic introduction to some of the experimental approaches used in enzymology.

The basic features of most of the mechanisms are known; for example, the positions of the catalytic groups on the enzyme have generally been determined, as has the overall chemical route of the reaction—although in certain cases either or both are obscure. The better-understood examples are often those in which covalent intermediates have been detected and characterized. Besides delineating the chemical mechanism, the presence of a covalent intermediate provides considerable additional structural information: it immediately locates the position of the substrate relative to a catalytic group on the enzyme.

The crucial problem in the structural work is the determination of the structure of the enzyme–substrate complex. Without this, it is not possible to obtain the fine details of the reaction, such as whether there is distortion of the enzyme or substrate, and precisely where the substrate is located relative to the catalytic groups. The structure of the enzyme by itself provides only a framework on which to hang hypotheses. Unfortunately, as described in Chapter 1, it is generally not possible to solve directly the structures of productively bound enzyme–substrate complexes, so substrate analogues and model building must be used as a substitute. This has the disadvantage that strain and distortion effects may be overlooked. Sometimes, as with ribonuclease, the structures of the analogues are very similar to that of the real substrate. In a few cases—trypsin and horse liver alcohol dehydrogenase, for example—it is possible to solve the structures of the productive complexes because of a favorable equilibrium between the substrate and the product.

It is important that we begin on a note of caution. The nature of science is such that experimentalists push their experimental data to the limits of reliability, and sometimes beyond this point. All experiments, including those from x-ray crystallography, are subject to *interpretation*: one crystallographer may interpret a particular feature in the electron density map as being significant, whereas another may consider it an artifact of statistical noise. The worst interpreters of all tend to be authors who interpret other people's interpretations!

A. The dehydrogenases

The dehydrogenases discussed in this section catalyze the oxidation of alcohols to carbonyl compounds. They utilize either NAD^+ or $NADP^+$ as coenzymes. The complex of the enzyme and coenzyme is termed the *holo*enzyme; the free enzyme is called the *apo*enzyme. Some dehydrogenases are specific for just one of the coenzymes; a few use both. The reactions are readily reversible, so that carbonyl compounds may be reduced by NADH or NADPH. The rates of reaction in either direction are conveniently measured by the appearance or disappearance of the reduced coenzyme, since it has a characteristic ultraviolet absorbance at 340 nm. The reduced coenzymes also fluoresce when they are excited at 340 nm, which provides an even more sensitive means of assay.

The chemistry of the reduction of NAD^+ has been solved most elegantly (Chapter 8, section B1).[1] Oxidation of the alcohol involves the removal of two hydrogen atoms. One is transferred directly to the 4 position of the nicotinamide ring of the NAD^+, and the other is released as a proton (equation 15·1).[2,3] It is generally thought that the hydrogen is transferred as a hydride ion H^-, but a radical intermediate cannot be ruled out. For convenience, we shall assume that the mechanism is the hydride transfer.

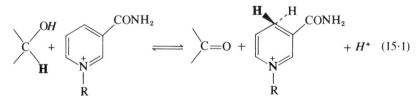

$$ \text{(15·1)} $$

The transfer is also stereospecific. Through the use of deuterated substrates, it has been found that some dehydrogenases will transfer to one side of the ring, and other enzymes to the opposite side (structures 15·2). The enzymes have been classified as "A" or "B" on this basis

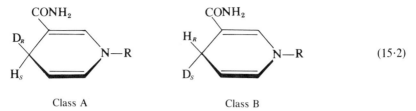

$$\text{Class A} \qquad\qquad \text{Class B} \tag{15·2}$$

(Table 15·1). Class A transfers the pro-R hydrogen from NADH, whereas Class B transfers the pro-S. The crystal structures of dehydrogenases solved so far show that NAD^+ is bound to the Class A enzymes with the nicotinamide ring in the anti conformation about the glycosidic bond, as is consistent with NMR studies in solution,[4] and that the Class B enzymes have the ring bound in the syn conformation.

A rationale for the two classes has been proposed, based on the observation that Class A enzymes catalyze the reduction of the more reactive carbonyls, while Class B is associated with the less reactive ones.[5] There is speculation that NADH is a weaker reducing agent when it is in the syn conformation (Class A) than when it is anti (Class B). This would tend to balance the equilibrium constant between enzyme-bound oxidized and reduced reagents, i.e., the internal equilibrium constant of Chapter 12, section E2, with the consequent rate advantages discussed in that section.

The structures of several dehydrogenases have now been solved. The earlier work on these has been reviewed in depth in the literature, as have their physical and kinetic properties.[6-10] Some generalizations can be made. As was discussed at the end of Chapter 1, the subunits may be divided into two domains: a catalytic domain, which can be quite variable in structure, and a nucleotide-binding domain, which is formed from a similar overall folding of the polypeptide chain for all the dehydrogenases. The detailed geometry of the nucleotide-binding domain varies consid-

TABLE 15·1. *The coenzyme specificity of some dehydrogenases*

Dehydrogenase	Coenzyme required	Stereospecificity class
Glutamate	NAD^+ or $NADP^+$	B
Glucose 6-phosphate	$NADP^+$	B
3-Glycerol phosphate	NAD^+	B
Glyceraldehyde 3-phosphate	NAD^+	B
Malate (soluble)	NAD^+	A
Alcohol	NAD^+	A
Lactate	NAD^+	A
Isocitrate	$NADP^+$	A

erably from one enzyme to another. However, the coenzyme binds in a similar extended, open conformation in all cases (Figure 15·1). The most significant variation concerns which side of the nicotinamide ring faces the substrate, with the side in the Class A enzymes being opposite to that in the Class B.

1. The alcohol dehydrogenases[7]

The alcohol dehydrogenases are zinc metalloenzymes of broad specificity. They oxidize a wide range of aliphatic and aromatic alcohols to their corresponding aldehydes and ketones, using NAD^+ as a coenzyme (see equation 15·1). The two most studied enzymes are those from yeast and horse liver. The crystal structures of the horse liver apo- and holoenzymes have been solved at 2.4 and 2.9 Å (0.24 and 0.29 nm), respectively.[11,12] The molecule is a symmetrical dimer, composed of two identical chains

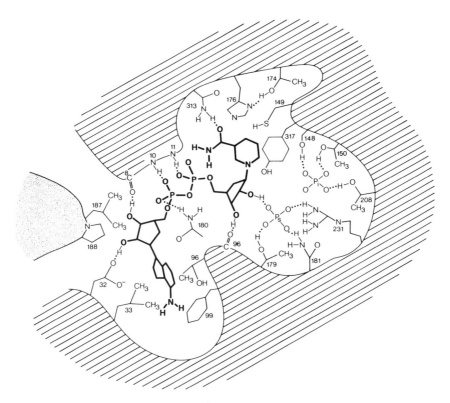

FIGURE 15·1. The binding of NAD^+ to glyceraldehyde 3-phosphate dehydrogenase from *Bacillus stearothermophilus*. [From G. Biesecker, J. I. Harris, J. C. Thierry, J. E. Walker, and A. J. Wonacott, *Nature, Lond.* **266**, 328 (1977).]

of M_r 40 000. Each chain contains one binding site for NAD^+ but two sites for Zn^{2+}. Only one of the zinc ions is directly concerned with catalysis. The yeast enzyme, on the other hand, is a tetramer of M_r 145 000, and each chain binds one NAD^+ and one Zn^{2+}. Despite these differences, sequence data indicate a good deal of homology between the two enzymes.[13] It is often assumed that the same overall reaction mechanism holds for both enzymes, although details, such as the rate-determining step, the pH dependence, and the Hammett plots, differ between the two.

The liver enzyme exists in two forms, E and S, which differ only by some six residues in their amino acid composition.[13] Only the S is active toward 3-β-hydroxysteroids, but both forms are active toward ethanol. None of the known amino acid differences is located in the subunit interfaces. Accordingly, E and S chains combine in statistical ratios to form EE, SS, and ES dimers. These different species are termed *isozymes*, which means that they are multiple molecular forms of the same enzyme. When it is isolated from liver, the enzyme consists of about 40 to 60% of the EE dimer, the remainder being SS and ES.

a. The structure of the active site of liver alcohol dehydrogenase[7,14]
The Zn^{2+} ion sits at the bottom of a hydrophobic pocket formed at the junction of the catalytic and nucleotide-binding domains (Figure 15·2). It is ligated by the sulfur atoms of Cys-46 and Cys-174, and by a nitrogen atom of His-67. The fourth ligand is an ionizable water molecule that is hydrogen-bonded to the hydroxyl group of Ser-48. It is known from the

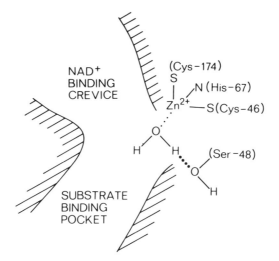

FIGURE 15·2. Sketch of the active site of horse liver alcohol dehydrogenase. (Courtesy of C.-I. Brändén.)

pH dependence of the binding of NAD^+,[15] from the direct determination of the proton release on the binding of NAD^+,[16] and from other kinetic procedures[17] that the apoenzyme has a functional group of $pK_a \sim 9.2$ that is perturbed to a pK_a of ~ 7.6 in the holoenzyme. Crystallographic studies suggest that this is the ionization of the zinc-bound water molecule. The nicotinamide ring of the NAD^+ is bound close to the zinc ion at the bottom of the pocket. The 2'-hydroxyl of the ribose ring is located between the hydroxyl of Ser-48 and the imidazole ring of His-51, within hydrogen-bonding distance of both.

b. The structure of the enzyme–substrate (ternary) complex
The structure of the *reactive* ternary complex composed of the enzyme, NAD^+, and 4-bromobenzyl alcohol has been solved at 2.9-Å resolution.[14] This was possible because there is a favorable equilibrium between this complex and the enzyme-bound ternary complex reaction product of NADH and 4-bromobenzaldehyde. The structure of the unreactive prod-uctlike complex composed of the enzyme, the coenzyme analogue H_2NADH (i.e., NAD^+ in which the nicotinamide ring has been reduced

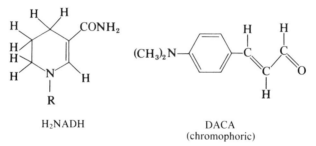

H_2NADH DACA
                                        (chromophoric)

to 1,4,5,6-tetrahydronicotinamide), and *trans*-4-(*N,N*-dimethylamino)-cinnamaldehyde (DACA) has also been solved at the same resolution.[18] In both examples, the relevant oxygen atom of the substrate is directly coordinated to the zinc ion, displacing the bound water molecule so that the metal remains tetracoordinated (Figure 15·3). The hydrophobic side chain of the substrate binds in the hydrophobic "barrel" of the pocket. Both observations are consistent with earlier solution studies using model building,[19] spectroscopy,[20] and kinetics:[21] the binding of the carbonyl oxygen of the chromophoric cinnamaldehyde (DACA) to a positively charged center in the enzyme was indicated by a characteristic shift in the spectrum. (The binding of substituted benzaldehydes is enhanced by electron-donating substituents.)

A major point of contention is the ionization state of the bound alcohol: Is it bound as the alcoholate anion[7,22] or as the neutral alcohol? The position of the proton cannot, of course, be located in the x-ray diffraction studies. Evidence in support of the alcoholate anion being

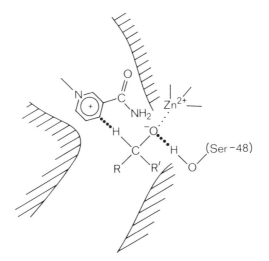

FIGURE 15·3. A proposed model for the productively bound ternary complex of horse liver alcohol dehydrogenase. It is suggested that the ionized alcohol displaces the zinc-bound water molecule shown in Figure 15·2. (Courtesy of C.-I. Brändén.)

bound comes from the pH dependence of the binding of substituted alcohols.[23] The complexes of the holoenzyme with various alcohols have the following pK_a values: trifluorethanol, 4.3; 2,2-dichloroethanol, 4.5; 2-chloroethanol, 5.4; ethanol, 6.4. These values are some 8 to 9 units below the values of the alcohols in solution. If they represent the ionization of the alcohols in the ternary complexes, then complexing to the enzyme has dramatically lowered the pK_a's. This does seem to have happened, since there is a very good correlation of the pK_a's of the ternary complexes with those of the alcohols in solution (the Brönsted coefficient = 0.6). Kinetic isotope effects are also consistent with the hydride ion transfer taking place from an alcoholate ion (equation 15·3).[24] On the other hand,

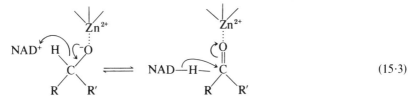

$$(15·3)$$

^{19}F NMR evidence on the binding of trifluoroethanol between pH 6 and 9 suggests—based on the magnitude and constancy of the chemical shift—that the alcohol is bound as the neutral species.[25] However, the proposed pK_a of the alcohol, 4.3,[23] is below this region and so could have been missed.

Interestingly, although the substrate is bound in the pocket in a water-free environment, there is a system of hydrogen bonds that can shuttle the proton from the bound alcohol to the surface (equation 15·4).[17] This

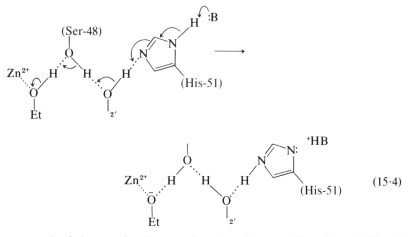

$$(15\cdot4)$$

is composed of the previously mentioned residues of Ser-48 and His-51 (which is on the surface), and the bridging 2′-hydroxyl of the coenzyme ribose ring.

c. The kinetic mechanism

The steady state and stopped-flow kinetic studies on the horse liver enzyme are now considered "classic" experiments. They have shown that the oxidation of alcohols is an ordered mechanism, with the coenzyme binding first and the dissociation of the enzyme–NADH complex being rate-determining.[15,26,27] Both the transient state and steady state methods have detected that the initially formed enzyme–NAD$^+$ complex isomerizes to a second complex.[27,28] In the reverse reaction, the reduction of aromatic aldehydes involves rate-determining dissociation of the enzyme–alcohol complex,[27,29] whereas the reduction of acetaldehyde is limited by the chemical step of hydride transfer.

A possible structural basis for the ordered kinetic mechanism of coenzyme binding before alcohol is the need for the 2′-hydroxyl of the ribose in the proton shuttle of equation 15·4.[14] Comparison of the structures of apo- and holoenzymes reveals extensive conformational changes on coenzyme binding.[12] The induced conformational changes are probably related to the kinetically detected isomerization, and may be another factor that contributes to ordered binding.

The enzyme–product complexes of the yeast enzyme dissociate rapidly so that the chemical steps are rate-determining.[30] This permits the measurement of kinetic isotope effects on the chemical steps of this reaction from the steady state kinetics. It is found that the oxidation of deuterated alcohols RCD_2OH and the reduction of benzaldehydes by deuterated NADH (i.e., NADD) are significantly slower than the reactions with the normal isotope (k_H/k_D = 3 to 5).[21,30] This shows that hydride (or deuteride) transfer occurs in the rate-determining step of the reaction. The rate constants of the hydride transfer steps for the horse liver enzyme have been measured from pre–steady state kinetics and found to give the same isotope effects.[31,32] More recent kinetic and kinetic isotope effect data are reviewed in reference 33.

d. Site-site interactions
Although the liver enzyme binds two moles of NAD^+ with equal affinity, it has been claimed that only one site of the enzyme reacts during the turnover of the enzyme.[29] Structural[14] and NMR[25] work indicate that both subunits are equivalent when substrates are bound. Kinetic data indicate that both sites do react simultaneously.[25,31,34-37] Stopped-flow studies show that the reaction course is biphasic during the turnover of the enzyme, an observation that at first sight appears to confirm that the two sites are nonequivalent. But this is just a natural consequence of the kinetics of an ordered ternary complex mechanism involving a single site:[24,36,37] the appearance or disappearance of the chromophore during the transient state is linked to several equilibria, so the complex kinetics of Chapter 4, sections D2 and D6, apply. This analysis stems from earlier studies on lactate dehydrogenase; these are discussed in the next section.

2. L-Lactate dehydrogenase[6,8,38]
L-Lactate dehydrogenase catalyzes the reversible oxidation of L-lactate to pyruvate, using NAD^+ as a coenzyme (equation 15·5). The enzyme,

$$H{-}\underset{\underset{CO_2^-}{|}}{\overset{\overset{CH_3}{|}}{C}}{-}OH + NAD^+ \rightleftharpoons \underset{\underset{CO_2^-}{|}}{\overset{\overset{CH_3}{|}}{C}}{=}O + NADH + H^+ \qquad (15·5)$$

which has been isolated from many species, is a tetramer of M_r 140 000. The two forms of the enzyme—the H_4, predominating in heart muscle, and the M_4, predominating in skeletal muscle—give rise to a family of isozymes.[39,40] The amino acid composition of the polypeptide chain that constitutes the M_4 form is significantly different from that of the H_4 form, and the two have different kinetic properties. Despite this, the sites for the association of the subunits must be very similar, since the hybrids M_3H, M_2H_2, and MH_3 occur in the expected statistical distribution.[41] There are no subunit interactions in catalysis, so the kinetic properties of, say, the H_3M form are identical to those expected for a 3:1 mixture of H_4 and M_4. The crystal structure of the apoenzyme from dogfish has been solved at 2.0-A resolution, and its complex with substrate analogues at 2.7 to 2.8 Å.[42–44] The molecule is symmetrical and its subunits are structurally equivalent.

a. The structure of the enzyme–substrate complex

The structures of the ternary complexes have been deduced from crystallographic studies of the binding of NAD-pyruvate,[43] a covalent analogue of pyruvate plus NADH,[45] and, subsequently, from the binding of the more realistic analogue, S-lac-NAD$^+$.[44]

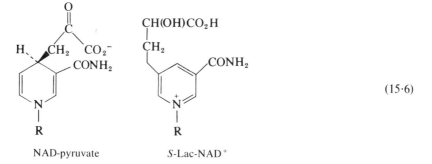

NAD-pyruvate   S-Lac-NAD$^+$

(15·6)

It seems most likely that the active ternary complexes are of the form shown in mechanism 15·7. The carboxylate group of the substrate forms

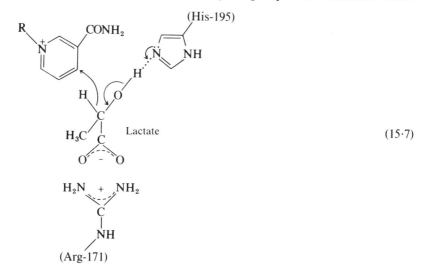

(15·7)

a salt bridge with the side chain of Arg-171. The hydroxyl group of lactate forms a hydrogen bond with the unprotonated imidazole ring of His-195. (When pyruvate is the substrate, its carbonyl group forms a hydrogen bond with the protonated imidazole.) As well as orienting the substrate,

His-195 acts as an acid-base catalyst, removing the proton from lactate during oxidation (mechanism 15·7). (In the reaction with pyruvate, His-195 stabilizes the negative charge that develops on the oxygen of pyruvate during reduction.)

The NAD-pyruvate is formed by the enzyme from NAD^+ and the enol form of pyruvate[46] (the keto tautomer is the substrate for the normal reaction).[47] The mechanism for the reaction is presumably catalyzed by His-195 in a manner similar to that of the oxidation of lactate (mechanism 15·8).

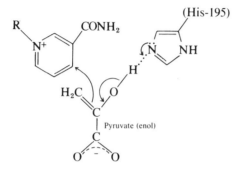

$$(15·8)$$

b. The kinetic mechanism[8,48]

The enzyme binds lactate or pyruvate only in the presence of coenzyme.[49] The mechanism is therefore ordered, with the coenzyme binding first. Pertinent to this is the observation from the crystallographic studies that coenzyme binding induces a conformational change in which residues 98 to 114 of the chain move through a relatively large distance to close over the active site in the ternary complex.[42,43] Rapid reaction studies show that the mechanism differs somewhat from that of horse liver alcohol dehydrogenase, although the dissociation of the enzyme–NADH complex is rate-determining in both reactions.[48,50] With alcohol dehydrogenase, the dissociation of the aldehyde from the ternary complex E·RCHO·NADH is faster than the rate of hydride transfer, so that the two ternary complexes do not have time to equilibrate. However, the hydride transfer steps are very rapid in the reactions of lactate dehydrogenase and the dissociation of pyruvate is slow, so that the two ternary complexes equilibrate. Furthermore, the equilibrium position favors lactate and NAD^+ at neutral pH. When lactate is mixed with the holoenzyme at pH 7, 20% of the bound NAD^+ is reduced during the first millisecond of reaction, as the equilibrium between [E·NAD$^+$·lactate] and [E·NADH·pyruvate] is rapidly attained. As the pyruvate dissociates, the equilibrium is displaced toward products, so that all four bound NAD^+ molecules are reduced. There is then the slower dissociation of NADH, the rate-limiting step in the steady state (equation 15·9). This behavior

$$E \cdot NAD^+ \cdot lactate \xrightleftharpoons{fast} E \cdot NADH \cdot pyruvate \xrightarrow{slow} E \cdot NADH \xrightarrow{slowest} E \qquad (15 \cdot 9)$$
$$\qquad\qquad\qquad\qquad\qquad\qquad\;\downarrow\qquad\qquad\;\downarrow$$
$$\qquad\qquad\qquad\qquad\qquad\quad Pyruvate \qquad NADH$$

could easily be mistaken for half-of-the-sites reactivity, but all four sites appear to be independent.[8,48,50] Coenzymes also bind independently to each site.

The catalytically important His-195 is unusually reactive toward diethylpyrocarbonate. This enabled the pK_a in both the apo and holoenzymes to be determined directly from the pH dependence of the rate of modification (the $pK_a = 6.7$).[51] There is evidence that lactate binds preferentially to the holoenzyme containing the un-ionized histidine, whereas pyruvate binds preferentially to the enzyme–NADH complex containing protonated histidine.

3. Malate dehydrogenase[9]

Malate dehydrogenase catalyzes the reversible oxidation of malate to oxaloacetate, using NAD^+ as a coenzyme (equation 15·10). The crystal structure of the soluble, or cytoplasmic, enzyme has been solved at 2.5-

$$
\begin{array}{c}
CO_2^- \\
| \\
H-C-OH \\
| \\
CH_2CO_2^-
\end{array}
+ NAD^+
\rightleftharpoons
\begin{array}{c}
CO_2^- \\
| \\
C=O \\
| \\
CH_2CO_2^-
\end{array}
+ NADH + H^+
\qquad (15 \cdot 10)
$$

Å resolution.[52] The amino acid sequence of the cytoplasmic enzyme has not yet been directly determined. However, the electron density of the crystal structure has been carefully examined to fit it to individual amino acids with some success.[53,54] The overall folding of the polypeptide chain is very similar to that of lactate dehydrogenase, although the malate enzyme is only a dimer of M_r 70 000. The x-ray analysis of its sequence suggests that the two are extensively homologous, with very similar active sites. Kinetic experiments show that malate dehydrogenase also has a catalytically essential histidine residue that may be modified with diethylpyrocarbonate.[55] Two moles of NADH or NAD^+ are bound with equal affinity.[56] The apparent negative cooperativity of coenzyme binding found for one enzyme preparation may have been due to the presence of two forms of the enzyme.[57,58]

4. Glyceraldehyde 3-phosphate dehydrogenase[59,60]

Glyceraldehyde 3-phosphate dehydrogenase, a tetrameric enzyme of M_r 150 000 containing four identical chains, catalyzes the reversible oxidative phosphorylation of glyceraldehyde 3-phosphate to 1,3-diphosphoglycerate, using NAD^+ as a coenzyme (equation 15·11). The reaction pathway

$$\underset{\substack{|\\ \mathrm{CH_2OPO_3^{2-}}}}{\overset{\substack{\mathrm{H}\diagdown\quad\diagup\mathrm{O}\\ \mathrm{C}\\ \|}}{\mathrm{H-C-OH}}} + \mathrm{NAD^+} + \mathrm{HPO_4^{2-}} \rightleftharpoons \underset{\substack{|\\ \mathrm{CH_2OPO_3^{2-}}}}{\overset{\substack{\mathrm{O}\diagdown\quad\diagup\mathrm{OPO_3^{2-}}\\ \mathrm{C}\\ \|}}{\mathrm{H-C-OH}}} + \mathrm{NADH} + \mathrm{H^+}$$

$$(15\cdot11)$$

consists of a series of reactions. The currently accepted mechanism (equations 15·12 to 15·16), which was first proposed in 1953,[61] is supported by extensive pre–steady state[62] and steady state[63,64] kinetic studies.

$$\mathrm{NAD^+\!\cdot\! E-SH} + \mathrm{RCHO} \rightleftharpoons \mathrm{NAD^+\!\cdot\! E-S-}\underset{\substack{|\\ \mathrm{H}}}{\overset{\substack{\mathrm{OH}\\ |}}{\mathrm{C}}}\mathrm{-R} \qquad (15\cdot12)$$

$$\mathrm{NAD^+\!\cdot\! E-S-}\underset{\substack{|\\ \mathrm{-H}}}{\overset{\substack{\mathrm{OH}\\ |}}{\mathrm{C}}}\mathrm{-R} \rightleftharpoons \mathrm{NADH\cdot E-S-}\overset{\substack{\mathrm{O}\\ \|}}{\mathrm{C}}\diagdown_{\mathrm{R}} + \mathrm{H^+} \qquad (15\cdot13)$$

$$\mathrm{NADH\cdot E-S-}\overset{\substack{\mathrm{O}\\ \|}}{\mathrm{C}}\diagdown_{\mathrm{R}} \rightleftharpoons \mathrm{E-S-}\overset{\substack{\mathrm{O}\\ \|}}{\mathrm{C}}\diagdown_{\mathrm{R}} + \mathrm{NADH} \qquad (15\cdot14)$$

$$\mathrm{E-S-}\overset{\substack{\mathrm{O}\\ \|}}{\mathrm{C}}\diagdown_{\mathrm{R}} + \mathrm{NAD^+} \rightleftharpoons \mathrm{NAD^+\!\cdot\! E-S-}\overset{\substack{\mathrm{O}\\ \|}}{\mathrm{C}}\diagdown_{\mathrm{R}} \qquad (15\cdot15)$$

$$\mathrm{NAD^+\!\cdot\! E-S-}\overset{\substack{\mathrm{O}\\ \|}}{\mathrm{C}}\diagdown_{\mathrm{R}} + \mathrm{HPO_4^{2-}} \rightleftharpoons \mathrm{NAD^+\!\cdot\! E-SH} + \mathrm{RC}\diagup^{\mathrm{OPO_3^{2-}}}_{\diagdown\mathrm{O}}$$

$$(15\cdot16)$$

The enzyme has a reactive cysteine residue that is readily acylated by acyl phosphates to form a thioester (the reverse of reaction 15·16).[65] The first step in the reaction sequence is the formation of a hemithioacetal between the cysteine and the substrate. This has the effect of converting the carbonyl group, which is not easy to oxidize directly, into an alcohol that is readily dehydrogenated by the usual procedure (reaction 15·13). The thioester that is formed in reaction 15·13 reacts with orthophosphate to give the acylphosphate (15·16). However, the acyl transfer is very slow unless NAD^+ is bound to the enzyme.[66,67] The replacement of NADH by NAD^+ in reactions 15·14 and 15·15 is therefore a necessary part of the reaction sequence. It is of interest that the dissociation of the complex of the acylenzyme and NADH (15·14) is the rate-determining step in the sequence at saturating reagent concentrations at high pH.[66] A consequence of this replacement of NADH by NAD^+ before the release of acylphosphate is that the free apoenzyme does not take part in the reaction. Also, because acylation of the enzyme by the diphosphate is activated by NAD^+, the holoenzyme initiates the reductive dephosphorylation of 1,3-diphosphoglycerate.

The Michaelis complexes of the holoenzyme with the aldehyde or the diphosphoglycerate, and the acylenzyme with orthophosphate, are not included in the scheme because their dissociation constants are too high for their accumulation.

a. The structures of the enzyme–substrate complexes
The crystal structures of the holoenzymes from lobster[68] and *Bacillus stearothermophilus*[69] have been solved at 2.9 and 2.7 Å, respectively; the human holoenzyme[70] has been solved at low resolution. The lobster apoenzyme has been solved at 3.0-Å resolution.[71] The structures of the enzyme–substrate complexes have been deduced from model-building experiments on the lobster and bacterial enzymes.[68,69] The following description is a composite of these; we use the specific details of the bacterial enzyme for convenience.

Two binding sites for sulfate ions were identified at the active site of the enzyme that had been crystallized from ammonium sulfate (Figure 15·1).[72] A chemically and stereochemically reasonable model for the course of the reaction may be constructed by assuming that these are the binding sites for the phosphate residue of the substrate and the nucleophilic phosphate in the deacylation reaction (15·16). The aldehyde group of the substrate can form a hemithioacetal with Cys-149 when the 3-phosphate is placed to make hydrogen bonds with the hydroxyl of Thr-179, the positively charged side chain of Arg-231, and the 2′-hydroxyl of the ribose ring that is attached to the nicotinamide of NAD^+ (Figure 15·1). The C-2 hydroxyl of the substrate can then form a hydrogen bond with Ser-148, while the C-1 hydroxyl forms one with a nitrogen of His-176. These interactions orient the substrate so that the C-1 hydrogen points toward the C-4 position of the nicotinamide ring, less than 3 Å away. In this mode of binding, the dehydrogenation reaction may take place as described earlier for lactate dehydrogenase, with His-176 as the general-base catalyst (reaction 15·17).

The transition state for the attack of orthophosphate on the thioester can be stabilized by hydrogen bonds to the attacking phosphate from the hydroxyl of Ser-148, the hydroxyl of Thr-150, and the C-2 hydroxyl of

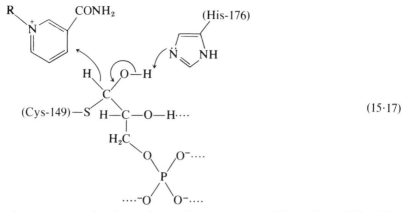

(15·17)

the substrate, and also from the amido nitrogens of Cys-149 and Thr-150. The presence of this specific binding site for phosphate explains why the thioester is phosphorolyzed rather than hydrolyzed. The sulfur atom of the thioester is close enough to the C-4 carbon of the nicotinamide ring of NAD^+ to be polarized by its positive charge. This perhaps explains the activation of the acyl transfer reactions on NAD^+ binding.

b. The symmetry of the enzyme and the cooperativity of ligand binding
There is considerable controversy in the literature about the symmetry of the dehydrogenase, the cooperativity of ligand binding, and half-of-the-sites vs. full-site reactivity.[73-80] The binding of NAD^+ to the enzyme is definitely cooperative; there is strong negative cooperativity in the binding to the rabbit muscle and bacterial enzymes,[73,77] although there is positive cooperativity in the binding to the yeast enzyme at some temperatures.[81] Glyceraldehyde 3-phosphate, on the other hand, binds independently to all four subunits.[80] Half-of-the-sites reactivity is found for the reactions of artificial substrates only; 1,3-diphosphoglycerate, for example, acylates all four reactive cysteines with a single rate constant.[66,75,77,80] It has been suggested from some of the kinetic and binding studies that the enzyme exists as a "dimer of dimers," having two pairs of structurally different subunits.[79,82] The interpretation of the electron density of the high-resolution crystal structure of the lobster holoenzyme supports this view,[68] but the more recent study on the bacterial enzyme strongly suggests that all four subunits are structurally identical and that the enzyme has precise 222 symmetry.[69] The bacterial apoenzyme also has this symmetry,[69] and the structure of the lobster apoenzyme was solved by assuming this symmetry.[71] A comparison of the bacterial apo- and holoenzymes shows that the binding of NAD^+ causes a large movement in the coenzyme domain, and that this contracts the volume of the molecule, but no such movement is seen in the lobster system. The structural origins of the negative cooperativity have yet to be elucidated.

5. Some generalizations about dehydrogenases

The structural studies have given a clear and chemically satisfying description of the stereochemical and catalytic requirements of the hydride transfer reaction. In three of the examples, there is an acid-base catalyst that forms a hydrogen bond with the carbonyl or alcohol group of the substrate, helps orient it correctly, and stabilizes the transition state for the reaction (equation 15·18). Liver alcohol dehydrogenase is similar, with

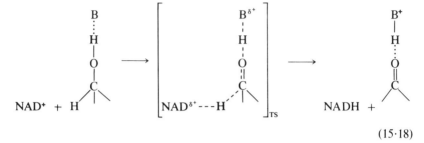

$$(15·18)$$

the His-51/Ser-48 shuttle and/or the Zn^{2+} ion taking the place of BH^+ in stabilization of the transition state and orientation of the substrate.

A consequence of the direction of the hydrogen bonding is that the alcohol binds preferentially to the basic form (B) of the catalyst, whereas the aldehyde binds preferentially to the acidic form (BH^+). The pK_a of B is lowered in the $E·NAD^+·RCH_2OH$ ternary complex and raised in the $E·NADH·RCHO$ complex. This means that the proton that is produced during the oxidation does not leave the ternary complex but is taken up by the catalytic group, and vice versa. The proton escapes into solution only when there is a change in substrate binding.[28,48,75,83]

The specificities of the enzymes are also nicely explained: The enantiomers of the substrates of L-lactate and D-glyceraldehyde 3-phosphate dehydrogenases cannot be productively bound; the hydrophobic pocket of alcohol dehydrogenase will not bind the charged side chains of lactate; etc. However, we do not know if conformational changes occur during catalysis or if there is strain.

A general kinetic feature is that NADH usually binds more tightly than NAD^+. The structural features responsible for this are not clear, although the charged nicotinamide ring is clearly more hydrophilic than the reduced form in NADH. The tight binding causes the dissociation of the enzyme–NADH complexes to be largely rate-determining at saturating concentrations of reagents at physiological pH. Furthermore, although the equilibrium constant for the oxidation reaction in solution greatly favors NAD^+ and alcohol, the tighter binding of the NADH causes the equilibrium constant for the enzyme-bound reagents to be more favorable: it was seen that the equilibrium constant between the two ternary complexes in the reactions of lactate dehydrogenase is not far from 1.

B. The proteases

The proteases may be conveniently classified according to their activities and functional groups. The serine proteases are endopeptidases that have a reactive serine residue and pH optima around neutrality. The carboxyl or aspartyl (formerly called acid) proteases are endopeptidases that have catalytically important carboxylates and pH optima at low pH (apart from chymosin, whose activity extends to neutral pH). The thiol proteases are endopeptidases that differ from the serine proteases by having reactive cysteine residues. The zinc proteases are metalloenzymes that function at neutral pH. Except for leucine aminopeptidase, which is of M_r ~250 000, the proteases are small monomeric enzymes of M_r 15 000 to 35 000, readily amenable to kinetic and structural study. Because of this, they are among the best-understood enzymes. Although the different classes catalyze the same reaction, they utilize different mechanisms. Some are well understood and have chemical models; the others are more obscure.

The notation of A. Berger and I. Schechter (Chapter 1, equation 1·6) is used throughout this section to describe the binding subsites. (The scissile bond of the peptide substrate sits across the S_1 and S_1' subsites with its C-terminal side occupying the S_1' to S_n' subsites, and its N-terminal side occupying S_1 to S_n.)

1. The serine proteases

These enzymes have been discussed in various parts of this text. Some major topics are: the enzymes as a family, specificity (Chapter 1, section C); the structures of the active site, the enzyme–substrate complex, the acylenzyme, and the enzyme–product complex (Chapter 1, section D2); proof of the reaction pathway, reaction kinetics (Chapter 7, section B); the pH dependence of catalysis and the state of ionization of the active site (Chapter 5, sections F and G2a); the utilization of binding energy to increase k_{cat} (Chapter 12, section B); transition state stabilization, specific solvation of the transition state (Chapter 12, section D5c); possible stereoelectric effects (Chapter 8, section F2). Following is a summary of these discussions.

The hydrolysis of ester or amide substrates catalyzed by the serine proteases involves an acylenzyme intermediate in which the hydroxyl group of Ser-195 is acylated by the substrate. The formation of the acylenzyme is the slow step in the reaction of amide substrates at saturating concentrations, but the acylenzyme often accumulates in the hydrolysis of esters. The attack of Ser-195 on the carboxyl group of the substrate probably forms a high-energy tetrahedral intermediate (equation 15·19), but there is no direct evidence for it.

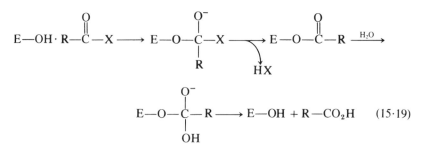

$$E-O-\underset{\underset{OH}{|}}{\overset{\overset{O^-}{|}}{C}}-R \longrightarrow E-OH + R-CO_2H \qquad (15\cdot19)$$

There is more direct experimental evidence about the mechanism of catalysis and the structures of the intermediates in the reactions of the serine proteases than there is about any other enzyme or family of enzymes. One of the major reasons for the structural knowledge is that it is possible to solve the crystal structures of the co-crystallized complexes of trypsin and some naturally occurring polypeptide inhibitors that mimic substrates (Chapter 1, section D). We know from these studies that the active site of the enzyme is complementary in structure to the transition state of the reaction, a structure that is very close to the tetrahedral adduct of Ser-195 and the carbonyl carbon of the substrate. Furthermore, the structure of the enzyme is not distorted when it binds the substrate. NMR studies on the binding of small peptides show that these are also not distorted on being bound. (The high-resolution study on the crystal structure of the complex between the pancreatic trypsin inhibitor and trypsin shows clearly that the reactive peptide bond is distorted toward its structure in the tetrahedral intermediate. However, this bond is distorted *before* the binding to the enzyme, the inhibitor being "designed" to bind as tightly as possible to the enzyme; i.e., it is a natural transition state analogue.)

a. The hydrogen-bond network at the active site
It has long been thought that the imidazole base of His-57 increases the nucleophilicity of the hydroxyl of Ser-195 by acting as a general-base catalyst (mechanism 15·20): The activity falls off at low pH according to

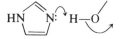 $\qquad\qquad\qquad\qquad\qquad\qquad\qquad (15\cdot20)$

the ionization of a base of $\sim pK_a$ 7, a characteristic value for a histidine residue; His-57 is modified by the affinity label *tos*-L-phenylalanine chloromethyl ketone, with an irreversible loss of enzymatic activity (Chapter 7, section G).[84] It came as a complete surprise when the crystallographers found that the carboxylate of Asp-102 is also involved at the active site to give a catalytic triad (structure 15·21) which was dubbed the "charge relay system."[85] Although the carboxyl group is completely buried in the

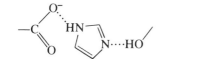

$$(15\cdot21)$$

interior of the protein, it is surrounded by polar residues and buried water molecules. Similar hydrogen-bond networks have subsequently been found at the active sites of all other serine proteases. In protease A from *Streptomyces griseus*, for example, the buried carboxylate ion is the recipient of four hydrogen bonds in an environment that has been described as very polar.[86]

A controversy arose about which residue is responsible for the ionization of the hydrogen-bond network. Kinetic data suggested that the group ionizing with a pK_a of 6.8 in chymotrypsin is the histidine,[87] and that the buried aspartate is ionized at neutrality, with a pK_a of <2.[88,89] Subsequent NMR and IR evidence purported to show that the group ionizing with pK_a 6.8 was Asp-102. However, additional NMR work[90,91] and the direct determination by neutron diffraction[92] of the position of the proton between Asp-102 and His-57 has proved that the histidine is the ionizing residue and that Asp-102 has an unusually low pK_a.

The role of the buried Asp-102 is an intriguing question. It may be there primarily to raise the pK_a of the histidine through the negative charge and also to confine its location.[88] An interesting experiment that is now feasible by using site-directed mutagenesis is to replace Asp-102 by an asparagine and then measure the catalytic activity. The author's guess is that the pK_a of His-57 would be about 5, and that the value of k_{cat} would drop by a factor of 10. (This is based on a lowering of the pK_a of the catalytic base by ~2 pH units, combined with a Brönsted β value of 0.5 for general-base catalysis.)[88]

A second query concerns whether or not there is a hydrogen bond between His-57 (N_ε) and Ser-195 (O_γ). The x-ray crystallographic criterion for the existence of a hydrogen bond between two such atoms is that their interatomic distance is about 2.8 to 3 Å (Chapter 11, section A3). The distance between the two relevant atoms is 3.7 Å in subtilisin[93,94] and 3.8 Å in γ-chymotrypsin[95] (see the next section for the different forms of chymotrypsin), distances that are far too long. But in the most recent studies, the distance between the O_γ and the N_ε is 3.0 Å in trypsin, and 2.9 Å in kallikrein.[96] In elastase, the binding of an inhibitor induces the histidine and serine to be aligned so that the distance is 2.9 Å.[97]

b. The structure and the reactivity of the substrate
The structural requirements for a substrate to be reactive have been determined by measuring the values of k_{cat} and K_M for a wide range of ester substrates, and the association constants of reversible inhibitors.[98] The inherently high reactivity of esters causes relatively poor ester substrates

to be hydrolyzed at a measurable rate. Thus esters have been most useful for working out the steric requirements of the acyl portion of the substrate. Amides and peptides are so unreactive that the only ones amenable to study are the derivatives of the specific substrates phenylalanine, tyrosine, and tryptophan. The kinetic studies may now be combined with those from x-ray diffraction.

1. *The deacylation step.* Listed in Table 15·2 are data for the deacylation of various acylenzymes. (Further values for amino acids were given in Table 7·3.) The table shows that the most reactive derivative is that of acetyl-L-phenylalanine. As was discussed in Chapter 1, chymotrypsin has a well-defined binding pocket for the aromatic side chain of the amino acid, and a hydrogen-bonding site (the $C=O$ of Ser-214) for the NH of the CH_3CONH- of the substrate (Chapter 1, Figure 1·12). When the $C_6H_5CH_2-$ and CH_3CONH- groups of acetyl-phenylalanine are replaced by hydrogen atoms to give the simple acetyl group, the deacylation rate drops by a factor of 10^4 (although 10 units of this factor is caused by the inductive effect of the CH_3CONH- group, as shown by the hydroxide ion–catalyzed rate constants listed in the last column of Table 15·2). Interestingly, it is seen in Table 15·2 that *both* the aromatic ring *and* the acylamino group are required for high reactivity. Acetyl-glycine-chymotrypsin deacylates only 12 times faster than acetyl-chymotrypsin, and the increase is seen from the hydroxide ion–catalyzed rate constants to be caused solely by the inductive effect of the acylamino group rather than by any binding effect. Similarly, β-phenylpropionyl-chymotrypsin deacylates only 17.8 times faster than acetyl-chymotrypsin. The fact that both the aromatic ring and the acylamino group are required for high reactivity has been nicely accounted for by x-ray diffraction studies. As was described in Chapter 1 (Figure 1·12), the carbonyl oxygen of a polypeptide substrate sits between the backbone NH groups of Ser-195

TABLE 15·2. *Structural requirements in the deacylation of acylchymotrypsins (at 25°C)[a]*

Acylchymotrypsin R— (RCO₂E)	k_{cat} (s^{-1}) (for deacylation)	k_{OH^-} $(s^{-1} M^{-1})$ (for hydrolysis of RCO_2CH_3)
CH_3-	0.01	0.19
$C_6H_5CH_2CH_2-$	0.178	0.15
$CH_2(NHCOCH_3)-$	0.12	2.48
L-$C_6H_5CH_2CH(NHCOCH_3)-$	111	1.94

[a] From A. Dupaix, J.-J. Bechet, and C. Roucous, *Biochem. Biophys. Res. Commun.* **41**, 464 (1970); I. V. Berezin, N. F. Kazanskaya, and A. A. Klyosov, *FEBS Lett.* **15**, 121 (1971) (see Table 7·3).

and Gly-193. This mode of binding has been found for the specific acyl-enzyme carbobenzoxy-L-alanine-elastase.[99] However, it has been shown that the carbonyl oxygen of the nonspecific acylenzyme indolylacryloyl-chymotrypsin is not productively bound in this manner.[100] Instead, there is a water molecule forming a hydrogen-bonded bridge between the carbonyl oxygen and the catalytic nitrogen atom of His-57 (Figure 15·4). For reaction to occur, the carbonyl oxygen must swing into the hydrogen-bonding site between Ser-195 and Gly-193, and the bound water molecule must attack the carbonyl carbon. Thus, the acylamino portion and the aromatic ring are together required to anchor the carbonyl group in the productive mode. If either of the anchors is missing, the carbonyl oxygen takes up a nonproductive binding mode.

2. *The acylation step.* It was pointed out in Chapter 12 that the binding energies of the S_2, S_3, S_4, and S_5 subsites often increase k_{cat} for the

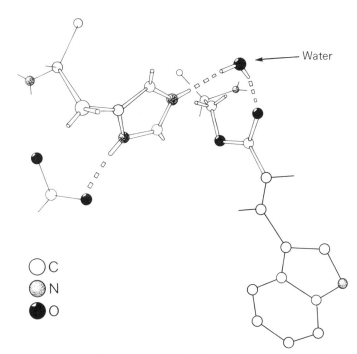

$\bigcirc$ C
$\bullet$ N
$\bullet$ O

FIGURE 15·4. The crystal structure of indolylacryloyl-chymotrypsin. [From R. Henderson, *J. Molec. Biol.* **54**, 341 (1970).] Note that the carbonyl oxygen of this nonspecific acylenzyme is not bound between the NH groups of Ser-195 and Gly-193, but is nonproductively linked to His-57 by a hydrogen-bonded water molecule. This is the acylenzyme that was found to deacylate at the same rate in solution and in the crystal (Chapter 1). [G. L. Rossi and S. A. Bernhard, *J. Molec. Biol.* **49**, 85 (1970).]

hydrolysis of polypeptide substrates rather than lower K_M. The reason
is not known. The binding energy of the S'_1 site is also used to increase
k_{cat} rather than give tighter binding.[101] This appears to result from a binding mode that cannot be fully occupied in the enzyme–substrate complex,
but that is accessible after the bond changes on formation of the unstable
tetrahedral intermediate.[102] It is also likely that the carbonyl oxygen atom
of the substrate forms better hydrogen bonds with the backbone $=$NH
groups of Ser-195 and Gly-193 (Figure 1·12) after reaction has occurred
to form the tetrahedral intermediate. Recent evidence supporting this hypothesis has come from studies on the hydrolysis of thionesters (substrates in which the carbonyl oxygen atom is replaced by a sulfur atom).
Thionesters are far less reactive than normal esters in the enzymatic reactions. This is attributed to the fact that the C$=$S bond length and the
optimal S$\cdots$HN$=$bond distances for the thionesters are longer than those
for the ester counterparts. The differential hydrogen bonding of the substrate and tetrahedral intermediate is thus altered and the transition state
stabilization is upset.[103,104]

c. The tetrahedral intermediate

The tetrahedral intermediate does not accumulate during the course of
the reaction.[89,105] The seemingly overwhelming mass of evidence that had
been gathered to "prove" that the intermediate accumulates was shown
convincingly to be an interesting set of artifacts.[106] Nor does the intermediate accumulate in the complexes of serine proteases with their naturally occurring polypeptide inhibitors.[107,108] The evidence for the intermediate is thus circumstantial. An analogous tetrahedral structure has
been observed, however, in the complexes of serine proteases with inhibitors that contain an aldehyde group instead of the peptide or ester
moiety. The *S. griseus* A protease forms such a covalent complex with
Ac-Pro-Ala-Pro-phenylalaninal[109] (equation 15·22) and with the naturally

$$\text{Ac-Pro-Ala-Pro-NHCH(CH}_2\text{-Phe)CHO} + \text{HO}\text{—(Ser-195)} \longrightarrow$$

$$\text{Ac-Pro-Ala-Pro-NHCH(CH}_2\text{-Phe)C}\overset{\displaystyle /\text{OH}}{\underset{\displaystyle \backslash\text{H}}{-}}\text{O}\text{—(Ser-195)} \qquad (15\cdot22)$$

occurring inhibitor chymostatin.[110] The oxygen atom is seen to occupy
its proposed position between the $=$NH groups of [the equivalent of] Gly-
193 and Ser-195. The lack of accumulation of the intermediate (i.e., its
presumed existence at steady state levels only) in the hydrolysis of peptides lends further justification to the assumption that the transition state
for the reaction resembles the intermediate.

d. A description of the reaction mechanism
The kinetic and structural data may be combined to give the following qualitative description of the mechanism of acylation of chymotrypsin by a good polypeptide substrate.[101]

The substrate binds in the specificity pocket of the enzyme, with the N-acylamino hydrogen binding to the carbonyl group of Ser-214. Any further residues in the N-acylamino chain bind in the subsites that are available. The reactive carbonyl group sits with its oxygen between the backbone NH groups of Ser-195 and Gly-193 (Figure 1·12). However, it is possible that the hydrogen bond between the oxygen and Gly-193 is long and weak. The first chemical step in the reaction is the attack of the hydroxyl of Ser-195 on the carbonyl carbon of the substrate to form the tetrahedral intermediate. During this, the proton on the hydroxyl is transferred to the imidazole of His-57. As the bond between Ser-195 and the carbonyl carbon is formed, the C=O bond lengthens to become a single bond. The oxygen, bearing a negative charge, moves closer to the NH of Gly-193, forming a shorter and stronger hydrogen bond. The transition state is stabilized relative to the Michaelis complex because of the better interactions of the leaving group with the S_1' subsite and the better hydrogen bond with Gly-193. The tetrahedral intermediate collapses to form the acylenzyme and expel the leaving group. The leaving group cannot bind in the S_1' site in the acylenzyme, as this would force the amino group to be too close to the carbonyl carbon. Thus, in the reverse reaction— the attack of the leaving group on the acylenzyme—the energy of binding to the S_1' site is realized only in the transition state. Deacylation takes place by the charge relay system activating the attack of water. Another tetrahedral intermediate is formed, and then it collapses to expel Ser-195 and give the enzyme–product complex.

Despite this detailed knowledge, many important questions still remain unanswered. For example, we do not know how the binding energies of the subsites for the N-acylamino chain are sometimes used to increase k_{cat} rather than decrease K_M (Table 12·1). We do not know the contribution to catalysis of the buried Asp-102 in the charge relay system: What would be the activity of chymotrypsin in which the aspartate is converted to asparagine?

e. The zymogens
Some of the serine proteases are stored in the pancreas as inactive precursors that may be activated by proteolysis. Trypsinogen, for example, is converted to trypsin by the removal of the N-terminal hexapeptide on the cleavage of the bond between Lys-6 and Ile-7 by enterokinase. Chymotrypsinogen is activated by the tryptic cleavage of the bond between Arg-15 and Ile-16. (In this case, further proteolysis by the chymotrypsin

that is released during the activation leads to the different forms of the enzyme—Figure 15·5.)

The mechanism of the activation and the reasons for the inactivity of chymotrypsinogen have been nicely explained by comparison of the crystal structures of the enzyme and the zymogen.[111-113] The zymogen has the charge relay system, and it ionizes in the same manner as it does in the enzyme.[88,114] However, the activity of the zymogen is extremely low, being devoid of proteolytic activity, and is only as reactive toward synthetic substrates as a solution of imidazole.[115,116] The reason for this is that the substrate-binding pocket is not properly formed in the zymogen, and the important NH group of Gly-193 points in the wrong direction for

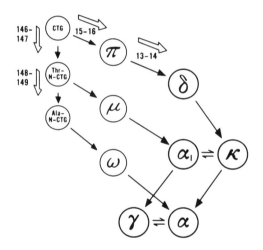

FIGURE 15·5. Activation of chymotrypsinogen (CTG). It was seen in Chapter 1, Figure 1·1, that CTG is susceptible to tryptic cleavage at the (Arg-15)—(Ile-16) bond, and to chymotryptic cleavages at (Tyr-146)—(Thr-147), (Ile-13)—(Ser-14), and (Asn-148)—(Ala-149) (the latter two because chymotrypsin has relatively broad specificity). In the rapid activation of chymotrypsinogen ([CTG]/[trypsin] ≈ 30), there is sufficient trypsin present to activate all the CTG before the accumulated chymotrypsin autolyzes or cleaves the CTG. The pathway of activation is CTG → π-chymotrypsin → δ-chymotrypsin → κ-chymotrypsin (+ α_1-) → α-chymotrypsin (+ γ-). The κ and α_1 forms are two different conformational states of the same primary structure, as are the α and γ. During the slow activation ([CTG]/[trypsin] ≈ 10^4), the small fraction of trypsin activates the zymogen slowly, allowing the chymotrypsin that is initially formed to cleave the unactivated zymogen to form neochymotrypsinogens (N-CTG's). Alpha$_1$- and α- chymotrypsin are produced from the N-CTG's via the generation of μ- and ω-chymotrypsin by tryptic cleavages. [From S. K. Sharma and T. R. Hopkins, *Biochemistry* **18**, 1008 (1979); *Bioorganic Chem.* **10**, 357 (1981).]

forming a hydrogen bond with the substrate.[113] This is an important lesson about enzyme catalysis. Enzyme catalysis often depends not on the presence of an unusually reactive catalytic group on the enzyme, but rather on the correct alignment of the substrate and ordinary catalytic groups.

The conformational change that forms the binding pocket and rotates Gly-193 results from a movement of Ile-16 as its α-ammonium group forms a salt bridge with the buried carboxylate of Asp-194. The activation process may be mimicked and studied by the effects of pH on the salt bridge.[117] This deprotonates at high pH and is destabilized so that the enzyme takes up a zymogen-like conformation. The energy difference between the two conformations is small and their equilibrium is delicately balanced.[117]

In trypsinogen, a region of the protein at the binding pocket is disordered, indicating conformational flexibility.[118,119] On activation or on the addition of a small peptide that can bind to the buried Asp-194, this region takes on a well-defined structure.

2. The thiol proteases

The thiol proteases are widely distributed in nature. The plant enzymes papain (from papaya), ficin (from figs), bromelain (from pineapple), and actinidin (from kiwi fruit or Chinese gooseberry) are members of a structurally homologous family. They are not homologous with the bacterial thiol protease clostripain (from *Clostridium histolyticum*) and streptococcal proteinase (from hemolytic streptococci). Perhaps the two groups will be found to be related in the same way as the mammalian and the bacterial serine proteases are. In mammals, the thiol proteases cathepsin B1 and B2 are found packaged in lysozomes.

a. Papain[120–123]

This enzyme is composed of a single polypeptide of 212 amino acids and M_r 23 406.[124] Kinetic studies have shown that the active site can accommodate seven amino acids, four on the acyl side of the cleaved bond (S_4 to S_1) and three on the amino side (S_1' to S_3').[125] Unlike the serine proteases that have S_1 as the primary specificity site, papain is specific for hydrophobic amino acids in the S_2 site. There is also a specificity for isoleucine or tryptophan in the S_1' site.[126] Esters, and presumably peptides, are hydrolyzed through an acylenzyme pathway in the same manner as the serine proteases are, except that Cys-25 is acylated.[127–130] A plot of k_{cat}/K_M against pH follows a bell-shaped curve with optimal activity at about pH 6. This is caused by the ionization of His-159 and Cys-25 with pK_a values of 4.2 and 8.2. If the histidine is denoted by "Im" and the cysteine by "RSH," the ionic form [RSH·HIm$^+$] is inactive at low pH, whereas the ionic form [RS$^-$·Im] is inactive at high pH. The catalytically active

form at neutral pH is one of the tautomers [RSH·Im] or [RS⁻·HIm⁺]:
one cannot distinguish between two ionic states bearing the same net
charge by examining a pH dependence (the principle of kinetic equivalence, Chapter 2, section F). The pH dependence of k_{cat} for deacylation
follows the ionization of a base of pK_a ~4. This may be attributed to His-
159, since the cysteine is blocked in the acylenzyme. The reaction mechanism is as shown in equation 15·23.

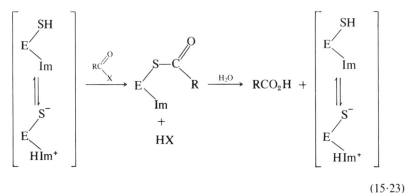

$$(15·23)$$

The crystal structures of papain[131–133] and actinidin[134] have been
solved at high resolution. Their polypeptide chains are virtually superimposable. Unfortunately, the active-site sulfydryl becomes oxidized
(probably to SO_2^-) during the exposure to x-rays.

    1. *The structure of the active site of papain.*[131–133,135] It is seen in the
crystal structure that the molecule is formed from two domains with a
deep cleft between them. The binding site for the substrate straddles this
cleft; although Cys-25 and His-159 are in close contact, they are on opposite sides of the cleft. The fairly deep pocket of the S_2 site for the
hydrophobic amino acids is lined with the hydrophobic side chains of Tyr-
67, Pro-68, and Trp-69 of one domain, and those of Phe-207, Ala-160,
Val-133, and Val-157 of the other. It was once thought that the pK_a of
4.2 that is found in the pH-activity profile is due to the ionization of a
carboxylate side chain, since they usually ionize in this region. But the
carboxyl group nearest to Cys-25, that of Asp-158, is 7.5 Å away.[122] This
is too distant for the carboxyl to act as an acid-base catalyst, unlike the
imidazole ring of His-159, which is correctly placed. The low pK_a of the
histidine is presumably due to its being partly buried in a hydrophobic
region. There is no equivalent to the charge relay system of the serine
proteases; the imidazole ring of His-159 does not interact with a buried
carboxylate.

    Model-building studies nicely explain the observed stereospecificity
of the enzyme.[135] D-Amino acid residues cannot be accommodated in the

subsites because of steric interference with the bulk of the enzyme. The enzyme is not an exopeptidase, since the free carboxylate of the substrate would be only 3 to 4 Å from the carboxylate of Asp-158, with consequent electrostatic repulsion. These studies also suggest a strain mechanism. The leaving group of the substrate appears to be forced against the α-CH_2 group of His-159 in the enzyme–substrate complex, but this interaction is relieved on formation of the tetrahedral intermediate. In support of this hypothesis it has been shown that substrate analogues that have a sterically small group in the leaving group position bind considerably more tightly than those with bulkier residues.[122,135]

The specificity for large hydrophobic residues in the S_2 subsite is manifested in increased values of k_{cat} rather than in tighter binding. G. Lowe and Y. Yuthavong have suggested that the binding of a residue such as phenylalanine in the S_2 site forces the cleft to open somewhat and increase the strain at the active site.[135] An outward movement of the walls of the cleft has subsequently been found in the crystal structure of the enzyme, which is inhibited by the chloromethyl ketone derivative of N-benzyloxycarbonyl-L-phenylalanine-L-alanine.[133] The crystal structure also shows that there is a binding site for the carbonyl oxygen of the scissile peptide bond. This comprises the backbone NH group of Cys-25, which is analogous to the binding site in the serine proteases; but the other hydrogen bond is to the side-chain —NH_2 of Gln-19 (Chapter 2, Figure 2·6).

2. *The protonic state of the active site and the reaction mechanism.* There has been a controversy about the position of the proton in the His-159/Cys-25 pair. Current evidence favors the ion pair $[RS^- \cdot HIm^+]$ as the dominant tautomeric form in the active enzyme.[123,136] It has also been noted in simple reactions in solution that the combination of a base B and a thiol RSH reacts as RS^-/BH^+; there is no known example of general-base catalysis of the attack of a thiol.[137] The acylation step in the reactions of papain and its family of homologous enzymes thus involves the attack of the —S^- ion of Cys-25 on the substrate to generate a tetrahedral intermediate.

The rate-determining step in the hydrolysis of amides and anilides appears to be the general-acid-catalyzed breakdown of the tetrahedral intermediate (structure 15·24).[138,139] The evidence for this is that (1) k_{cat} and k_{cat}/K_M are higher for substrates containing the more basic anilines ($\rho = -1.04$),[138] which shows that the nitrogen of the substrate becomes protonated during the transition state (the better the base, the easier the protonation); and (2) k_{cat}/K_M for the hydrolysis of benzoyl-L-arginine amide is 2.4% greater for a substrate containing ^{14}N in the leaving group than for a substrate containing ^{15}N ($k_{^{14}N}/k_{^{15}N} = 1.024$), a value that would be expected for the nearly complete cleavage of a C—N bond.[139] (Note:

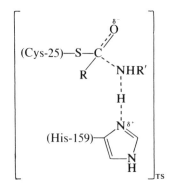

$$(15 \cdot 24)$$

Rate-determining breakdown does not mean that the intermediate accumulates, but merely that it reverts to starting materials faster than it proceeds to products.)

The balance of the evidence favors the deacylation of the acylenzyme intermediate being catalyzed by His-159 acting as a general base, although the possibility that Asp-158 is the catalytic base cannot be absolutely ruled out.[140]

3. The zinc proteases

The zinc proteases are another widespread group of enzymes. The most studied member is the digestive enzyme bovine pancreatic carboxypeptidase A, which is a metalloenzyme containing one atom of zinc bound to its single polypeptide chain of 307 amino acids and M_r 34 472.[141–144] It is an exopeptidase, which catalyzes the hydrolysis of C-terminal amino acids from polypeptide substrates, and is specific for the large hydrophobic amino acids such as phenylalanine. The closely related carboxypeptidase B catalyzes the hydrolysis of C-terminal lysine and arginine residues. The two enzymes are structurally almost identical, except that the B form has an aspartate residue that binds the positively charged side chain of the substrate.[145]

Other well-known zinc proteases are: collagenase; angiotensin-converting enzyme (important in regulating blood pressure); thermolysin, a bacterial endopeptidase of M_r 34 600 containing 316 residues in its single polypeptide chain;[146] and the Zn^{2+} G protease from *Streptomyces albus*, a D-alanyl-D-alanine carboxypeptidase that catalyzes carboxypeptidation and transpeptidation in cell wall metabolism.[147]

As described in Chapter 1, section C1b, the pancreatic carboxypeptidases are not homologous with thermolysin (and the G protease), but their active sites have evolved convergently to have striking similarities.

a. The structure of the active site of carboxypeptidase A
The crystal structure of the enzyme has been solved at 2.0-Å resolution.[148] The active site consists of a shallow groove on the surface of the enzyme

leading to a deep pocket, lined with aliphatic and polar side chains and parts of the polypeptide chain, for binding the C-terminal amino acid. The catalytically important zinc ion is ligated by the basic side chains of Glu-72, His-196, and His-69. In about 20% of the molecules, the phenolic oxygen of Tyr-248 is a fourth ligand.[149] The side chain of this residue is conformationally very mobile, a matter leading to much discussion and debate in the literature, concerning the similarities of the solution and the crystal structure.[143] The phenolic side chain may rotate about its C_α—C_β bond, and ~80% of its electron density in the map of the crystal structure is found in an orientation on the surface of the molecule pointing into solution.[149] This residue is also so mobile in carboxypeptidase B that its position cannot be determined in the crystal structure.[145] The fourth ligand of the zinc ion is normally a water molecule or a hydroxide ion.

A model for the binding of substrates has been extrapolated from the structure of the complex of the enzyme and glycyl-L-tyrosine, which had been solved by the difference Fourier method at 2.0-Å resolution (Figure 15·6).[141,148] The dipeptide is hydrolyzed only very slowly and is presumably bound nonproductively. It is reasoned that the slow hydrolysis rate is caused by the free amino group of the substrate binding to the carboxylate of Glu-270 via the intervening water molecule. This prevents the carboxylate from acting as a general base or nucleophile in the reaction (see below). The remaining features of the complex are used in the construction of the model for productive binding: the aromatic side chain binds in the binding pocket; the carboxylate ion of the C-terminus forms a salt linkage with Arg-145; the carbonyl oxygen of the scissile bond becomes the fourth ligand of the zinc ion; and the phenolic oxygen of Tyr-248 is within about 3 Å of the scissile bond, the side chain having rotated through about 120° from its predominant orientation in the free

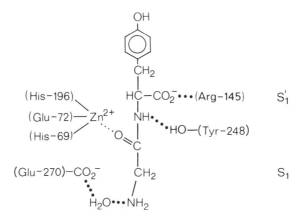

FIGURE 15·6. The (partly) nonproductively bound complex of a dipeptide (glycyl-L-tyrosine) and carboxypeptidase A. (Courtesy of W. N. Lipscomb.)

enzyme. Using this as a basis, the polypeptide chain may be extended to give the structure in Figure 15·7.

b. The structure of the active site of thermolysin[146,150–153]
Thermolysin differs from the carboxypeptidases in being an endopeptidase rather than an exopeptidase. This is manifested in the nature of the binding site: instead of the deep pocket of the carboxypeptidases, thermolysin has an open extended cleft that can bind the polypeptide chain of the substrate on both sides of the scissile bond. There are also important differences in the catalytic apparatus:

1. The catalytic glutamate residue (Glu-143) is located at the bottom of a narrow cleft, where it is bound to a water molecule and cannot approach the substrate as readily as can Glu-270 in carboxypeptidase.
2. There is no tyrosine residue corresponding to Tyr-248. Instead, a histidine residue, His-231, is suitably located to be a general-acid-base catalyst in the reaction.

The crystal structure of the complex between the enzyme and a tight-binding inhibitor, phosphoramidon (structure 15·25), has been solved.[151]

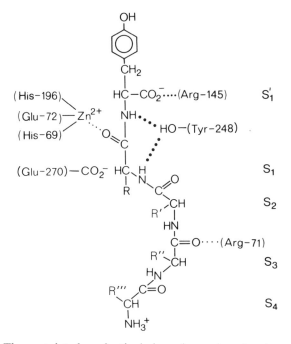

FIGURE 15·7. The postulated productively bound complex of carboxypeptidase A and a polypeptide substrate. (Courtesy of W. N. Lipscomb.)

$$(15 \cdot 25)$$

The phosphoryl group mimics a carbonyl group to which a water molecule has been added to generate a tetrahedral intermediate. It is found that a phosphoryl oxygen displaces the molecule of water that is bound to the Zn^{2+} ion and becomes its fourth ligand, at a distance of ~ 2.0 Å. The hydroxyl oxygen bonded to the phosphorus atom is 2.6 Å from an oxygen atom of Glu-143, and is thus hydrogen-bonded to it. The phosphoramide nitrogen is 4.1 Å from a nitrogen atom of His-231.

A likely reaction mechanism is immediately suggested by this structure, since it appears to be a transition state analogue for the Glu-143 general-base-catalyzed attack of a water molecule on a carbonyl group that has been polarized by binding to the zinc ion. Furthermore, His-231 is suitably placed to be a proton donor to the leaving group (equation 15·26).

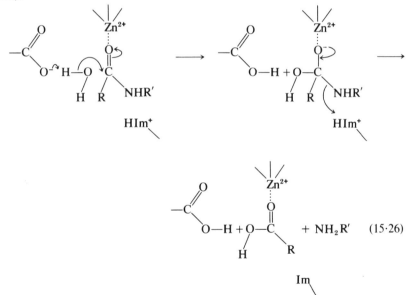

$$(15 \cdot 26)$$

An alternative catalytic role for Glu-143 is that it acts as a nucleophile (equation 15·27). Indeed, it is possible to alkylate it with active-site-directed α-chloro ketone irreversible inhibitors.[153] But the steric inaccessibility of the carboxylate group leads to such distortion of the enzyme on the formation of covalent bonds that nucleophilic attack on a substrate is most unlikely.[153]

$$E{-}CO_2^- + RCONHCH(R')CO_2^- \longrightarrow E{-}C\overset{O}{\underset{O-C\underset{R}{\overset{O}{\diagup}}}{\diagup}} \xrightarrow{H_2O} E{-}CO_2^- + RCO_2$$

$$+$$
$$NH_2CH(R')CO_2^-$$

c. The reaction mechanism of carboxypeptidase

An important difference between thermolysin and carboxypeptidase leads to the major uncertainty in the mechanism of carboxypeptidase. This difference is that the catalytic carboxylate of carboxypeptidase is far more sterically accessible. The crucial question is whether or not the carboxy-peptidase-catalyzed hydrolysis of peptides proceeds via general-base catalysis, as in equation 15·26, or via nucleophilic catalysis, as in 15·27. Early kinetic work concentrated on establishing the participation of the various groups in catalysis.

There is little doubt that the zinc ion acts as an electrophilic catalyst to polarize the carbonyl group and stabilize the negative charge that develops on the oxygen (Chapter 2, section B7).[154] The ionized carboxylate of Glu-270 is implicated in catalysis from the pH-rate profile. The activity follows a bell-shaped curve with pH; there is an optimum at pH 7.5, due to the basic form of a group of pK_a 6 and the acidic form of a group of pK_a 9.1 in the free enzyme.[155,156] The lower pK_a is that of Glu-270; the higher pK_a has not yet been assigned unambiguously. The hydroxyl of Tyr-248 probably acts as a general acid in the reaction.[141] There is no direct evidence of this, but Tyr-248 is certainly essential for the peptidase activity. Modification of the tyrosine side chain by acetylation destroys the peptidase activity, whereas nitration significantly alters the pH dependence of the peptidase activity.[157,158]

Carboxypeptidase acts as an esterase toward synthetic ester substrates based on L-β-phenyllactate [e.g., *O*-(*trans-p*-chlorocinnamoyl)-L-β-phenyllactate (structure 15·28)]. There is little doubt that the hydrolysis of esters takes place via nucleophilic catalysis. The mixed anhydride

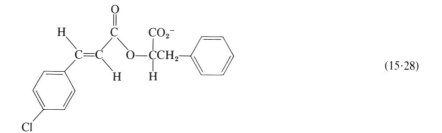

$$(15·28)$$

formed from the cinnamoyl group of structure 15·28 and Glu-270 via reaction 15·27 has been trapped by cryoenzymological procedures, and a
spin-labeled derivative has been characterized by ESR and NMR spectroscopy.[159,160] It has been suggested that in this reaction the Zn^{2+} ion
expands its coordination number to 5, retaining a metal-bound hydroxide
ion that acts as a catalyst in the deacylation step.[161]

The mixed anhydride acylenzyme does not accumulate in the hydrolysis of peptide substrates. It has been proposed that peptide hydrolysis also involves nucleophilic catalysis but that the rate-determining
step is acylation of the enzyme[160] (a situation analogous to that in the
chymotrypsin-catalyzed hydrolysis of esters and peptides). There are,
however, differences between the esterase and the peptidase activities
of carboxypeptidase:

1. Nitration of Tyr-248 significantly perturbs the pH dependence of the
   peptidase activity, so that there is a dramatic decrease in activity at
   neutrality, but enhances the esterase activity.[157,158]
2. Replacement of the zinc by mercury, cadmium, or lead destroys the
   peptidase activity but retains the esterase activity.[162]
3. Rapid-scanning low-temperature stopped-flow spectrophotometry suggests that different intermediates are formed in the esterase and peptidase reactions.[163]

These differences do not necessarily imply a different mechanism for the
two activities; they may just reflect the different catalytic requirements
of ester and peptide hydrolysis. For example, it is likely that the —OH
group of Tyr-248 acts as a general-acid catalyst in the peptidase activity,
by protonating the —NH of the leaving group as it is expelled from a
tetrahedral intermediate (whether generated by general-base catalysis, as
in equation 15·26, or by nucleophilic catalysis, as in 15·27). But it is well
known from studies of ester hydrolysis in solution that the tetrahedral
intermediate generated from the attack of a nucleophile on an ester can
expel an alcoholate ion without general-acid catalysis.

Evidence against the mixed anhydride mechanism in peptide hydrolysis has come from isotope exchange experiments.[164] If the hydrolysis
occurs by the anhydride route of equation 15·27, then the synthesis of
the peptide in the reverse reaction requires the initial formation of the
anhydride from the enzyme and RCO_2^-. The enzyme should thus be able
to catalyze the exchange of ^{18}O between the substrate and water (equation
15·29). However, exchange does not occur in the absence of added free
amino acid $NH_3^+CH(R')CO_2^-$, and occurs in its presence via the re-
synthesis of the peptide. The exchange is not stimulated by the analogue
$HOCH(R')CO_2^-$. This rules out the anhydride pathway *unless* the added
amino acid is an activator of the exchange reaction (and its hydroxyl
analogue is not), or unless the $H_2{}^{18}O$ released in equation 15·29 does not

$$RC^{18}O_2^- + E{-}CO_2^- \xrightarrow[H_2{}^{18}O]{} E{-}CO_2CR^{18}O \xrightarrow[H_2O]{} RCO^{18}O^- + E{-}CO_2^-$$

$$(15{\cdot}29)$$

exchange with the medium but remains attached to the enzyme. The mechanistic situation is thus unresolved.

d. The zymogen

Procarboxypeptidase A is activated by the removal of a peptide of some 64 residues from the N-terminus by trypsin.[165] This zymogen has significant catalytic activity. As well as catalyzing the hydrolysis of small esters and peptides,[166,167] procarboxypeptidase removes the C-terminal leucine from lysozyme only seven times more slowly than does carboxypeptidase. Also, the zymogen hydrolyzes Bz-Gly-L-Phe with $k_{cat} = 3$ s^{-1} and K_M = 2.7 mM, compared with values of 120 s^{-1} and 1.9 mM for the reaction of the enzyme.[166] In contrast to the situation in chymotrypsinogen, the binding site clearly pre-exists in procarboxypeptidase, and the catalytic apparatus must be nearly complete.

4. The carboxyl (aspartyl) proteases[168–171]

The carboxyl proteases are so called because they have two catalytically essential aspartate residues. They were formerly called acid proteases because most of them are active at low pH. The best-known member of the family is pepsin, which has the distinction of being the first enzyme to be named (in 1825, by T. Schwann). Other members are chymosin (rennin), cathepsin D, *Rhizopus*-pepsin (from *Rhizopus chinensis*), penicillinopepsin (from *Penicillium janthinellum*), and the enzyme from *Endothia parasitica*. All carboxyl proteases that have been sequenced so far are homologous in regions constituting about 30% of the molecule;[172] all are of M_r ~35 000; and all contain the two essential aspartyl residues. The carboxyl proteases are all specific for peptide bonds located between large hydrophobic residues, and they are inhibited by a common class of inhibitors (pepstatin). Three-dimensional crystal structures of four carboxyl proteases have been reported and have been shown to be very similar.[173–175] Most of the kinetic work has been performed on pepsin and penicillinopepsin. Their reaction mechanism is the most obscure of all the proteases, and there are no simple chemical models for guidance. The following discussion refers to both enzymes; they are similar both structurally and kinetically.

a. Pepsin

Pepsin consists of a single polypeptide chain of molecular weight 34 644 and 327 amino acid residues.[176,177] Ser-68 is phosphorylated, but this phosphate may be removed without significantly altering the catalytic properties of the enzyme.[178] As in other acid proteases, the active site is an extended area that can accommodate at least four or five, and maybe as many as seven, substrate residues.[179,180] The enzyme has a preference for hydrophobic amino acids on either side of the scissile bond. A statistical survey of the bond cleavages in proteins shows that there is a specificity for leucine, phenylalanine, tryptophan, and glutamate (!) in the S_1 site, and for tryptophan, tyrosine, isoleucine, and phenylalanine in the S_1' site.[180] Pepsin rarely catalyzes the hydrolysis of esters, the exceptions being esters of L-β-phenyllactic acid and some sulfite esters.

There are two catalytically active residues in pepsin: Asp-32 and Asp-215. Their ionizations are seen in the pH-activity profile, which has an optimum at pH 2 to 3, and which depends upon the acidic form of a group of $pK_a \sim 4.5$ and the basic form of a group of $pK_a \sim 1.1$.[181,182] The pK_a values have been assigned from the reactions of irreversible inhibitors that are designed to react specifically with ionized or un-ionized carboxyl groups. Diazo compounds—such as N-diazoacetyl-L-phenylalanine methyl ester, which reacts with un-ionized carboxyls—react specifically with Asp-215 up to pH 5 or so (equation 15·30).[183–185] Epoxides, which react specifically with ionized carboxyls, modify Asp-32 (equation 15·31).

$$(\text{Asp-215})\text{—CO}_2\text{H} + \text{N}_2\text{CHCONHCH(Ph)CO}_2\text{CH}_3 \longrightarrow$$

$$(\text{Asp-215})\text{—CO}_2\text{CH}_2\text{CONHCH(Ph)CO}_2\text{CH}_3 + \text{N}_2 \quad (15\cdot30)$$

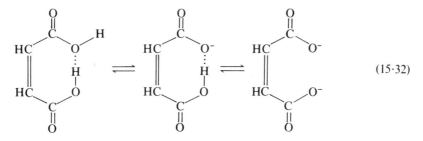

$$(\text{Asp-32})\text{—CO}_2{}^- + \overset{O}{\overset{\diagup\diagdown}{\text{CH}_2\text{—CHR}}} \longrightarrow (\text{Asp-32})\text{—CO}_2\text{CH}_2\overset{\text{OH}}{\overset{|}{\text{CHR}}} \quad (15\cdot31)$$

The pH dependence of the rate of modification shows that the pK_a of Asp-32 is less than 3.[186] It is seen in the high-resolution crystal structures that the carboxyl groups of the two aspartate residues are hydrogen-bonded to each other. This is similar to the ionization of maleic acid, which has pK_a values of 1.9 and 6.2 (equation 15·32).

$$(15\cdot32)$$

1. *The chemical reaction mechanism of pepsin.* Evidence accumulated over the years appeared to support the idea that both an *acyl*-enzyme and an *amino*enzyme were formed during the reaction. Some of

the evidence, such as the incorporation of ^{18}O into the enzyme during the reaction performed in $H_2\,^{18}O$, was simply incorrect, but other evidence, principally from transpeptidation studies, may be reinterpreted and shown to be consistent with the *absence* of these intermediates. The interpretation of these experiments is quite instructive and will be discussed next. Current evidence favors general-acid-base catalysis with no intermediates.

Transpeptidation experiments. The pepsin-catalyzed hydrolysis of Leu-Tyr-Leu gives the product Leu-Leu, which can be formed from the acyl transfer shown in equation 15·33.[187,188]

$$E—OH + Leu*-Tyr-Leu \longrightarrow (Leu*-CO)—O—E \xrightarrow{Leu\text{-}Tyr\text{-}Leu}$$

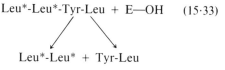

$$Leu*-Leu*-Tyr-Leu + E—OH \quad (15·33)$$

$$Leu*-Leu* + Tyr-Leu$$

This experiment has been extended by using the double-labeled substrate $[^{14}C]$Leu-Tyr-$[^3H]$Leu to show that simultaneous amino and acyl transfer could take place. It is found that both $[^3H]$Leu-$[^3H]$Leu and $[^{14}C]$Leu-$[^{14}C]$Leu are formed.[189] The ^{14}C-labeled product, which predominates by a factor of 3 or 4, could come from the acyl transfer route, whereas the ^3H-labeled product could arise from the $[^{14}C]$Leu-Tyr-$[^3H]$Leu-$[^3H]$Leu produced from an aminoenzyme by mechanism 15·34, where E-$[^3H]$(NH-Leu) is $[^3H]$Leu bound to E by the NH group of the amino acid.

$$E—OH + [^{14}C]Leu-Tyr-[^3H]Leu \longrightarrow E—[^3H](NH-Leu) \xrightarrow{Leu\text{-}Tyr\text{-}Leu}$$

$$E—OH + [^{14}C]Leu-Tyr-[^3H]Leu-[^3H]Leu \quad (15·34)$$

$$[^{14}C]Leu-Tyr + [^3H]Leu-[^3H]Leu$$

It is clear that this mechanistic interpretation is not unique, because there is no direct evidence for the proposed intermediates. For example, alternative possible explanations are: (1) The amino group in the "aminoenzyme" may not be covalently bound, but may merely be "activated" by the enzyme; and (2) similarly, the acyl transfer reaction of equation 15·33 could occur by the direct attack of Leu-Tyr-Leu on the enzyme-bound Leu-Tyr-Leu. However, M. S. Silver and S. L. T. James[190,191] have proposed a further interpretation, based on the observation that small peptides stimulate the pepsin-catalyzed hydrolysis of other peptides by being first synthesized into larger peptides in a condensation reaction that is the reverse of the hydrolytic step; e.g., equations 15·35. The idea of the condensation of two small peptides to give a larger

$$Z\text{-Ala-Leu} + \text{Phe-Trp-NH}_2 \rightleftharpoons Z\text{-Ala-Leu-Phe-Trp-NH}_2$$
$$Z\text{-Ala-Leu-Phe-Trp-NH}_2 \rightarrow \text{Trp-NH}_2 + Z\text{-Ala-Leu-Phe} \qquad (15\cdot35)$$
$$Z\text{-Ala-Leu-Phe} \rightarrow Z\text{-Ala-Leu} + \text{Phe}$$

peptide at a rate that is relatively fast compared with hydrolysis of the small peptides is quite reasonable when the following is considered: pepsin has an extended active site so that long peptides react far more rapidly than short ones; and the equilibrium constant for the hydrolysis of a peptide bond is close to 1 M. Therefore, at high concentrations of small peptides that are the reaction products of a reactive larger peptide, the rate of condensation of the small peptides should be comparable with the rate of hydrolysis of the larger peptide, and faster than the rate of hydrolysis of the small peptides. The reaction products in equations 15·33 and 15·34 may be generated by the prior condensation of two molecules of Leu-Tyr-Leu, followed by various modes of hydrolysis and condensation.[191]

Isotope incorporation experiments have provided direct evidence against covalent intermediates between a carboxyl group in the enzyme and the substrate.[192,193] An aminoenzyme can definitely be excluded: if a carboxyl group in the enzyme forms an amide bond with an NH group in a substrate to give E—CO—NHR, then hydrolysis of the aminoenzyme in $H_2{}^{18}O$ will lead to incorporation of ^{18}O into that carboxyl group—and no such incorporation is found. The following experiments on the transpeptidation reaction of Leu-Tyr-NH$_2$ to give Leu-Leu-Tyr-NH$_2$ argue against an acylenzyme. If transpeptidation occurs, as in equation 15·33, by the formation of an acylenzyme that then reacts with the amino group of Leu-Tyr-NH$_2$, there will be no incorporation of ^{18}O from solvent $H_2{}^{18}O$ into the resultant larger peptide. But it is found that such incorporation does occur at a high rate. This is most readily explained by the reaction involving the general-base-catalyzed attack of solvent water on the Leu-Tyr-NH$_2$ to generate Leu-CO^{18}OH, which acts as the acceptor. Furthermore, there is also exchange of some ^{18}O into unreacted Leu-Tyr-NH$_2$. This could result from the formation of a tetrahedral intermediate in which the two oxygens become equivalent and a fraction of the intermediate reverts to starting material. Clearly, however, there are complications from the transpeptidation possibilities suggested by Silver and James.[190,191]

The naturally occurring inhibitor pepstatin (structure 15·36) binds very tightly to carboxyl proteases: K_i with porcine pepsin is 4.5×10^{-11} M.[194] The statine residue has a tetrahedral carbon replacing the normal carboxyl carbon, and so perhaps is an analogue of a tetrahedral intermediate that the protein binds tightly.[195] The crystal structure of pepstatin

bound to the *Rhizopus chinensis* enzyme has been solved, and it is found
that the [middle] statine binds next to the active-site carboxyls.[196] A pepstatin analogue containing a keto group in the reactive position is found
from ^{13}C NMR experiments to be bound as a tetrahedral adduct.[197] It is
possible that the enzyme binds the hydrated form, $=C(OH)_2$, since this
is a transition state analogue of the general-base-catalyzed attack of a
water molecule on a carboxyl group (cf. phosphoramidon, structure
15·25).

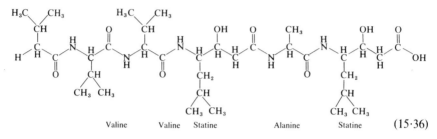

| Valine | Valine | Statine | Alanine | Statine | (15·36) |

2. *The zymogen.* Pepsin is formed from pepsinogen by the proteolysis of 44 residues from the N-terminus. The zymogen is stable at neutral
pH, but below pH 5 it rapidly and spontaneously activates. The activation
process takes place by two separate routes, a pepsin-catalyzed and an
intramolecularly catalyzed process. There is much evidence that pepsinogen may activate itself in a unimolecular process, the active site cleav-
ing the N-terminus of its own polypeptide chain.[198–204] Perhaps the neatest
demonstration of this intriguing phenomenon is the autoactivation of pepsinogen that is covalently bound to a sepharose resin.[200] The molecules
are immobilized and, in general, are not in contact with each other. Yet
on exposure to pH 2, the pepsinogen spontaneously activates. The two
routes for activation compete, with the bimolecular activation dominating
at higher zymogen concentrations and above pH 2.5, and the intramolecular activation dominating at low pH. The result of this spontaneous
activation of the pure zymogen is that the majority of pepsinogen molecules will be active 10 s after mixing with the hydrochloric acid in the
stomach.

C. Ribonuclease [205–207]

Bovine pancreatic ribonuclease catalyzes the hydrolysis of RNA by a
two-step process in which a cyclic phosphate intermediate is formed
(equation 15·37). The cyclization step is usually much faster than the
subsequent hydrolysis, so the intermediate may be readily isolated. DNA
is not hydrolyzed, as it lacks the 2'-hydroxyl group that is essential for
this reaction. There is a strong specificity for the base B on the 3' side
of the substrate to be a pyrimidine—uracil or cytosine.

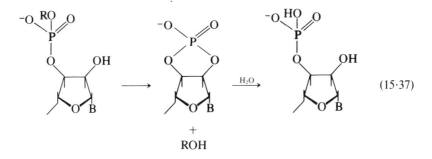

$$+$$
$$ROH$$

$$(15·37)$$

The enzyme consists of a single polypeptide chain of M_r 13 680 and 124 amino acid residues.[208,209] The bond between Ala-20 and Ser-21 may be cleaved by subtilisin. Interestingly, the peptide remains attached to the rest of the protein by noncovalent bonds. The modified protein, called ribonuclease S, and the native protein, now termed ribonuclease A, have identical catalytic activities. Because of its small size, its availability, and its ruggedness, ribonuclease is very amenable to physical and chemical study. It was the first enzyme to be sequenced.[208] The crystal structures of both forms of the enzyme were solved at 2.0-Å resolution several years-ago.[210,211] Furthermore, because the catalytic activity depends on the ionizations of two histidine residues, the enzyme has been extensively studied by NMR (the imidazole rings of histidines are easily studied by this method—Chapter 5).

The currently accepted chemical mechanism for the reaction was de-duced by an inspired piece of chemical intuition before the crystal struc-ture was solved.[212] It was found that the pH-activity curve is bell-shaped, with optimal rates around neutrality. The pH dependence of k_{cat}/K_M shows that the rate depends upon the ionization of a base of pK_a 5.22 and an acid of pK_a 6.78 in the free enzyme, whereas the pH dependence of k_{cat} shows that these are perturbed to pK_a values of 6.3 and 8.1 in the enzyme–substrate complex. It was proposed that the reaction is catalyzed by concerted general-acid–general-base catalysis by two histidine resi-dues, later identified as His-12 and His-119 (reactions 15·38 and 15·39).

In the cyclization step, His-12 acts as a general-base catalyst and His-119 acts as a general acid to protonate the leaving group. Their catalytic roles are reversed in the hydrolysis step: His-119 activates the attack of water by general-base catalysis and His-12 is the acid catalyst, protonating the leaving group. This reversal of roles is quite logical. Reaction 15·39 is essentially the reverse of 15·38, except that HOH replaces ROH. It is expected from the principle of microscopic reversibility that a group re-acting as a general acid in one direction will react as a general base in the opposite direction.

The chemistry and stereochemistry of the reactions were extensively discussed in Chapter 8, sections E1 and E3. There is an in-line mechanism

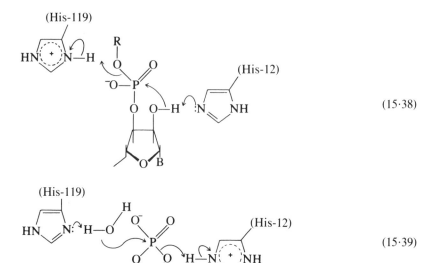

$$(15 \cdot 38)$$

$$(15 \cdot 39)$$

that generates a pentacovalent intermediate, with the attacking nucleophile and leaving group occupying the apical positions of the trigonal bipyramid.

1. The structures of the enzyme and the enzyme–substrate complexes

Ribonuclease has a well-defined binding cleft for the substrate. In it are located His-12, His-119, and the side chains of Lys-7, Lys-41, and Lys-66. The structure of the enzyme–substrate complex for the cyclization step was first deduced from the crystal structure of the enzyme and the substrate analogue UpcA (structure 15·40),[213] the phosphonate analogue of UpA. It is a very good analogue, differing from the real substrate only in that a —CH$_2$— group replaces an oxygen atom, so that the structure of its complex with ribonuclease should be close to that of a productively bound enzyme–substrate complex. It was found that His-119 is within hydrogen-bonding distance of the leaving group, and His-12 within hydrogen-bonding distance of the 2′-hydroxyl of the pyrimidine ribose. The pK_a values of His-12 and His-119 have been determined by NMR measurements to be 5.8 and 6.2, respectively, at 40° C.[214] A considerable fraction of each is in the suitable ionic state at physiological pH for the general-acid–general-base catalysis shown in equations 15·38 and 15·39 to occur.

The structures of the native enzyme and its complexes with several inhibitors have since been obtained at higher resolution in other laboratories, to afford a more complete description of the enzyme–substrate interactions.[215–219] Particularly noteworthy are the lysine residues 7, 41, and 66. That these are an important part of the catalytic machinery has

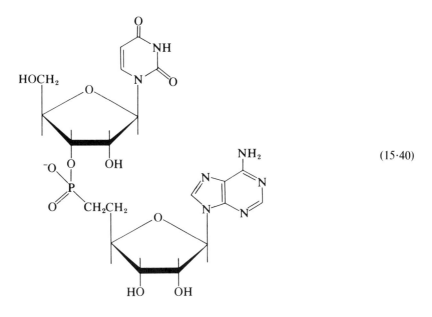

(15·40)

been deduced from their conservation in evolution (they have been found in all homologous ribonucleases that have been sequenced), and from their loss of activity when they are acetylated. Lys-41 is particularly important.[206,215] The lysine side chains are very mobile in the free enzyme, but their mobilities are much decreased on the binding of nucleotide substrate analogues. Lys-41 interacts directly with the phosphate moiety and is thought to stabilize the pentacovalent intermediate. Another residue that has been conserved through evolution is Thr-45.[206] This residue is responsible for the specificity of the enzyme for pyrimidines (on the 3′ side). Its backbone NH group and —OH side chain are able to form complementary hydrogen bonds with both uracil and cytosine (structures 15·41).

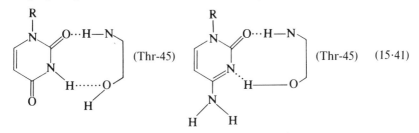

An attempt has been made to obtain the crystal structures of all the enzyme–substrate, enzyme–intermediate, enzyme–product, and enzyme–transition state complexes by using a combination of methods: conventional x-ray analysis of the binding of substrate and transition state analogues, cryoenzymology, and model building, all based on refinement

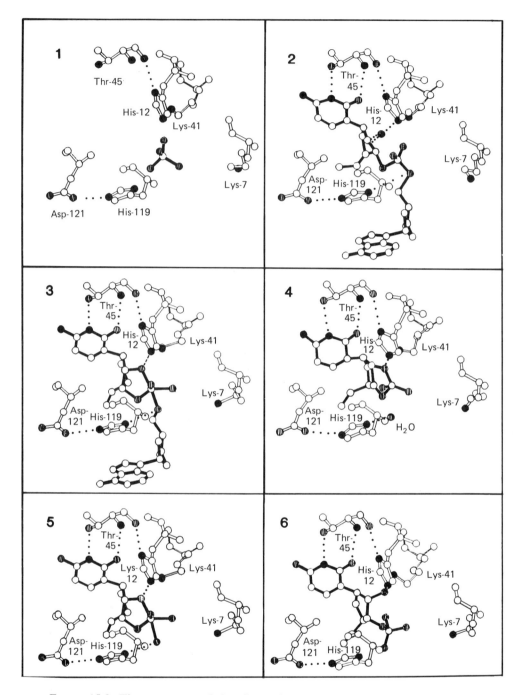

FIGURE 15·8. The structures of the ribonuclease active sites (frame 1) and the complexes with a substrate (frame 2), transition states (frames 3 and 5), the intermediate (frame 4), and the product (frame 6) (see text). [Modified from W. A. Gilbert, A. L. Fink, and G. A. Petsko (in press).]

to 1.5 to 1.9 Å.[219] These are summarized in a series of "stop-action" photographs in Figure 15·8. Frame 1 shows the free enzyme, containing a bound sulfate ion under the conditions of crystallization. Frame 2 shows the complex of the enzyme and 2'-deoxy-CpA. This is a good substrate analogue, lacking only the 2'-hydroxyl, which can be filled in by model building. His-12 is in position to act as a general-base catalyst, and His-119 as a general acid. Thr-45 binds the pyrimidine. It is suggested here that Lys-41 remains disordered, but other studies on enzyme–inhibitor complexes have indicated convincingly that Lys-41 becomes less mobile and bonds with the phosphate.[217] Frame 3 shows the first transition state, obtained by model building. Lys-41 moves to stabilize the pentacovalent intermediate electrostatically. Frame 4 shows the enzyme–cyclic phosphate intermediate complex. This was obtained by using a flow cell at −70°C and flowing cytidine 2',3'-monophosphate through the crystal. Under these conditions, the reaction rate is so slow that the complex is maintained as the cyclic phosphodiester. A water molecule replaces the 5'-hydroxyl that was bonded to His-119. The histidine residue activates the water by general-base catalysis to act as a nucleophile and form the next transition state, frame 5. Frame 5 shows the actual structure of the complex with the transition state analogue, uridine vanadate (structure 15·42), which mimics the pentacovalent intermediate formed by the attack of OH⁻ on uridine 3',5'-cyclic phosphate.[220] The transition state analogue binds 10^4 times more tightly than the corresponding 3',5'-cyclic phosphodiester or 3'-monophosphate product. The increased negative charge on the equatorial oxygens is stabilized by ionic interactions. His-12 (the H^+ form) is able to protonate the 2'-hydroxyl group to perform the general-acid catalysis of its expulsion. The stereochemical evidence implies that Lys-41 and His-12 must be in these positions, but a combined neutron and x-ray diffraction study on the uridine vanadate complex has been used to argue that the two residues are interchanged, with His-12 hydrogen-bonding to an equatorial oxygen and Lys-41 to the 2'-hydroxyl. Frame 6 shows the stable enzyme–product complex with uridine 3'-monophosphate. During the collapse of the previous pentacovalent intermediate, the phosphorus atom moves through about 1 Å. The concurrent movement of the oxygen atoms weakens the ionic interactions with Lys-41. These high-resolution structures will clearly facilitate the molecular mechanics calculations that are already being performed on this reaction.[221,222]

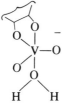

(15·42)

D. Staphylococcal nuclease[223–225]

Staphylococcal nuclease is a phosphodiesterase that cleaves DNA and RNA to form 3'-phosphomononucleosides (equation 15·43). The enzyme consists of a single polypeptide chain of M_r 16 900 and 149 amino acid residues. The structures of the enzyme and its complex with thymidine 3',5'-diphosphate were solved some years ago at 2.0-Å resolution.[226,227] But the studies did not elucidate the mechanism of the enzyme. After the success of the experiments on ribonuclease, our ignorance about staphylococcal nuclease provides a salutary lesson: that structural knowledge does not automatically solve the problems of mechanism and function. A subsequent crystal structure solved at 1.5-Å resolution has, however, given the following clues concerning the mechanism.[228]

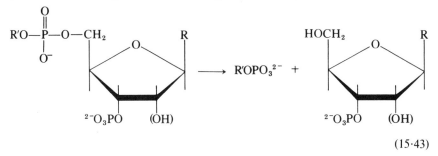

$$(15·43)$$

The 5'-phosphate group of thymidine 3',5'-diphosphate in the enzyme–inhibitor complex presumably occupies the binding site for the phosphate of a diester substrate. It is seen to be hydrogen-bonded to the positively charged side chains of Arg-35 and Arg-87, which will activate it to nucleophilic attack. There is also a calcium ion in the active site, bound by the carboxylate groups of Asp-21 and Asp-40. Calcium ions are known to be essential for activity. The calcium ion appears to coordinate with the phosphate and activate it further to nucleophilic attack. Glu-43

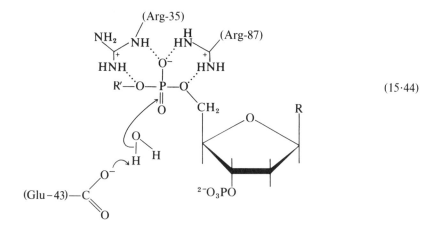

$$(15·44)$$

is positioned to act as a general-base catalyst for the attack of H_2O (mechanism 15·44). Two further observations have yet to be explained. The pH optimum of the enzyme activity is rather high (the pH–k_{cat}/K_M profile is bell-shaped and depends on pK_a values of 8.4 and 9.2).[229] Two moles of calcium ions bind to the enzyme–inhibitor complex,[230] but only one mole is seen in the crystal structure.

E. Lysozyme[231–233]

Hen egg white lysozyme is a small protein of M_r 14 500 and 129 amino acid residues. This enzyme was introduced in Chapter 1, where it was pointed out that examination of the crystal structure of the enzyme stimulated most of the solution studies. A mechanism proposed for the enzymatic reaction was based on the structure of the active site and on ideas from physical organic chemistry.[234,235] This mechanism consisted of the following points:

1. There are six subsites—labeled A, B, C, D, E, and F—for binding the glucopyranose rings of the substrate.
2. The scissile bond lies between sites D and E, close to the carboxyl groups of Glu-35 and Asp-52.
3. The reaction proceeds via a carbonium ion intermediate (more strictly, a carboxonium ion) which is stabilized by the ionized carboxylate of Asp-52 (equation 15·45).

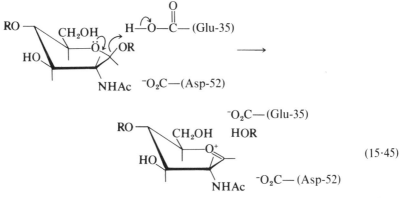

$$(15·45)$$

4. The expulsion of the alcohol is general-acid-catalyzed by the un-ionized carboxyl of Glu-35.
5. Furthermore, as discussed in Chapter 12, sections B2a and D5c, the sugar ring in site D is distorted to the sofa conformation expected for a carbonium ion.
6. Small polysaccharides avoid the strain in the D subsite by binding the A, B, and C sites.

We shall now see how all of these points, apart from the role of distortion in site D, have been experimentally verified. (The conformation in D was originally called a "half-chair," but "sofa" is more appropriate.)[236]

1. The carbonium ion

Alternatives to the carbonium ion mechanism are the direct attack of water on the substrate, and the nucleophilic attack of Asp-52 on the C-1 carbon to give an ester intermediate. The single-displacement reaction has been ruled out by showing that the reaction proceeds with retention of configuration (Chapter 7, section C3; Chapter 8, section C2).[237-239] The carbonium ion or S_N1 mechanism has been substantiated by secondary isotope effects, using substrates containing either deuterium or tritium instead of hydrogen attached to the C-1 carbon.[240,241] For example, k_H/k_D for structures 15·46 is 1.11, compared with values of 1.14 found

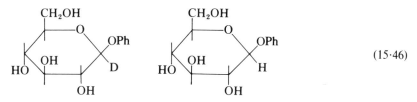

(15·46)

for a carbonium ion reaction and 1.03 for a bimolecular displacement (S_N2) in simple chemical models. The $^{16}O/^{18}O$ kinetic isotope effect for the leaving group oxygen atom indicates high C—O bond fission in the transition state.[242]

2. Electrostatic and general-acid catalysis

The pH dependence of k_{cat}/K_M shows that the reaction rate is dependent on an acid of $pK_a \sim 6$ and a base of $pK_a \sim 4$ in the free enzyme.[243] Asp-52 may be specifically blocked by esterification with triethoxonium fluoroborate (a reaction that requires an ionized carboxylate). The difference titration between the modified and native proteins shows that the pK_a of Asp-52 is 4.5 whereas that of Glu-35 is 5.9.[244] The reaction rate therefore depends on the ionization of the two residues in the manner predicted by the mechanism.

An observation consistent with general-acid catalysis by Glu-35 comes from a study of the reverse reaction. It is found that the rate of reaction of alcohols with the carbonium ion intermediate is virtually independent of their pK_a's. This is consistent with the general-base-catalyzed attack of the alcohol on the ion; hence, by the principle of microscopic reversibility, the expulsion of the alcoholate ion from the glycoside is general-acid-catalyzed.[245]

It was emphasized by C. A. Vernon in the initial formulation of the lysozyme mechanism that the electrostatic stabilization of the carboxonium ion by Asp-52 is the most important catalytic factor.[235] Theoretical calculations suggest that the activation energy for k_{cat} is lowered by ~32 kJ/mol (~8 kcal/mol) by electrostatic stabilization (Chapter 2, Figure 2·7).[246,247] A high value for this is indicated by experiments showing that the chemical conversion of the carboxyl group of Asp-52 to —CH$_2$OH abolishes the enzymatic activity.[248]

3. Binding energies of the subsites

It was originally proposed that nonbonded interactions between the enzyme and the sugar ring in site D distort it to the sofa conformation of the carbonium ion. Various workers have searched for weak binding in this site by estimating the binding energies of the individual sites from binding and kinetic measurements.[231,233] Some estimates for these are given in Table 15·3. Although the values are not precise, it is clear that there is a repulsive energy against the binding of NAM in site D. Also, as (NAG)$_4$ is found to bind about equally in sites A, B, and C, on the one hand, and A, B, C, and D, on the other, there is no net binding energy for NAG in site D.[249] The positions of binding of small substrates have been located from their interactions with probes, such as the dye Biebrich Scarlet[250] and the lanthanide ion,[249] which have been bound in the cleft; these substrates have been shown to be predominantly nonproductively bound. This would appear to provide very strong evidence for the strain mechanism. However, the evidence has been reviewed by M. F. Levitt,[251] who, as discussed in Chapter 12, section D5c, finds by calculation that it is very unlikely that the sugar ring in site D is distorted.[246,251] Also, transition state analogues that have the sugar ring in site D chemically modified to resemble the carbonium ion bind no more than 100 times more

TABLE 15·3. *Binding energies of subsites in hen egg white lysozyme^a*

Site	Residue binding^b	Binding energy kJ/mol	Binding energy kcal/mol
A	NAG	−8	−2
B	NAG	−12	−3
	NAM	−16	−4
C	NAG	−20	−5
D	NAM	+12	+3
	NAG	0	0
E	NAG	−16	−4
F	NAG	−8	−2

^{a} See also M. Schindler, Y. Assaf, N. Sharon, and D. M. Chipman, *Biochemistry* **16**, 423 (1977).
^{b} NAG = *N*-acetylglucosamine; NAM = *N*-acetylmuramic acid.

tightly than substrates containing the unmodified ring (Chapter 12, section B2).[249,254] Furthermore, it is suggested that the poor binding of the ring in site D is due to the displacement by the ring of two water molecules that are bound to the carboxylate of Asp-52.

Subsequent binding studies have shown that there is little strain associated with binding in the D site,[252-254] and the most recent crystallographic data agree with this.[255] However, the idea of strain in the D site has been raised again in crystallographic studies on the enzyme encoded by bacteriophage T4[256] (whose evolutionary relationship with the hen egg white lysozyme was discussed in Chapter 1). But, as was stressed in Chapter 12, strain does not necessarily mean that the substrate is distorted when it binds to the enzyme; strain only implies that the transition state is stabilized with respect to the substrate.

The deductions from this remarkable example of x-ray crystallography have not only stood the test of time, but have been neatly confirmed by solution studies, except that the emphasis of the strain mechanism is now on the electrostatic component rather than on that due to distortion.

To end on a note of caution, it has been found that the binding of small substrates is a two-step process with a step involving the isomerization of the enzyme–substrate complex.[257,258] It is not clear what is involved in this so-called conformational change and how it fits into the above scheme, but it could be related to the small structural changes that are observed when (NAG)₃ is bound to the crystalline enzyme.

F. Carbonic anhydrase[259,260]

The carbonic anhydrases catalyze the hydration of carbon dioxide and the dehydration of bicarbonate (equation 15·47). The crystal structures

$$CO_2 + H_2O \rightleftharpoons HCO_3^- + H^+ \qquad\qquad (15\cdot47)$$

of the human B and C isozymes have been solved at high resolution.[261,262] They are both metalloenzymes containing 1 mol of tightly bound zinc per single polypeptide chain of $M_r \sim 29\,000$. The C enzyme, which is the more active, has 259 amino acid residues in the chain, 1 less than the B form.[263,264] The sequences of the two are about 60% homologous and their tertiary structures are quite similar. The Zn^{2+} ion is ligated by the imidazole rings of three histidine residues at the bottom of the active-site cavity, some 12 Å from the surface. The fourth ligand is probably a water molecule or a hydroxide ion, depending on the pH.[261,262]

The C enzyme is an extremely efficient catalyst. In the hydration reaction, k_{cat} is 10^6 s^{-1} and K_M for CO_2 is 8.3 mM, whereas in the dehydration reaction, k_{cat} is 6×10^5 s^{-1} and K_M for HCO_3^- is 32 mM.[265] The catalytic activity depends on the ionization of a group of pK_a 7 in the free enzyme.[265] The hydration reaction depends on this group being in the basic form, and dehydration, in the acidic form. The turnover numbers of the reactions are far higher than the rate constants for the transfer of protons between water and a group of pK_a 7 (about 2×10^3 s^{-1}— Chapter 4, Table 4·2). The enzymes will also catalyze the hydration of aldehydes and the hydrolysis of esters in nonphysiological reactions.

There four questions in particular that have stimulated recent studies:

1. What is the ionizing group of pK_a 7 responsible for activity?
2. Is the bound CO_2 directly coordinated to the Zn^{2+} ion?
3. What is the bound substrate in the dehydration reaction: HCO_3^- or H_2CO_3?
4. Are the high turnover numbers compatible with the rates of proton transfer?

The classic mechanism invokes a zinc-bound water molecule that ionizes with a pK_a of 7 to give the nucleophilic zinc-bound hydroxide ion at high pH.[260,266] The zinc-bound hydroxide is known to be nucleophilic and to ionize in this region (Chapter 2, section B7b). Many other proposals have been made,[267–269] but the only other groups at the active site that can ionize with this pK_a are the histidines. However, their ionizations have been studied by NMR and found to be inconsistent with the activity-linked ionization.[270] Furthermore, measurements on nuclear quadrupole interactions in the cadmium-substituted enzyme are consistent with the ionization of a metal-bound water molecule with the same pK_a as is found in the pH-activity profile.[271]

The IR absorbance maximum of CO_2 bound to carbonic anhydrase is at a very similar frequency to that of the free ligand. The CO_2 is thus not polarized by the zinc ion and not directly bonded to it.[272] It has been suggested, however, that CO_2 was bound in a nonproductive mode in these studies, since N_2O was found to compete with CO_2 in the IR studies but not in kinetic studies.[273]

Another method of measuring ligand–metal distances is to substitute a paramagnetic ion for the zinc and measure its effects on the relaxation rates of NMR signals of the substrates. Paramagnetic ions cause rapid nuclear relaxation because of their high dipoles. The effect falls off rapidly with distance, and so enables internuclear distances to be calculated. Such a study on carbonic anhydrase in which Mn(II) had been substituted for Zn^{2+} and ^{13}C NMR measurements were made on $^{13}CO_2$ and $H^{13}CO_3^-$ showed that the bicarbonate ion is directly bonded to the metal, whereas the CO_2 is bonded very weakly, if at all.[274] (The Mn-substituted enzyme has 7% of the activity of the native enzyme.)

The simplest chemically reasonable mechanism is shown in equation 15·48.

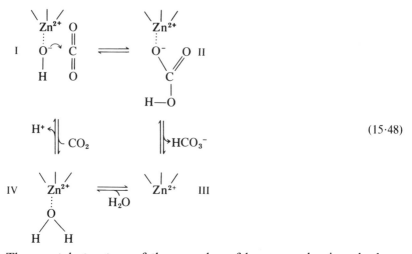

$$(15\cdot48)$$

The crystal structure of the complex of human carbonic anhydrase B with its competitive inhibitor imidazole has been solved.[275] The imidazole is bound in a hydrophobic pocket with a nitrogen atom directly coordinated to the zinc ion. Most interestingly, it appears to bind as a fifth ligand without displacing the zinc-bound H_2O or OH^-. Analogously, it is suggested that an oxygen atom of carbon dioxide could also bind directly, but loosely, to the zinc and not displace the bound water. The zinc ion could thus both orient the CO_2 and weakly polarize it, as well as providing a zinc-bound nucleophilic water molecule (reaction 15·49).

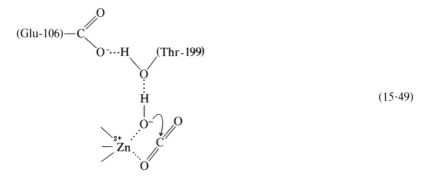

$$(15\cdot49)$$

The authors (reference 275) note that the bound water molecule is hydrogen-bonded to the hydroxyl of Thr-199, which is in turn hydrogen-bonded to the carboxylate of Glu-106. This adds a further twist to the problem of identification of the group ionizing with pK_a 7. It could perhaps be the carboxylate Glu-106, perturbed to a high value by its environment, rather than the zinc-bound water. If this is so, then the nucleophile is not the hydroxide ion as in reaction 15·49, but the un-ionized water molecule activated by general-base catalysis from Glu-106, transmitted by the intervening hydroxyl of Thr-199.

Any mechanism involving HCO_3^- as a substrate requires that at some stage in the reaction there be proton release from the enzyme to the medium. In equation 15·48, this occurs at the step IV $\rightarrow$ I. In an unbuffered solution, the rate of the proton transfer step would be too slow to regenerate the free enzyme. For example, from the rate constants in Chapter 4, Table 4·2, it is seen that the rate constant for the transfer of a proton from a base of pK_a 7 to water is 2.5×10^3 s^{-1}, and that the comparable rate constant for transfer to 10^{-7}-M hydroxide ions at pH 7 is 2.3×10^3 s^{-1}—values well below the turnover number of 5×10^5 s^{-1} at this pH. This cannot be circumvented by invoking the combination of the zinc ion with OH$^-$ instead of water at step III $\rightarrow$ IV, for similar reasons. The maximum value of a second-order rate constant is about 10^{10} to 10^{11} s^{-1} M^{-1} (Chapter 4, Part 2, Section A), which, combined with the hydroxide concentration of 10^{-7} M, gives a limit of 10^3 to 10^4 s^{-1} for this step at pH 7. A similar calculation indicates that it is unlikely that the undissociated H_2CO_3 is the substrate in the dehydration reaction, since its concentration is so low compared with HCO_3^- at pH 7.[266,276] (If this were the substrate, there would be no proton uptake or release in the reaction.) However, a simple calculation shows that low concentrations of buffers in solution account for the transfer rate constants (Table 4·2).[266,277] This has been verified by experiments in unbuffered solutions, where it has been found that the rate of the dehydration reaction is anomalously slow. The addition of 10-mM buffer restores the maximum reaction rate.[278,279]

G. Triosephosphate isomerase[280–282]

Triosephosphate isomerase catalyzes the interconversion of D-glyceraldehyde 3-phosphate and dihydroxyacetone phosphate (equation 15·50).

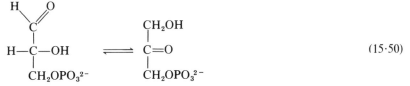

$$(15·50)$$

The crystal structures of the enzymes from chicken muscle[283] and yeast[284] have been solved at 2.5- and 3-Å resolution, respectively. Both are symmetrical dimers of M_r 53 000, and their identical chains contain 247 and 248 amino acid residues, respectively. The two enzymes have very similar tertiary structures: the secondary structures are nearly superimposable, the differences being confined to the external, irregular loops in the chain; the active sites have identical amino acids in the same positions.[284] There is no evidence for cooperativity of substrate binding during catalysis, but the dimer is the functional unit.

1. Deuterium and tritium tracer experiments and the mechanism of aldose–ketose isomerases

One of the most important techniques for studying the mechanism of aldose–ketose isomerases, which transfer a hydrogen between two carbon atoms, is the use of isotopically labeled hydrogen. The experiments have increased in complexity and sophistication over the years, and are best understood by a historical approach.

a. Exchange of protons with the medium: The enediol intermediate[285–288]

The first experiments performed in aqueous solutions enriched with deuterium or tritium showed that up to 1 mol of isotope is stereospecifically incorporated into the products of the reactions of mannose 6-phosphate, glucose 6-phosphate, ribose 5-phosphate, and triosephosphate isomerases (equation 15·51). This type of experiment has always been interpreted as ruling out a direct hydride transfer and instead showing that protons that can exchange with the medium are formed. It was suggested that an enediol intermediate, $RC(OH)=C(OH)H$, is formed.[286]

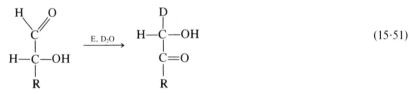

$$(15·51)$$

b. Detection of intramolecular proton transfer: The syn-enediol and a single base[289]

The next development was the discovery that there is also an intramolecular transfer of hydrogen (tritium) between the two carbons (equation 15·52). The simplest interpretation was that the tritium from the C-2 of the substrate is transferred to a catalyzing base, and that some of the tritium exchanges with protons from the water before it is transferred to the C-1 carbon, whereas the rest is incorporated into the C-1 position before it has time to escape into solution. Two important conclusions were drawn:

$$(15·52)$$

1. The transfer is mediated by a single base. (If a base removed the proton from C-2 and a separate acid residue transferred its proton to C-1, there would be no transfer.)
2. Because the proton is transferred by a single base, it must return to the *same* face of the enediol from which it is removed. The stereochemistry of the products shows that the enediol is syn (Chapter 8, section B3). On the basis of this, I. A. Rose proposed mechanism 15·53.[281]

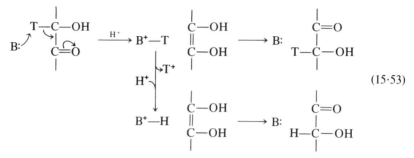

$$(15·53)$$

c. Construction of the minimal Gibbs energy profile for the reaction[290]
The next development in the use of isotopes in these reactions was the complete steady state analysis of the exchange of tritium from tritiated water into the product and the remaining substrate, the exchange of tritium from the tritiated substrate into the solvent and the products, and the primary kinetic isotope effects on the reaction rate. From this, it was possible to determine all the rate constants for mechanism 15·53 and to construct the Gibbs energy profile for the reaction, based on the assumption of the minimum number of intermediate states in the reaction (equation 15·54, where D = dihydroxyacetone phosphate, X = the enediol, and G = glyceraldehyde 3-phosphate).

$$E + D = E·D = E·X = E·G = E + G \qquad\qquad (15·54)$$

d. Refinement of the Gibbs energy profile[291,292]
Other intermediates in addition to those in equation 15·54 have been detected through direct observation of enzyme-bound intermediates by using high concentrations of enzyme (Chapter 7). The equilibrium between E·D, E·X, and E·G in equation 15·54 was directly measured by quenching the enzyme-bound reagents with acid and assaying the components.[291] The ratio of 41:5:54 found for [E·D]:[E·X]:[E·G] by this procedure differs from the ratio of about 20:1 found for [E·D]:[E·G] in the steady state study.[290] This measurement was the least accurate in the steady state study and so the two results are not irreconcilable. It has been suggested from further partitioning experiments that there are two additional intermediates in the reaction scheme: E′·D, produced by a

conformational change in the enzyme before enolization; and E'·G, a conformer preceding E·G (equation 15·55).[292] There is substantial spec-

$$E + D = E·D = E'·D = E'·X = E'·G = E·G = E + G \qquad (15·55)$$

troscopic evidence that substrate binding causes conformational changes in triosephosphate isomerase.[293,294] The release of glyceraldehyde 3-phosphate is the slowest step in the forward direction in the steady state.

One important point to emerge from the measurements of the energetics is that, independent of the precise number of intermediates, the intermediates do not substantially accumulate at the substrate concentrations encountered *in vivo*, and the equilibrium between the major complexes is reasonably balanced.[290] Furthermore, the value of k_{cat}/K_M for the favorable direction is close to the diffusion-controlled limit (Chapter 4, Table 4·4). The criteria of Chapter 12, sections C3 and E2, for "evolutionary" perfection are thus obeyed.[290]

2. The structure of the enzyme–substrate complex

With triosephosphate isomerase, nature has provided a rare opportunity to directly study an enzyme–substrate complex by x-ray diffraction methods, because the enzyme catalyzes a simple reversible reaction. The crystal structure of dihydroxyacetone phosphate bound to the chicken muscle enzyme has been solved at 6-Å resolution,[295] and the yeast enzyme at 3.5 Å.[296] There are problems, however, that cannot be resolved at these resolutions. Whereas the equilibrium constant between the two triosephosphates greatly favors the dihydroxyacetone phosphate in solution†, the results discussed in the last section suggest that the equilibrium constant is close to 1 for the enzyme-bound reagents.[291]

It is known from chemical evidence that the catalytic base in the reaction is Glu-165: this residue is modified by affinity labels, and the pH dependence of the reaction shows that the base has a pK_a of 3.9.[297–302] There is also evidence for an electrophile that polarizes the carbonyl oxygen of dihydroxyacetone phosphate: its rate of reduction by borohydride is increased 8-fold on its binding to the enzyme,[303] and its IR

---

† The aldehyde and ketone substrates are partly hydrated in solution. At 25°C, ~55% of the dihydroxyacetone phosphate and ~3.3% of the glyceraldehyde- 3-phosphate are in the unhydrated keto forms, which are the substrates for enzymatic-reactions. The equilibrium

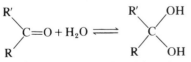

constant between the unhydrated keto forms is ~330:1 in favor of the dihydroxyacetone phosphate. [D. R. Trentham, C. H. McMurray, and C. I. Pogson, *Biochem. J.* **114**, 19 (1969); S. J. Reynolds, D. W. Yates, and C. I. Pogson, *Biochem. J.* **122**, 285 (1971); B. D. Peczon and H. O. Spivey, *Biochemistry* **11**, 2209 (1972).]

stretching frequency is reduced by 19 cm^{-1}.[304] The crystal structure of the complex reveals, despite any ambiguity about the precise composition of the enzyme-bound reagents, that the substrate is placed in juxtaposition with Glu-165 so that it can shuttle the protons from C-1 to C-2 at the *re-re* face of the *syn*-enediol. His-95 is in a position to interact with the oxygens on C-1 and C-2. Lys-13 may also interact with the C-2 oxygen and/or the bridging oxygen of the phosphate ester. There is a large change in the positions of residues 168 to 177 in the yeast enzyme on substrate binding. These residues, which are part of a mobile external loop in the free enzyme, become much less mobile in the complex. It has been suggested that the reduction in mobility weakens binding by a large entropic factor.[296]

H. Epilogue

The traditional way of viewing enzyme catalysis was formulated from chemical studies done long before the determination of the crystal structure of any enzyme. Enzyme catalysis was attributed to a combination of the following factors: general-acid-base catalysis, metal ion catalysis, nucleophilic catalysis, electrostatic catalysis, propinquity effects (a combination of an intramolecular reaction and correct orientation), strain (distortion of the substrate), and induced fit (distortion of the enzyme). All of these factors have now been verified to some extent.

General-acid-base catalysis is the most ubiquitous, being found in the reactions of the dehydrogenases, the serine proteases (and most probably the thiol proteases and carboxypeptidase), ribonuclease, lysozyme, and the aldose–ketose isomerases. Metal ion catalysis, in its classic form of stabilizing an anion, is found in the mechanisms of carboxypeptidase and possibly alcohol dehydrogenase. In the form of a metal-bound hydroxide ion, it is thought to provide a nucleophile in the reactions of carbonic anhydrase. Nucleophilic catalysis certainly occurs during the hydrolyses catalyzed by the thiol and serine proteases and also in enzymatic reactions in which Schiff bases are formed between the carbonyl group of the substrate and the side chain of a lysine residue. Transfer of a phosphoryl group is found between the substrates and histidine side chains in phosphoglycerate mutase. Electrostatic catalysis is important in the stabilization of the carbonium ion intermediate in the reactions of lysozyme. Propinquity effects in enzymatic reactions are not easily studied, and the best information has come from simple model systems and calculations, as was discussed in Chapter 2. The only direct evidence for distortion of a substrate on its binding to an enzyme is the IR work on the polarization of the carbonyl group of dihydroxyacetone phosphate on its binding to triosephosphate isomerase; but, as was discussed in Chapter 12, there are many examples of strain in the form of "transition state stabilization."

There are also several examples of the distortion of the enzyme on its binding of a substrate.

At the time of the first edition of this book, the possibility of dissecting the rate enhancement by a particular enzyme into individual contributions from all these factors seemed rather remote. Since then, however, the major advances both in theory (molecular dynamics and mechanics, and electrostatic calculations) and in experiment (site-directed mutagenesis) have brought this goal much closer.

References

1. G. Popják, *The Enzymes* **2**, 115 (1970).
2. H. F. Fisher, E. E. Conn, B. Vennesland, and F. H. Westheimer, *J. Biol. Chem.* **202**, 687 (1953).
3. M. E. Pullman, A. San Pietro, and S. P. Colowick, *J. Biol. Chem.* **206**, 129 (1954).
4. A. M. Gronenborn and G. M. Clore, *J. Molec. Biol.* **157**, 155 (1982).
5. S. A. Benner, *Experientia* **38**, 633 (1982).
6. M. G. Rossmann, A. Liljas, C.-I. Brändén, and L. J. Banaszak, *The Enzymes* **11**, 61 (1975).
7. C. -I. Brändén, H. Jornvall, H. Eklund, and B. Furugren, *The Enzymes* **11**, 104 (1975).
8. J. J. Holbrook, A. Liljas, S. J. Steindel, and M. G. Rossmann, *The Enzymes* **11**, 191 (1975).
9. L. J. Banaszak and R. A. Bradshaw, *The Enzymes* **11**, 369 (1975).
10. K. Dalziel, *The Enzymes* **11**, 2 (1975).
11. H. Eklund, B. Nordstrom, E. Zeppezauer, G. Söderlund, I. Ohlsson, T. Boiwe, B.-O. Söderberg, O. Tapia, C.-I. Brändén, and Å. Åkeson, *J. Molec. Biol.* **102**, 27 (1976).
12. H. Eklund, J.-P. Samama, L. Wallen, C.-I. Brändén, Å. Åkersen, and T. A. Jones, *J. Molec. Biol.* **146**, 561 (1981).
13. H. Eklund, C.-I. Brändén, and H. Jornvall, *J. Molec. Biol.* **102**, 61 (1976).
14. H. Eklund, B. V. Plapp, J.-P. Samama, and C.-I. Brändén, *J. Biol. Chem.* **257**, 14349 (1982).
15. K. Dalziel, *J. Biol. Chem.* **238**, 2850 (1963).
16. J. D. Shore, H. Gutfreund, R. L. Brooks, D. Santiago, and P. Santiago, *Biochemistry* **13**, 4185 (1974).
17. P. Andersson, J. Kvassman, A. Linström, B. Olden, and G. Pettersson, *Eur. J. Biochem.* **113**, 425 (1981).
18. E. Cedergren-Zeppezauer, J.-P. Samama, and H. Eklund, *Biochemistry* **21**, 4895 (1982).
19. H. Dutler and C.-I. Brändén, *Bioorg. Chem.* **10**, 1 (1981).
20. M. F. Dunn, F. F. Biellman, and G. Bruylant, *Biochemistry* **14**, 3176 (1975).
21. J. P. Klinman, *Biochemistry* **15**, 2018 (1976).
22. J. Kvassman and G. Pettersson, *Eur. J. Biochem.* **103**, 557, 565 (1980).
23. J. Kvassman, A. Larsson, and G. Pettersson, *Eur. J. Biochem.* **114**, 555 (1981).
24. P. F. Cook and W. W. Cleland, *Biochemistry* **20**, 1805 (1981).
25. D. C. Anderson and F. W. Dahlquist, *Biochemistry* **21**, 3569, 3578 (1982).
26. H. Theorell and B. Chance, *Acta Chem. Scand.* **5**, 1127 (1951).

27 C. C. Wratten and W. W. Cleland, *Biochemistry* **2**, 935 (1963); **4**, 2442 (1965).
28 J. D. Shore, H. Gutfreund, and D. Yates, *J. Biol. Chem.* **250**, 5276 (1975).
29 S. A. Bernhard, M. F. Dunn, P. L. Luisi, and P. Schack, *Biochemistry* **9**, 185 (1970).
30 J. P. Klinman, *J. Biol. Chem.* **247**, 7977 (1972).
31 J. D. Shore and H. Gutfreund, *Biochemistry* **9**, 4655 (1970).
32 R. L. Brooks and J. D. Shore, *Biochemistry* **10**, 3855 (1971).
33 J. P. Klinman, *C. R. C. Crit. Rev. Biochem.* **10**, 39 (1981).
34 M. Hadorn, V. A. John, F. K. Meier, and H. Dutler, *Eur. J. Biochem.* **54**, 65 (1975).
35 R. J. Kordal and S. M. Parsons, *Archs. Biochem. Biophys.* **104**, 439 (1979).
36 J. Kvassman and G. Pettersson, *Eur. J. Biochem.* **69**, 279 (1976).
37 C. F. Weidig, H. R. Halvorson, and J. D. Shore, *Biochemistry* **16**, 2916 (1977).
38 J. Everse and N. O. Kaplan, *Adv. Enzymol.* **37**, 61 (1973).
39 C. L. Markert and F. Moller, *Proc. Natn. Acad. Sci. USA* **45**, 753 (1959).
40 I. Fine, N. O. Kaplan, and D. Kuftinec, *Biochemistry* **2**, 116 (1963).
41 O. P. Chilson, L. A. Costello, and N. O. Kaplan, *Biochemistry* **4**, 271 (1965).
42 J. L. White, M. L. Hackert, M. Buehner, M. J. Adams, G. C. Ford, P. J. Lentz, Jr., I. E. Smiley, S. J. Steindel, and M. G. Rossmann, *J. Molec. Biol.* **102**, 759 (1976).
43 M. J. Adams, M. Buehner, K. Chandrasekhar, G. C. Ford, M. L. Hackert, A. Liljas, M. G. Rossmann, I. E. Smiley, W. S. Allison, J. Everse, N. O. Kaplan, and S. S. Taylor, *Proc. Natn. Acad. Sci. USA* **70**, 1968 (1973).
44 U. M. Grau, W. E. Trommer, and M. G. Rossmann, *J. Molec. Biol.* **150**, 289 (1981).
45 J. Everse, R. E. Barnett, C. J. R. Thorne, and N. O. Kaplan, *Archs. Biochem. Biophys.* **143**, 444 (1971).
46 C. J. Coulson and B. R. Rabin, *FEBS Lett.* **3**, 333 (1969).
47 J. H. Griffin and R. S. Criddle, *Biochemistry* **9**, 1195 (1970).
48 J. J. Holbrook and H. Gutfreund, *FEBS Lett.* **31**, 157 (1973).
49 W. B. Novoa and G. W. Schwert, *J. Biol. Chem.* **236**, 2150 (1961).
50 J. R. Whitaker, D. W. Yates, N. G. Bennett, J. J. Holbrook, and H. Gutfreund, *Biochem. J.* **139**, 677 (1974).
51 J. J. Holbrook and V. A. Ingram, *Biochem. J.* **131**, 729 (1973).
52 L. E. Webb, E. Hill, and L. J. Banaszak, *Biochemistry* **12**, 5101 (1973).
53 J. J. Birktoft, R. J. Fernley, R. A. Bradshaw, and L. J. Banaszak, *Proc. Natn. Acad. Sci. USA* **79**, 6166 (1982).
54 J. J. Birktoft and L. J. Banaszak, *J. Biol. Chem.* **258**, 472 (1983).
55 J. J. Holbrook, A. Lodola, and N. P. Illsley, *Biochem. J.* **139**, 797 (1974).
56 J. J. Holbrook and R. G. Wolfe, *Biochemistry* **11**, 2499 (1972).
57 M. Cassman and R. C. King, *Biochemistry* **11**, 4937 (1972).
58 M. Cassman and D. Vetterlein, *Biochemistry* **13**, 684 (1974).
59 J. I. Harris and M. Waters, *The Enzymes* **13**, 1 (1976).
60 K. Dalziel, N. V. McFerran, and A. J. Wonacott, *Phil. Trans. R. Soc.* **B293**, 105 (1981).
61 H. L. Segal and P. D. Boyer, *J. Biol. Chem.* **204**, 265 (1953).
62 P. J. Harrigan and D. R. Trentham, *Biochem. J.* **143**, 353 (1974).
63 R. G. Duggleby and D. T. Dennis, *J. Biol. Chem.* **249**, 167 (1974).
64 J.-C. Meunier and K. Dalziel, *Eur. J. Biochem.* **82**, 483 (1978).
65 I. Krimsky and E. Racker, *Science, N.Y.* **122**, 319 (1955).
66 D. R. Trentham, *Biochem. J.* **122**, 59, 71 (1971).

67 L. D. Byers and D. E. Koshland, *Biochemistry* **14**, 3661 (1975).
68 D. Moras, K. W. Olsen, M. N. Sabesan, M. Buehner, G. C. Ford, and M. G. Rossmann, *J. Biol. Chem.* **250**, 9137 (1975).
69 G. Biesecker, J. I. Harris, J. C. Thierry, J. E. Walker, and A. J. Wonacott, *Nature, Lond.* **266**, 328 (1977).
70 H. C. Watson, E. Duée, and W. D. Mercer, *Nature New Biology, Lond.* **240**, 130 (1972).
71 M. R. N. Murthy, R. M. Garavito, J. E. Johnson, and M. G. Rossmann, *J. Molec. Biol.* **138**, 859 (1980).
72 M. Buehner, G. C. Ford, D. Moras, K. W. Olsen, and M. G. Rossmann, *J. Molec. Biol.* **90**, 25 (1974).
73 A. Conway and D. E. Koshland, Jr., *Biochemistry* **7**, 4011 (1968).
74 B. D. Peczon and H. O. Spivey, *Biochemistry* **11**, 2209 (1972).
75 P. J. Harrigan and D. R. Trentham, *Biochem. J.* **135**, 695 (1973).
76 F. Seydoux, S. A. Bernhard, O. Pfenninger, M. Payne, and O. P. Malhotra, *Biochemistry* **12**, 4290 (1973).
77 J. Schlessinger and A. Levitzki, *J. Molec. Biol.* **82**, 547 (1974).
78 A. Levitzki, *J. Molec. Biol.* **90**, 451 (1974).
79 F. Seydoux and S. A. Bernhard, *Bioorg. Chem.* **1**, 161 (1974).
80 N. Kelemen, N. Kellershohn, and F. Seydoux, *Eur. J. Biochem.* **57**, 69 (1975).
81 L. S. Gennis, *Proc. Natn. Acad. Sci. USA* **73**, 3928 (1976).
82 J. Bode, M. Blumenstein, and M. A. Raftery, *Biochemistry* **14**, 1146 (1975).
83 M. Dunn, *Biochemistry* **13**, 1146 (1974).
84 G. Schoellman and E. Shaw, *Biochemistry* **2**, 252 (1963).
85 D. M. Blow, J. J. Birktoft, and B. S. Hartley, *Nature, Lond.* **221**, 337 (1970).
86 G. D. Brayer, L. T. J. Delbaere, and M. N. G. James, *J. Molec. Biol.* **124**, 261 (1978).
87 T. Inagami and J. M. Sturtevant, *Biochim. Biophys. Acta* **38**, 64 (1960).
88 A. R. Fersht and J. Sperling, *J. Molec. Biol.* **74**, 137 (1973).
89 A. R. Fersht and M. Renard, *Biochemistry* **13**, 1416 (1974).
90 M. A. Porubcan, W. M. Westler, I. B. Ibañez, and J. L. Markley, *Biochemistry* **18**, 4108 (1979).
91 W. W. Bachovchin, R. Kaiser, J. H. Richards, and J. D. Roberts, *Proc. Natn. Acad. Sci. USA* **78**, 7323 (1981).
92 A. A. Kossiakoff and S. A. Spencer, *Biochemistry* **20**, 6462 (1981).
93 D. A. Matthews, R. A. Alden, J. J. Birktoft, S. T. Freer, and T. Kraut, *J. Biol. Chem.* **252**, 8875 (1977).
94 J. Kraut, *Ann. Rev. Biochem.* **46**, 331 (1977).
95 G. H. Cohen, E. W. Silverton, and D. R. Davies, *J. Molec. Biol.* **148**, 449 (1981).
96 W. Bode, Z. Chen, K. Bartels, C. Kutzbach, G. Schmidt-Kastner, and H. Bartunik, *J. Molec. Biol.* **164**, 237 (1983).
97 D. L. Hughes, L. C. Sieker, J. Bieth, and J.-L. Dimicoli, *J. Molec. Biol.* **162**, 645 (1982).
98 G. E. Hein and C. Niemann, *J. Am. Chem. Soc.* **84**, 4495 (1962).
99 T. Alber, G. A. Petsko, and D. Tsernoglou, *Nature, Lond,* **263**, 297 (1976).
100 R. Henderson, *J. Molec. Biol.* **54**, 341 (1970).
101 A. R. Fersht, D. M. Blow, and J. Fastrez, *Biochemistry* **12**, 2035 (1973).
102 S. A. Bizzozero, W. K. Baumann, and H. Dutler, *Eur. J. Biochem.* **122**, 251 (1982).

103   P. Campbell, N. T. Nashed, B. A. Lapinskas, and J. Gurrieri, *J. Biol. Chem.* **258**, 59 (1983).
104   B. Asboth and L. Polgar, *Biochemistry* **22**, 117 (1983).
105   J. Fastrez and A. R. Fersht, *Biochemistry* **12**, 1067 (1973).
106   J. L. Markley, F. Travers, and C. Balny, *Eur. J. Biochem.* **120**, 477 (1981).
107   M. Hunkapiller, M. D. Forgac, E. H. Yu, and J. H. Richards, *Biochem. Biophys. Res. Commun.* **87**, 25 (1979).
108   M. Fujinaga, R. J. Read, A. Sielecki, W. Ardelt, M. Laskowski, Jr., and M. N. G. James, *Proc. Natn. Acad. Sci. USA* **79**, 4868 (1982).
109   G. D. Brayer, L. T. J. Delbaere, M. N. G. James, C.-A. Bauer, and R. C. Thompson, *Proc. Natn. Acad. Sci. USA* **76**, 96 (1979).
110   L. T. J. Delbaere and G. D. Brayer, *J. Molec. Biol.* **139**, 45 (1980).
111   S. T. Freer, J. Kraut, J. D. Robertus, H. T. Wright, and Ng. H. Xuong, *Biochemistry* **9**, 1997 (1970).
112   H. T. Wright, *J. Molec. Biol.* **79**, 1, 13 (1973).
113   J. J. Birktoft, J. Kraut, and S. T. Freer, *Biochemistry* **15**, 4481 (1976).
114   G. Robillard and R. G. Shulman, *J. Molec. Biol.* **86**, 519 (1974).
115   A. R. Fersht, *FEBS Lett.* **29**, 283 (1973).
116   A. Gertler, K. A. Walsh, and H. Neurath, *Biochemistry* **13**, 1302 (1974).
117   A. R. Fersht, *J. Molec. Biol.* **64**, 497 (1972).
118   H. Fehlhammer, W. Bode, and R. Huber, *J. Molec. Biol.* **111**, 415 (1977).
119   W. Bode, *J. Molec. Biol.* **127**, 357 (1979).
120   A. N. Glazer and E. L. Smith, *The Enzymes* **3**, 501 (1971).
121   J. Drenth, J. N. Jansonius, R. Koekoek, and B. G. Wolthers, *Adv. Protein Chem.* **25**, 79 (1971).
122   G. Lowe, *Tetrahedron* **32**, 291 (1976).
123   L. Polgar and P. Halasz, *Biochem. J.* **107**, 1 (1982).
124   R. E. J. Mitchell, I. M. Chaiken, and E. L. Smith, *J. Biol. Chem.* **245**, 3485 (1970).
125   A. Berger and I. Schechter, *Phil. Trans. R. Soc.* **B257**, 249 (1970).
126   M. R. Alecio, M. L. Dann, and G. Lowe, *Biochem. J.* **141**, 495 (1974).
167   A. Stockell and E. L. Smith, *J. Biol. Chem.* **227**, 1 (1957).
128   G. Lowe and A. Williams, *Biochem. J.* **96**, 189, 199 (1965).
129   P. M. Hinkle and J. F. Kirsch, *Biochemistry* **10**, 2717 (1971).
130   L. J. Brubacher and M. L. Bender, *J. Am. Chem. Soc.* **88**, 5871 (1966).
131   J. Drenth, J. N. Jansonius, and B. G. Wolthers, *J. Molec. Biol.* **24**, 449 (1967).
132   J. Drenth, J. N. Jansonius, R. Koekoek, H. M. Swen, and B. G. Wolthers, *Nature, Lond.* **218**, 929 (1968).
133   J. Drenth, K. H. Kalk, and H. M. Swen, *Biochemistry* **15**, 3731 (1976).
134   E. N. Baker, *J. Molec. Biol.* **141**, 441 (1981).
135   G. Lowe and Y. Yuthavong, *Biochem. J.* **124**, 107 (1971).
136   F. A. Johnson, S. D. Lewis, and J. A. Shafer, *Biochemistry* **20**, 44, 48, 52 (1981).
137   A. R. Fersht, *J. Am. Chem. Soc.* **93**, 3504 (1971).
138   G. Lowe and Y. Yuthavong, *Biochem. J.* **124**, 117 (1971).
139   M. H. O'Leary, M. Urberg, and A. P. Young, *Biochemistry* **13**, 2077 (1974).
140   V. I. Zannis and J. F. Kirsch, *Biochemistry* **17**, 2669 (1978).
141   J. A. Hartsuck and W. N. Lipscomb, *The Enzymes* **3**, 1 (1971).
142   F. A. Quiocho and W. N. Lipscomb, *Adv. Protein Chem.* **25**, 1 (1971).
143   W. N. Lipscomb, *Proc. Natn. Acad. Sci. USA* **77**, 3875 (1980).

144 R. A. Bradshaw, L. H. Ericsson, K. A. Walsh, and H. Neurath, *Proc. Natn. Acad. Sci. USA* **63**, 1389 (1969).
145 M. F. Schmid and J. R. Herriott, *J. Molec. Biol.* **103**, 175 (1976).
146 M. A. Holmes and B. W. Matthews, *J. Molec. Biol.* **160**, 623 (1982).
147 O. Dideberg, P. Charlier, G. Dive, B. Joris, J. M. Frère, and J. M. Ghuysen, *Nature, Lond.* **299**, 469 (1982).
148 G. N. Reeke, J. A. Hartsuck, M. L. Ludwig, F. A. Quiocho, T. A. Steitz, and W. N. Lipscomb, *Proc. Natn. Acad. Sci. USA* **58**, 2220 (1967).
149 W. N. Lipscomb, *Proc. Natn. Acad. Sci. USA* **70**, 3797 (1973).
150 W. R. Kester and B. W. Matthews, *J. Biol. Chem.* **252**, 7704 (1977).
151 L. H. Weaver, W. R. Kester, and B. W. Matthews, *J. Molec. Biol.* **114**, 119 (1977).
152 M. A. Holmes and B. W. Matthews, *Biochemistry* **20**, 6912 (1981).
153 M. A. Holmes, D. E. Tronrud, and B. W. Matthews, *Biochemistry* **22**, 236 (1983).
154 B. L. Vallee, J. F. Riordan, and J. E. Coleman, *Proc. Natn. Acad. Sci. USA* **49**, 109 (1963).
155 D. S. Auld and B. L. Vallee, *Biochemistry* **10**, 2892 (1971).
156 J. W. Bunting and S. S.-T. Chu, *Biochemistry* **15**, 3237 (1976).
157 J. F. Riordan, M. Sokolovsky, and B. L. Vallee, *Biochemistry* **6**, 358 (1967).
158 W. L. Mock and J.-T. Chen, *Archs. Biochem. Biophys.* **203**, 542 (1980).
159 M. W. Makinen, K. Yamamura, and E. T. Kaiser, *Proc. Natn. Acad. Sci. USA* **73**, 3882 (1976).
160 L. C. Kuo, J. M. Fukuyama, and M. W. Makinen, *J. Molec. Biol.* **163**, 63 (1983).
161 L. C. Kuo and M. W. Makinen, *J. Biol. Chem.* **257**, 24 (1982).
162 J. E. Coleman and B. L. Vallee, *J. Biol. Chem.* **236**, 2244 (1961).
163 K. F. Geoghegan, A. Galdes, R. A. Martinelli, B. Holmquist, D. S. Auld, and B. Vallee, *Biochemistry* **22**, 2255 (1983).
164 R. Breslow and D. L. Wernick, *Proc. Natn. Acad. Sci. USA* **74**, 1303 (1977).
165 J. H. Freisheim, K. A. Walsh, and H. Neurath, *Biochemistry* **6**, 3010, 3020 (1967).
166 J. R. Uren and H. Neurath, *Biochemistry* **13**, 3512 (1974).
167 T. J. Bazzone and B. L. Vallee, *Biochemistry* **15**, 818 (1976).
168 J. S. Fruton, *The Enzymes* **3**, 119 (1971).
169 G. E. Clement, *Progr. Bioorg. Chem.* **2**, 177 (1973).
170 J. S. Fruton, *Adv. Enzymol.* **44**, 1 (1976).
171 *Acid proteinases, structure, function and biology* (J. Tang, Ed.), Plenum (1983).
172 B. Foltmann and V. B. Pedersen, in reference 171, p. 3.
173 M. N. G. James and A. R. Sielecki, *J. Molec. Biol.* **163**, 299 (1983).
174 E. Subramanian, I. D. A. Swan, M. Liu, D. R. Davies, J. A. Jenkins, I. J. Tickle, and T. L. Blundell, *Proc. Natn. Acad. Sci. USA* **74**, 556 (1977).
175 N. S. Andreeva, A. A. Federov, A. A. Gutshina, R. R. Riskulov, N. E. Schutzkever, and M. G. Safro, *Molec. Biol. (USSR)* **12**, 922 (1978).
176 P. Sepulveda, J. Marciniszyn Jr., D. Liu, and J. Tang, *J. Biol. Chem.* **250**, 5082 (1975).
177 L. Moravek and V. Kostka, *FEBS Lett.* **43**, 207 (1974).
178 G. E. Clement, J. Rooney, D. Zakheim, and J. Eastman, *J. Am. Chem. Soc.* **92**, 186 (1970).
179 P. S. Sampath-Kumar and J. S. Fruton, *Proc. Natn. Acad. Sci. USA* **71**, 1070 (1974).

180   A. A. Zinchenko, L. D. Rumsh, and V. K. Antonov, *Bioorg. Chem. (USSR)* **2**, 803 (1976).
181   J. L. Denburg, R. Nelson, and M. S. Silver, *J. Am. Chem. Soc.* **90**, 479 (1968).
182   A. J. Cornish-Bowden and J. R. Knowles, *Biochem. J.* **113**, 353 (1969).
183   G. R. Delpierre and J. S. Fruton, *Proc. Natn. Acad. Sci. USA* **54**, 1161 (1965); **56**, 1817 (1966).
184   R. L. Lundblad and W. H. Stein, *J. Biol. Chem.* **244**, 154 (1969).
185   R. S. Bayliss, J. R. Knowles, and G. B. Wybrandt, *Biochem. J.* **113**, 377 (1969).
186   J. A. Hartsuck and J. Tang, *J. Biol. Chem.* **247**, 2575 (1972).
187   M. Takahashi and T. Hofmann, *Biochem. J.* **127**, 35P (1972); **147**, 549 (1975).
188   M. Takahashi, T. T. Wang, and T. Hofmann, *Biochem. Biophys. Res. Commun.* **57**, 39 (1974); T. T. Wang and T. Hofmann, *Biochem. J.* **153**, 691 (1976).
189   A. K. Newmark and J. R. Knowles, *J. Am. Chem. Soc.* **97**, 3557 (1975).
190   M. S. Silver and S. L. T. James, *Biochemistry* **20**, 3177 (1980).
191   M. S. Silver and S. L. T. James, *Biochemistry* **20**, 3183 (1980).
192   V. K. Antonov, L. M. Ginodman, Y. V. Kapitannikov, T. N. Barshevskaya, A. G. Gurova, and L. D. Rumsh, *FEBS Lett.* **88**, 87 (1978).
193   V. K. Antonov, L. M. Ginodman, L. D. Rumsh, Y. V. Kapitannikov, T. N. Barshevskaya, L. B. Yavashev, A. G. Gurova, and L. I. Volkova, *Eur. J. Biochem.* **117**, 195 (1981).
194   R. J. Workman and D. W. Burkitt, *Archs. Biochem. Biophys.* **194**, 157 (1979).
195   J. Marciniszyn, J. A. Hartsuck, and J. Tang, *J. Biol. Chem.* **251**, 7088 (1976).
196   R. Bott, E. Subramanian, and D. R. Davies, *Biochemistry* **21**, 6956 (1982).
197   D. H. Rich, M. S. Bernatowicz, and P. G. Schmidt, *J. Am. Chem. Soc.* **104**, 3535 (1982).
198   M. Bustin, M. C. Lin, W. H. Stein, and S. Moore, *J. Biol. Chem.* **245**, 846 (1970).
199   J. Tang, *Biochem. Biophys. Res. Commun.* **41**, 697 (1970).
200   M. Bustin and A. Conway-Jacobs, *J. Biol. Chem.* **246**, 615 (1971).
201   J. Al-Janabi, J. A. Hartsuck, and J. Tang, *J. Biol. Chem.* **247**, 4628 (1972).
202   P. McPhie, *J. Biol. Chem.* **247**, 4277 (1972).
203   C. G. Sanny, J. A. Hartsuck, and J. Tang, *J. Biol. Chem.* **250**, 2635 (1975).
204   C. W. Dykes and J. Kay, *Biochem. J.* **153**, 141 (1976).
205   F. M. Richards and H. W. Wyckoff, *The Enzymes* **4**, 647 (1971).
206   P. Blackburn and S. Moore, *The Enzymes* **15**, 317 (1982).
207   F. M. Richards and H. W. Wyckoff, *Atlas of molecular structures in biology*, Clarendon Press, Oxford (1973).
208   C. H. W. Hirs, S. Moore, and W. H. Stein, *J. Biol. Chem.* **235**, 633 (1960).
209   D. G. Smyth, W. H. Stein, and S. Moore, *J. Biol. Chem.* **238**, 227 (1963).
210   G. Kartha, J. Bello, and D. Harker, *Nature, Lond.* **213**, 862 (1967).
211   H. W. Wyckoff, D. Tsernoglou, A. W Hanson, J. R. Knox, B. Lee, and F. M. Richards, *J. Biol. Chem.* **245**, 305 (1970).
212   D. Findlay, D. G. Herries, A. P. Mathias, B. R. Rabin, and C. A. Ross, *Nature, Lond.* **190**, 781 (1961).
213   F. M. Richards, H. W. Wyckoff, W. D. Carlson, N. M. Allewell, B. Lee, and Y. Mitsui, *Cold Spring Harbor Symp. Quant. Biol.* **36**, 35 (1971).
214   J. L. Markley, *Biochemistry* **14**, 3546 (1975).
215   A. G. Pavlovsky, S. N. Borisova, V. V. Borisov, I. V. Antonov, and M. Ya. Karpeisky, *FEBS Lett.* **92**, 258 (1978).

216 S. Y. Wodak, M. Y. Liu, and H. W. Wyckoff, *J. Molec. Biol.* **116**, 855 (1977).
217 N. Borkakoti, *Eur. J. Biochem.* **132**, 89 (1983).
218 A. Wlodawer, M. Miller, and L. Sjölin, *Proc. Natn. Acad. Sci. USA* **80**, 3628 (1983).
219 W. A. Gilbert, A. L. Fink, and G. A. Petsko, (in press).
220 R. N. Linquist, J. L. Lynn, and G. E. Leinhard, *J. Am. Chem. Soc.* **95**, 8762 (1973).
221 C. A. Deakyne and L. C. Allen, *J. Am. Chem. Soc.* **101**, 3951 (1979).
222 R. R. Holmes, J. A. Deiters, and J. C. Galluci, *J. Am. Chem. Soc.* **100**, 7393 (1978).
223 F. A. Cotton and E. E. Hazen, Jr., *The Enzymes* **4**, 153 (1971).
224 C. B. Anfinsen, P. Cuatrecasas, and H. Taniuchi, *The Enzymes* **4**, 177 (1971).
225 P. W. Tucker, E. E. Hazen, Jr., and F. A. Cotton, *Mol. Cell. Biochem.* **23**, 3, 67, 131 (1979).
226 A. Arnone, C. J. Bier, F. A. Cotton, V. W. Day, E. E. Hazen, Jr., D. C. Richardson, J. S. Richardson, and A. Yonath, *J. Biol. Chem.* **246**, 2302 (1971).
227 F. A. Cotton, C. J. Bier, V. W. Day, E. E. Hazen, Jr., and S. W. Larsen, *Cold Spring Harbor Symp. Quant. Biol.* **36**, 243 (1971).
228 F. A. Cotton, E. E. Hazen, Jr., and M. J. Legg, *Proc. Natn. Acad. Sci.. USA* **76**, 2551 (1979).
229 B. M. Dunn, C. Di Bello, and C. B. Anfinsen, *J. Biol. Chem.* **248**, 4769 (1973).
230 P. Cuatrecasas, S. Fuchs, and C. B. Anfinsen, *J. Biol. Chem.* **242**, 3063 (1967).
231 T. Imoto, L. N. Johnson, A. C. T. North, D. C. Phillips, and J. A. Rupley, *The Enzymes* **7**, 665 (1972).
232 B. Dunn and T. C. Bruice, *Adv. Enzymol.* **37**, 1 (1973).
233 D. M. Chipman and N. Sharon, *Science, N.Y.* **165**, 454 (1969).
234 C. C. F. Blake, L. N. Johnson, G. A. Mair, A. C. T. North, D. C. Phillips, and V. R. Sarma, *Proc. R. Soc.* **B167**, 378 (1967).
235 C. A. Vernon, *Proc. R. Soc.* **B167**, 389 (1967).
236 L. O. Ford, L. N. Johnson, P. A. Machin, D. C. Phillips, and R. Tjian, *J. Molec. Biol.* **88**, 349 (1974).
237 J. A. Rupley and V. Gates, *Proc. Natn. Acad. Sci. USA* **57**, 496 (1967).
238 M. A. Raftery and T. Rand-Meir, *Biochemistry* **7**, 3281 (1968).
239 U. Zehavi, J. J. Pollock, V. I. Teichberg, and N. Sharon, *Nature, Lond.* **219**, 1152 (1968).
240 F. W. Dahlquist, T. Rand-Meir, and M. A. Raftery, *Proc. Natn. Acad. Sci. USA* **61**, 119 (1968).
241 L. E. H. Smith, L. H. Mohr, and M. A. Raftery, *J. Am. Chem. Soc.* **95**, 7497 (1973).
242 S. Rosenberg and J. F. Kirsch, *Biochemistry* **20**, 3196 (1981).
243 J. J. Pollock, D. M. Chipman, and N. Sharon, *Archs. Biochem. Biophys.* **120**, 235 (1967).
244 S. M. Parsons and M. A. Raftery, *Biochemistry* **11**, 1623 (1972).
245 J. A. Rupley, V. Gates, and R. Bilbrey, *J. Am. Chem. Soc.* **90**, 5633 (1968).
246 A. Warshel and M. Levitt, *J. Molec. Biol.* **103**, 227 (1976).
247 A. Warshel, *Biochemistry* **20**, 3167 (1981).
248 Y. Eshdat, A. Dunn, and N. Sharon, *Proc. Natn. Acad. Sci. USA* **71**, 1658 (1974).
249 I. I. Secemski and G. E. Lienhard, *J. Biol. Chem.* **249**, 2932 (1974).

250 E. Holler, J. A. Rupley, and G. P. Hess, *Biochemistry* **14**, 1088, 2377 (1975).
251 M. Levitt, in *Peptides, polypeptides and proteins* (E. R. Blout, F. A. Bovey, M. Goodman, and N. Lotan, Eds.), Wiley, p. 99 (1974).
252 M. Schindler, Y. Assaf, N. Sharon, and D. M. Chipman, *Biochemistry* **16**, 423 (1977).
253 F. W. Ballardie, B. Capon, M. W. Cuthbert, and W. M. Dearie, *Bioorg. Chem.* **6**, 483 (1977).
254 M. Schindler and N. Sharon, *J. Biol. Chem.* **251**, 4330 (1976).
255 J. A. Kelly, A. R. Sielecki, B. D. Sykes, M. N. G. James, and D. C. Phillips, *Nature* **282**, 875 (1979).
256 W. F. Anderson, M. G. Grütter, S. J. Remington, L. C. Weaver, and B. W. Matthews, *J. Molec. Biol.* **147**, 523 (1981).
257 E. Holler, J. A. Rupley, and G. P. Hess, *Biochem. Biophys. Res. Commun.* **37**, 423 (1969).
258 J. H. Baldo, S. E. Halford, S. L. Patt, and B. D. Sykes, *Biochemistry* **14**, 1893 (1975).
259 S. Lindskog, L. E. Henderson, K. K. Kannan, A. Liljas, P. O. Nyman, and B. Strandberg, *The Enzymes* **5**, 587 (1971).
260 J. E. Coleman, *Progr. Bioorg. Chem.* **1**, 159 (1971).
261 A. Liljas, K. K. Kannan, P.-C. Bergsten, I. Waara, K. Fridborg, B. Strandberg, U. Carlbom, L. Järup, S. Lövgren, and M. Petef, *Nature New Biology, Lond.* **235**, 131 (1972).
262 K. K. Kannan, B. Notstrand, K. Fridborg, S. Lövgren, A. Ohlsson, and M. Petef, *Proc. Natn. Acad. Sci. USA* **72**, 51 (1975).
263 B. Andersson, P. O. Nyman, and L. Strid, *Biochem. Biophys. Res. Commun.* **48**, 670 (1972).
264 L. E. Henderson, D. Henriksson, and P. O. Nyman, *J. Biol. Chem.* **251**, 5457 (1976).
265 H. Steiner, B. H. Jonsson, and S. Lindskog, *Eur. J. Biochem.* **59**, 253 (1975).
266 S. Lindskog and J. E. Coleman, *Proc. Natn. Acad. Sci. USA* **70**, 2505 (1973).
267 S. H. Koenig and R. D. Brown III, *Proc. Natn. Acad. Sci. USA* **69**, 2422 (1972).
268 D. W. Appleton and B. Sarkar, *Proc. Natn. Acad. Sci. USA* **71**, 1686 (1974).
269 J. M. Pesando, *Biochemistry* **14**, 675, 681 (1975).
270 I. D. Campbell, S. Lindskog, and A. I. White, *J. Molec. Biol.* **90**, 469 (1974); **98**, 597 (1975).
271 R. Bauer, P. Limkilde, and J. T. Johansen, *Biochemistry* **15**, 334 (1976).
272 M. E. Riepe and J. H. Wang, *J. Biol. Chem.* **243**, 2779 (1968).
273 R. G. Khalifah, *J. Biol. Chem.* **246**, 2561 (1971).
274 J. J. Led, E. Neesgaard, and J. T. Johansen, *FEBS Lett.* **147**, 74 (1982).
275 K. K. Kannan, M. Petef, K. Fridborg, H. Cid-Dresdener, and S. Lövgren, *FEBS Lett.* **73**, 115 (1977).
276 R. H. Prince and P. Woolley, *Bioorg. Chem.* **2**, 337 (1973).
277 R. G. Khalifah, *Proc. Natn. Acad. Sci. USA* **70**, 1986 (1973).
278 D. N. Silvermann and C. K. Tu, *J. Am. Chem. Soc.* **97**, 2263 (1975).
279 C. K. Tu and D. N. Silverman, *J. Am. Chem. Soc.* **97**, 5935 (1975).
280 I. A. Rose, *The Enzymes* **2**, 281 (1970).
281 I. A. Rose, *Adv. Enzymol.* **43**, 491 (1975).
282 E. A. Noltmann, *The Enzymes* **6**, 271 (1972).
283 D. W. Banner, A. C. Bloomer, G. A. Petsko, D. C. Phillips, C. I. Pogson, I. A. Wilson, P. H. Corran, A. J. Furth, J. D. Milman, R. E. Offord, J. D. Priddle, and S. G. Waley, *Nature, Lond.* **255**, 609 (1975).
284 T. Alber, F. C. Hartman, and G. A. Petsko (in press).

285 S. V. Rieder and I. A. Rose, *J. Biol. Chem.* **234**, 1007 (1958).
286 Y. J. Topper, *J. Biol Chem.* **225**, 419 (1957).
287 B. Bloom and Y. J. Topper, *Nature, Lond.* **181**, 1128 (1958).
288 I. A. Rose and E. L. O'Connell, *Biochim. Biophys. Acta* **42**, 159 (1960).
289 I. A. Rose and E. L. O'Connell, *J. Biol. Chem.* **236**, 3086 (1961).
290 W. J. Albery and J. R. Knowles, *Biochemistry* **15**, 5588, 5627 (1976).
291 R. Iyengar and I. A. Rose, *Biochemistry* **20**, 1223, 1229 (1981).
292 I. A. Rose and R. Iyengar, *Biochemistry* **21**, 1591 (1982).
293 L. N. Johnson and R. G. Wolfenden, *J. Molec. Biol.* **47**, 93 (1970).
294 R. B. Jones and S. G. Waley, *Biochem. J.* **179**, 623 (1979).
295 D. C. Phillips, M. J. E. Sternberg, J. M. Thornton, and I. A. Wilson, *Biochem. Soc. Trans.* **5**, 642 (1977).
296 T. Alber and G. Petsko (in press).
297 F. C. Hartman, *Biochem. Biophys. Res. Commun.* **33**, 888 (1968); **39**, 384 (1970).
298 S. G. Waley, J. C. Miller, I. A. Rose, and E. L. O'Connell, *Nature, Lond.* **227**, 181 (1970).
299 S. De La Mare, A. F. W. Coulson, J. R. Knowles, J. D. Priddle, and R. E. Offord, *Biochem. J.* **129**, 321 (1972).
300 B. Plaut and J. R. Knowles, *Biochem. J.* **129**, 311 (1972).
301 K. J. Schray, E. L. O'Connell, and I. A. Rose, *J. Biol. Chem.* **248**, 2214 (1973).
302 F. C. Hartman, G. M. LaMuraglia, Y. Tomozawa, and R. Wolfenden, *Biochemistry* **14**, 5274 (1975).
303 M. R. Webb and J. R. Knowles, *Biochem. J.* **141**, 589 (1974).
304 J. G. Belasco and J. R. Knowles, *Biochemistry* **19**, 472 (1980).

Author Index

Aaviksaar, A. *96*
Abbott, S. J. *247*
Abeles, R. H. *96, 262, 345*
Abelson, J. *388*
Abraham, M. H. *310*
Adams, M. J. *445*
Adair, G. S. *291*
Adelman, R. C. *329*
Ahmed, A. I. *43*
Åkeson, Å. *444*
Alber, T. *43, 219, 430, 446, 451, 452*
Alberts, B. *368*
Alberty, R. A. *120, 174*
Albery, W. J. 331, *345, 452*
Alden, R. A. *43, 44, 346, 446*
Aldridge, W. N. *219*
Alecio, M. R. *447*
Al-Janabi, J. *449*
Allen, L. C. *450*
Allewell, N. W. *449*
Allison, W. S. *445*
Alworth, W. L. *345*
Ambrose, M. C. *45*
Anderson, D. C. *444*
Anderson, W. F. *43, 345, 451*
Andersson, B. *451*
Andersson, P. *444*
Andreeva, N. *448*
Anfinsen, C. B. *450*
Antonetti, A. *154*
Antonov, V. K. *449*
Appleton, D. W. *451*
Arai, K.-I. *45*
Ardelt, W. *247, 447*
Argos, P. *42, 43, 46*
Arigoni, D. *246*
Arnone, A. *279, 292, 450*
Arnott, M. S. *43*

Artymiuk, P. J. *43, 45*
Asboth, B. *447*
Ashford, J. A. *154*
Ashour, A.-L.E. *246*
Assaf, Y. *435, 451*
Atkins, G. L. *120, 192*
Atkinson, D. E. *268*
Atkinson, R. F. *95*
Atkinson, T. C. *388*
Atlas, D. *44*
Auld, D. S. *448*
Ault, B. *151, 154*
Axelrod, B. *210, 220*

Bachovin, W. W. *175, 446*
Bagshaw, C. R. *120*
Baker, B. R. *262*
Baker, E. N. *447*
Balaban, R. S. *154*
Bald, R. *310*
Baldo, J. H. *151, 154, 451*
Baldwin, A. N. *220, 367*
Baldwin, J. M. *276, 291*
Baldwin, R. L. 8, *42*
Ballardie, F. W. *451*
Balls, A. K. *219*
Balny, C. *447*
Banaszak, L. J. *43, 444, 445*
Banner, D. W. *451*
Baraniak, J. *247*
Barker, D. *387*
Barker, R. *329*
Barley, F. *95*
Barman, T. E. *219*
Barnard, E. A. *211, 220*
Barnett, D. R. *43*
Barnett, R. *95, 445*
Barrett, H. *210, 220*

453

Subject Index